DATE DUE

JUL 1 3 1998	
SEP 4 1998	
SEP 2 9 1998	

Biological Effects
of
Heavy Metals

Volume I

Editor

E. C. Foulkes

Professor
Department of Environmental Health
Kettering Laboratory
University of Cincinnati
Cincinnati, Ohio

CRC Press, Inc.
Boca Raton, FL

Library of Congress Cataloging in Publication Data

Foulkes, Ernest C., 1924–
    Biological effects of heavy metals/author, E.C. Foulkes.
      p. cm.
    Includes bibliographies and index.
    ISBN 0-8493-4241-4 (v.1).
    1. Heavy metals — Toxicology. I. Title.
    RA1231.M52F68   1990
    615.9'2531— dc20   89-15716

Direct all inquiries to CRC Press, Inc., 2000 Corporate Blvd., N.W., Boca Raton, Florida, 33431.

© 1990 by CRC Press, Inc.

International Standard Book Number 0-8493-4241-4 (v.1)

Library of Congress Card Number 89-15716
Printed in the United States

PREFACE

It is now part of general wisdom that heavy metals like lead, mercury, and cadmium are increasingly important toxic constituents of our environment. Appropriately, therefore, the toxicology of heavy metals provides the focus of extensive research. This interest was reflected in the formation of a Metals Specialty Section of the Society of Toxicology. When a series of volumes titled "Focus on Biological Effects of Heavy Metals" was first discussed in 1987, the members at that time of the Executive Committee of the section agreed to serve as the Editorial Advisory Committee. The present volume on the neurotoxicology of heavy metals is the first of this series, and we hope that members of the Metals Specialty Section will continue to be able to contribute to its success.

Not only are heavy metals of great toxicological significance, but their experimental use has constituted a powerful tool in elucidating the mechanisms of normal physiological function. This is well illustrated by the chapters contributed to this volume. The contributors are all established investigators in this field, and they were invited to write a chapter on the basis of their interest in basic mechanisms of metal action. I am grateful to them for the care, if not always the timeliness, with which they completed their task. This task was the critical appraisal of how metals interact with specific portions of the nervous system, with emphasis where appropriate, on experimental techniques. No attempt could be made either to cover the whole nervous system, to consider all neurotoxic metals, or to establish complete consensus between the independent authors.

Aspects of heavy metal neurotoxicology have been discussed in several recent reviews and volumes. The contents of the present volume are based on a symposium held at the Annual Meeting of the Society of Toxicology in Washington, D.C. in 1987 (Fund. Appl. Toxicol. 9:599-615, 1987), sponsored jointly by the Neurotoxicology and Metals Specialty Sections of the Society. On the molecular level there is strong evidence that a common denominator in the action of many toxicants is the interference with cellular Ca homeostasis; this is certainly true of the nervous system and the first three chapters deal with aspects of this problem. Drs. Chang and Fu concentrated on neuropathological effects of various metals, while Drs. Bierkamper and Buxton illustrated metal neurotoxicology by concentrating on a specific metal (tin). The last two chapters deal with effects of metals on a specific sensory system (the retina), and on neurobehavioral aspects of metal toxicology.

Of the people who have helped put this volume together, I want to single out my coworker, Ms. Annette Townsley, whose critical skills and organizing ability were invaluable. I hope the final work will encourage future research by raising questions, pointing to gaps in our knowledge and indicating directions in which further progress needs to be made.

E. C. Foulkes
October 1988

THE EDITOR

Ernest C. Foulkes, Ph.D., is a professor of Environmental Health and Physiology/ Biophysics at the University of Cincinnati, College of Medicine. He received his B.Sc. and M.Sc. degrees from the University of Sydney, NSW, in 1946 and 1947, and his doctorate from Oxford University in 1952.

After working on a fellowship at the May Institute of the Jewish Hospital in Cincinnati and as established investigator of the American Heart Association, he joined the University of Cincinnati in the Departments of Environmental Health and Physiology in 1965.

As a physiological toxicologist, Dr. Foulkes has worked for many years in the field of heavy metals, with special emphasis on their effects on biological membranes in kidney, intestine and other tissues. These interests have taken him to Universities in Europe and Japan as a visiting professor.

At present, work in his laboratory is supported mainly by the National Institute of Environmental Health Sciences, and continues to focus on the interaction between toxic heavy metals and epithelial membranes. This same focus led to work with the National Committee on Radiation Protection dealing with the toxicology of uranium compounds, and is also reflected in active involvement with the Metals Specialty Section of the Society of Toxicology (Section President, 1987—1988).

ADVISORY BOARD

CONTRIBUTORS

Gerald Audesirk
Associate Professor
Biology Department
University of Colorado/Denver
Denver, Colorado

George G. Bierkamper
(deceased)
Professor
Department of Pharmacology
School of Medicine
University of Nevada
Reno, Nevada

Robert Bornschein
Associate Professor
Department of Environmental Health
University of Cincinnati School of
 Medicine
Cincinnati, Ohio

Iain L. O. Buxton
Associate Professor
Department of Pharmacology
School of Medicine
University of Nevada
Reno, Nevada

Louis W. Chang
Professor
Departments of Pathology and
 Pharmacology/Toxicology
University of Arkansas Medical School
Little Rock, Arkansas

Gary P. Cooper
Professor
Department of Environmental Health
University of Cincinnati School of Medicine
Cincinnati, Ohio

Durisala Desaiah
Professor
Department of Neurology
University of Mississippi Medical Center
Jackson, Mississippi

Donald A. Fox
Associate Professor
College of Optometry
University of Houston
Houston, Texas

Casey S. Fu
Research Associate
Departments of Pathology and Nutritional
 Science
UCLA
Los Angeles, California

Shou-Ren Kuang
Attending Physician
Clinical Department
Guangdong Institute of Occupational
 Medicine
Guangzhou, People's Republic of China

Daniel J. Minnema
Assistant Professor
Department of Environmental Health
University of Cincinnati School
 of Medicine
Cincinnati, Ohio

TABLE OF CONTENTS

Chapter 1

EFFECTS OF HEAVY METALS ON NEURONAL CALCIUM CHANNELS

Gerald Audesirk

TABLE OF CONTENTS

I. INTRODUCTION

Calcium is an important messenger within the cytoplasm of all living cells. In most cells, the resting concentration of free calcium ions in the cytoplasm, $[Ca^{2+}]_{in}$, is kept very low, about 1 to 2×10^{-7} M. Changes in $[Ca^{2+}]_{in}$ modulate a wide variety of structural and biochemical events within cells, including cytoskeletal mobility, secretion, metabolic activity, and the opening or closing of ion channels in the plasma membrane. $[Ca^{2+}]_{in}$ may be rapidly changed either by the release of Ca^{2+} from intracellular stores, especially the endoplasmic reticulum or sarcoplasmic reticulum, or by the influx of Ca^{2+} through the plasma membrane from the extracellular fluid.

Most, if not all, excitable cells have calcium channels in their plasma membranes.[1-3] Since the extracellular fluid bathing cells has approximately millimolar Ca^{2+} concentrations, while the cytoplasm of "resting" cells contains less than micromolar concentrations, there is a large gradient favoring the movement of Ca^{2+} into the cell. Plasma membrane calcium channels, when opened by depolarization or by appropriate ligands, consequently allow an influx of Ca^{2+}. This influx transiently increases $[Ca^{2+}]_{in}$, triggering changes in cellular activity.

Calcium channels are blocked by micromolar to millimolar concentrations of many divalent and trivalent cations, including Cd^{2+}, Co^{2+}, La^{3+}, Mg^{2+}, Mn^{2+}, and Ni^{2+}.[1-4] In most electrophysiological research, the blockade of inward current into neurons by these cations has been used, first, as an indicator that the current in question most likely passes through calcium channels, and second, to study the role of Ca^{2+} influx on neuronal functions such as repetitive action potential activity, release of transmitters or hormones, or opening of other ion channels. Much less research has focused on the potential neurotoxicological effects of metal cations on calcium channel functioning. However, in view of the importance of Ca^{2+} as an intracellular messenger, even a partial blockade of calcium channels may lead to profound alterations in neuronal function. Therefore, the interactions of divalent cations, particularly heavy metals such as Cd^{2+} and Pb^{2+}, with calcium channels has potentially great significance for neurotoxicology.

In this review, I will focus on the effects of inorganic forms of heavy metals on voltage-dependent calcium channels in neuronal plasma membranes. Organic forms of heavy metals, such as triethyl lead or tributyl tin, will not be discussed. Similarly, ligand-gated calcium channels, such as those associated with N-methyl-D-aspartate receptors, will not be covered. Calcium channels in non-neuronal excitable tissues, in particular the heart, probably have structures that are similar to certain types of calcium channels in neurons and are also affected by heavy metals, but these will not be considered in detail.

II. VOLTAGE-DEPENDENT CALCIUM CHANNELS IN NEURONS

A. CALCIUM CHANNEL STRUCTURE

Based largely on studies of "L-type" calcium channels in cardiac muscle cells (see below), and to a lesser extent, of calcium channels in gastropod central neurons, the following model of calcium channel function has been developed.[5-7]

Calcium channels are transmembrane proteins composed of multiple subunits. These subunits are thought to form a pore through which calcium ions pass in single file. A voltage-sensitive "gate", probably on the cytoplasmic side of the channel, determines whether the pore is open. Which ions permeate through an open pore seems to be determined by the presence of at least two Ca^{2+} binding sites within the pore.

The affinity of the binding sites for various cations accounts for the selectivity of the channel for Ca^{2+}, the rates of permeation by Ca^{2+} and Ba^{2+}, and the blocking effects of many divalent cations. The rate of movement of cations through the open calcium channel depends in large measure on the affinity of the cations for the binding sites. Both monovalent and divalent cations form a continuum of "blocking" and "permeation". Any cation that occupies the pore at least temporarily blocks passage by other cations. Monovalent cations bind very weakly to the intrapore sites, are readily displaced by calcium, and do not normally constitute a major part of the current flow through the open channel. Ca^{2+}, Ba^{2+}, and Sr^{2+} bind relatively strongly to the sites; flow of these ions through the channel is probably due to electrostatic repulsion of two cations simultaneously inside the pore. Ba^{2+} binds somewhat more weakly to the intrapore sites than does Ca^{2+}. When Ba^{2+} is the major cation available, electrostatic repulsion between ions inside the channel readily displaces Ba^{2+} ions from the binding sites, allowing high current flow through the channel. When Ca^{2+} is the major cation available, its tighter binding within the pore leads to slower permeation. Blocking cations, such as Cd^{2+} or La^{3+}, apparently enter the pore and bind extremely tightly to the intrapore sites. Thus they tend to remain within the pore, neither permeating nor allowing other cations to pass through.

B. TYPES OF CALCIUM CHANNELS

Calcium channels exist in a variety of types. All open in response to depolarization and have substantial permeabilities to calcium. The model presented above is thought to apply to all channel types,[7] although channel types may differ in some respects, for example in the voltage sensitive gate and the affinity of the intrapore binding sites.

Calcium channel types have thus far been characterized by electrophysiological criteria, including the activation and inactivation voltages, the rate of inactivation of open channels during sustained depolarization, and the single-channel conductance, and by pharmacological criteria, notably their responses to dihydropyridines and cone-snail toxin[8-18] (see the reviews by McCleskey et al.[19] and Miller;[20] for an extensive list of examples of calcium channel types in vertebrate excitable cells, see Table 3 in Fox et al.[12]). The various channel types also probably differ in their functions.[13,20] Finally, calcium channel types appear to differ in their responses to heavy metals.[11,15,20]

Voltage and patch-clamping studies have revealed at least three types of calcium channels in the plasma membranes of vertebrate neurons, usually labeled L, T, and N[11,12,15,19] (see Figure 1 and Table 1). In brief, the L-type channels (Long lasting) are found in cardiac muscle cells and in many neurons. These channels open in response to depolarizations to relatively positive voltages, inactivate slowly during sustained depolarizations, require only relatively small polarizations to remove inactivation, have large single-channel conductances, and are sensitive to dihydropyridines and conotoxin GVIA. (Recent data suggest that there may be at least two subtypes of L-type channels, one subtype sensitive to conotoxin and one insensitive[10].) The T-type channels (Transient) open in response to smaller depolarizations and require much more negative potentials to remove inactivation. T-type channels inactivate rapidly during sustained depolarizations, have small single-channel conductances, and are insensitive to dihydropyridines and conotoxin GVIA. Like T-type channels, N-type channels (Neither T nor L) inactivate rapidly and are insensitive to dihydropyridines. However, they have intermediate single-channel conductances, require large depolarizations for opening, and are blocked by conotoxin GVIA.

Among the invertebrates, calcium channels in gastropod neurons[21-26] and squid giant synapse[27-29] have been most thoroughly studied. Although there is evidence for multiple calcium channels in gastropod neurons,[30] the best studied gastropod calcium channels are "L-like" channels (although they are not usually given this term in the literature). In general, gastropod "L-like" channels inactivate slowly during sustained depolarizations and require small polarizations to remove inactivation. However, they open in response to relatively small depolarizations and are apparently insensitive to conotoxin GVIA.[14] Several studies report that the gastropod "L-like" channels are sensitive to dihydropyridines,[31-33] but others have found no effect.[34,35] Calcium channels in squid giant synapse presynaptic terminals seem to be similar to gastropod neuronal channels, and are insensitive to dihydropyridines.

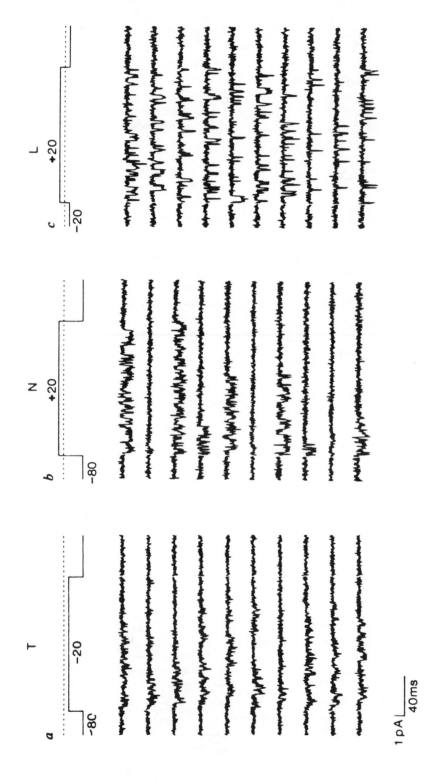

FIGURE 1. Three types of unitary calcium channels in cell-attached patch clamp recordings of chick dorsal root ganglion cells, with Ba²⁺ as the charge carrier. Note the differences in holding potential, step potential, unitary conductance, and persistence of channel opening for each channel type. See also Table 1. (Reprinted from Nowycky, et al., *Nature*, 316, 440, 1985. With permission.)

TABLE 1
Characteristics of Calcium Channel Types in Vertebrate Neurons[12,20]

Property	Channel type		
	T	N	L
Activation Range	>–70 mV	>–30 mV	>–10 mV
Inactivation Range	–100 to –60 mV	–120 to –30 mV	–60 to –10 mV
Inactivation Rate	Fast	Moderate	Slow
	($t_{1/2}$ 2F—50 ms)	($t_{1/2}$ 50—80 ms)	($t_{1/2}$ > 500 ms)
Single-channel conductance	8 pS	13 pS	25 pS
Sensitivity to Cd^{2+} block	low	high	high
Sensitivity to Ni^{2+} block	high	low	low
Sensitivity to dihydropyridines	low	low	high
Sensitivity to conotoxin GVIA	low	high	high

1. Functions of Calcium Channel Types

The neuronal functions served by the various channel types are not completely understood, and in fact the (presumably) same type of channel appears to have different functions in different cells. Nevertheless, some generalizations have emerged.

L-type channels — Calcium influx through L-type channels in cardiac cells produces the plateau phase of the action potential, carrying large amounts of calcium into the cells.[36] L-type channels appear to be involved in transmitter release from adrenal chromaffin cells[37] and PC12 cells[16] (see also Miller[20]).

T-type channels — Llinas and coworkers[38,39] have proposed that a low-voltage-activated, rapidly inactivating calcium channel much like the T-type channel may be involved in causing bursting action potential activity in several types of neurons of the mammalian central nervous system.

N-type channels — Several studies have suggested that the N-type channels may be involved in transmitter release.[13,20]

Molluscan "L-like" channels — At squid giant synapse, these channels control neurotransmitter release. In gastropod central neurons, their function is less well characterized; however, large amounts of calcium enter these neurons during action potentials (which often have plateaus much like vertebrate cardiac muscle cells), so calcium influx must play a major role in triggering a variety of intracellular events.

III. EFFECTS OF *IN VITRO* HEAVY METAL EXPOSURE ON CALCIUM CHANNELS

The focus of the discussion of the effects of heavy metals will be on calcium channels in four experimental systems: (1) depolarization-evoked transmitter release, especially by the motor neurons at the neuromuscular junction, by adrenal chromaffin cells, and by synaptosomes; (2) depolarization (high potassium) stimulated calcium uptake, especially in synaptosomes and adrenal chromaffin cells; (3) binding of radioactive dihydropyridines to synaptosomes and brain membrane preparations; and (4) current flow through calcium channels in neuronal cell bodies under voltage or patch clamp.

Each of these experimental systems involves changes in the functional state of voltage-dependent calcium channels, but they do not provide equally direct measures of calcium channel function. (1) Evoked transmitter release requires a series of intracellular steps following calcium entry; these steps may themselves be sensitive to the effects of heavy metals.[40-44] Appropriate precautions can be taken to minimize the likelihood that these later events contribute to the observed effects of heavy metals, especially with short-term exposure to low concentrations of metals. However, with higher concentrations and longer exposures, contributions by intracellular events become more likely, particularly with metals that permeate through calcium channels into the cytoplasm. (2) Measurements of the uptake of radioactive calcium eliminate contributions by exocytosis-related events, but nevertheless present some potential difficulties. First, the time resolution of even the fastest procedures is slow relative to the normal activation time of calcium channels; second, calcium-extrusion mechanisms quickly come into play as the intracellular calcium ion concentration increases, potentially counterbalancing some or all of the calcium influx, particularly in the slower assay systems. (3) Insofar as dihydropyridine binding sites are parts of calcium channels, binding studies provide information about the functional state of the channels. However, not all dihydropyridine binding sites may be related to calcium channels. (4) Voltage and patch clamping probably provide the most accurate assessment of the impact of heavy metals on the functional states of calcium channels themselves. Aside from the normal technical difficulties of these techniques, the multiplicity of calcium channel types (see above) and the rundown of calcium channel functioning in excised patches and during whole-cell internal perfusion make relatively long-term studies, e.g., tens of minutes to hours, difficult (see Byerly and Yazejian[45] and Chad and Eckert[31] for methods that slow the rundown of calcium currents in perfused cells). In this section, the effects of heavy metals on calcium channels as measured by each of these techniques will be discussed.

A. EFFECTS OF HEAVY METALS ON DEPOLARIZATION-EVOKED TRANSMITTER RELEASE

Another chapter in this volume deals with the effects of heavy metals on transmitter release. Further, this topic has been the subject of recent reviews by Cooper and colleagues.[46,47] This section will therefore be confined to a brief discussion of the implications of these data for calcium channel function. Unless otherwise noted, the data presented here are extracted from the review of Cooper et al.[46]

Most divalent and multivalent cations inhibit action potential-evoked release of neurotransmitter at neuromuscular junctions (usually frog, toad, rat, or mouse). In a survey of the literature, Cooper et al.[46] estimated the order of effectiveness of cations in blocking transmitter release (lowest effective concentration to highest effective concentration) as $Hg^{2+} > Pb^{2+} > Sn^{4+} > La^{3+} > Er^{3+} > Pr^{3+} > Cd^{2+} > Y^{3+} > Cr^{3+} > Be^{2+} > Zn^{2+} = Ni^{2+} > Co^{2+} = Mn^{2+} > Mg^{2+}$. Great differences in experimental protocol, concentrations used, and durations of exposure make comparisons difficult, but a few generalizations can be drawn.

First, the majority of cations tested are reversible inhibitors, at least when applied for short durations and at reasonably low concentrations. Some of these cations have been extensively studied, particularly Cd^{2+}, Co^{2+}, Mg^{2+}, Mn^{2+}, and Pb^{2+}. These five cations have all been found to be competitive inhibitors of Ca^{2+}. Mg^{2+} is at least an order of magnitude less potent in blocking transmitter release (or in blocking calcium currents in other experimental systems) than are the other competitive inhibitors. According to Lansman et al.,[6] this is probably due to an excessively slow rate of association of Mg^{2+} with the binding sites within calcium channels, perhaps because it is very slow to lose its waters of hydration and form a tight complex with the binding sites. The other competitive inhibitors presumably bind strongly to the intrapore sites, allowing them to block calcium flux through the channel at low concentrations. The binding must still be weak enough, however, that a reasonably short period of washout removes essentially all of the blocking cations and restores normal calcium influx. Even though it is not technically feasible to measure calcium currents directly at vertebrate presynaptic nerve terminals, it is fairly certain that the reversible blockers interfere with calcium channel functioning rather than (or in addition to) affecting later stages in the release process.

A few cations cause irreversible or only partly reversible blockade of evoked release, including Cr^{3+}, Y^{3+}, Sn^{4+}, and Hg^{2+}. A few others appear to be reversible at low concentrations but cause irreversible blockade at high concentrations or after long exposures (La^{3+}, Cd^{2+}, Pb^{2+}). In these cases, it is difficult to assess whether the cations interfere with calcium channel functioning or with later steps in release.

Spence et al.[48] found that synaptic transmission in rat spinal cord is inhibited by acute exposure to Pb^{2+} at concentrations at least as low as 20 μM. The amount of inhibition was very variable, somewhat independent of concentration, and reversible only with short exposures.

The depolarization-induced release of transmitters by adrenal chromaffin cells[49] and by synaptosomes[50-52] is inhibited by heavy metals. In rat brain synaptosomes, Cd^{2+} and Pb^{2+} reduce the high-potassium-stimulated release of acetylcholine, with K_i's of 4.5 μM and 16 μM, respectively. The lesser effectiveness of Pb^{2+} is probably due to the fact that lead appears to penetrate into the synaptosome interior, where it stimulates basal, spontaneous release; Cd^{2+}, on the other hand, appears unable to permeate into synaptosomes in any appreciable quantity.[50] In adrenal chromaffin cells, Pb^{2+} also penetrates readily through the plasma membrane (probably through calcium channels) into the cell interior,[53] but does not stimulate basal catecholamine release. Pb^{2+} inhibits the depolarization-evoked catecholamine release with an IC_{50} of 0.9 to 1.6 μM .[49]

B. EFFECTS OF HEAVY METALS ON DEPOLARIZATION-STIMULATED UPTAKE OF RADIOACTIVE CALCIUM

A long list of heavy metals inhibit the uptake of ^{45}Ca into synaptosomes, including Al^{3+} (K_i 600 μM);[54] Mg^{2+} (IC_{50} 3000 μM), Sr^{2+} (2600 μM), Ba^{2+} (1500 μM), Hg^{2+} (120 μM), Mn^{2+} (70 μM), Co^{2+} (60 μM), Ni^{2+} (40 μM), Cu^{2+} (40 μM), Zn^{2+} (30 μM), Y^{3+} (1 μM), Cd^{2+} (1 μM), Pb^{2+} (0.4 μM), and La^{3+} (0.3 μM) (all from Nachshen[55]); Cd^{2+} (K_i 2.2 μM) and Pb^{2+} (1.1 μM) (from Suszkiw et al.[50]); and Hg^{2+} (IC_{50} 50 μM).[56] Many heavy metals also reduce ^{45}Ca uptake into bovine adrenal medullary cells: Pb^{2+} (IC_{50} 2.1 μM);[49] Mg^{2+} (IC_{50} 980 μM), Mn^{2+} (725 μM), Co^{2+} (273 μM), Ni^{2+} (127 μM), La^{3+} (66 μM); Cd^{2+} (52 μM), and Zn^{2+} (24 μM) (all from Gandia et al.[57]). Even allowing for differences in experimental techniques, there appear to be great differences between synaptosomes and adrenal chromaffin cells both in the sequence of potencies of these metals in blocking Ca^{2+} uptake and in the absolute concentrations that produce a 50% inhibition. Calcium channels in synaptosomes are much more susceptible to block by La^{3+} and Cd^{2+}, and somewhat more susceptible to Co^{2+} and Pb^{2+}. Medullary cells are more susceptible to Mg^{2+}. Although its absolute blocking concentrations are similar in the two preparations, Zn^{2+} ranks as an extremely potent blocker in medullary cells, but only falls in the middle of the blocking sequence in synaptosomes.

It is interesting to note that calcium channels in adrenal medullary cells are quite different pharmacologically and electrophysiologically from calcium channels in synaptosomes and several other neuronal transmitter-releasing systems. The dihydropyridines are potent blockers of calcium channels in medullary cells.[57,58] Electrophysiological studies show that medullary calcium channels are slowly inactivating.[37,59] Thus the medullary calcium channels appear to be similar to the L-type channel. Calcium channels in synaptosomes are insensitive to dihydropyridine blocking compounds,[18,60-62] as measured both by calcium uptake and transmitter release. Transmitter release in several whole-cell preparations, for example in cultured sympathetic neurons,[13] is insensitive to dihydropyridines. Dihydropyridines also do not block transmitter release at the rat neuromuscular junction.[63] Based on these findings and on electrophysiological characteristics of calcium channels in sympathetic neurons, it has been proposed that N-type channels are most intimately involved in transmitter release from neurons.[13,20] Thus, the differential effects of heavy metals on synaptosomes and adrenal medullary cells suggest that L-type and N-type channels may differ not only in their pharmacology and electrophysiological characteristics, but also in their sensitivities to heavy metals.

C. EFFECTS OF HEAVY METALS ON NITRENDIPINE BINDING TO SYNAPTOSOMES

Metal cations affect the binding of [^{3}H]nitrendipine, a dihydropyridine calcium channel blocker, to rat brain membranes and synaptosomes. Gould et al.[64] found that *in vitro* exposure to divalent cations that can pass through calcium channels, including Ca^{2+}, Ba^{2+}, and Sr^{2+}, enhances [^{3}H]nitrendipine binding to rat brain membranes, at least up to cation concentrations of 10 mM. Mn^{2+}, which is slightly permeable, is somewhat less effective, but still stimulates binding at high concentrations. Cations that normally block current flow through calcium channels, including La^{3+}, Co^{2+}, and Cu^{2+}, slightly stimulate [^{3}H]nitrendipine binding at low concentrations (approximately 10 to 100 μM), but depress binding at higher concentrations. Rius et al.[65] studied metal effects on [^{3}H]nitrendipine binding to synaptosomes prepared from several areas of the rat brain. Although there were some differences noted among cortex, striatum, and hippocampus, their results agree that La^{3+} inhibits, while Mn^{2+} slightly stimulates, [^{3}H]nitrendipine binding. However, Pb^{2+}, which is a potent blocker of calcium channels, strongly enhances [^{3}H]nitrendipine binding, even at mM concentrations. Interestingly, Pb^{2+} is not only a strong blocker of calcium currents through calcium channels, but also appears to pass through at least some types of calcium channels.[53] (Note that nitrendipine, which is a potent blocker of L-type calcium channels, blocks neither the depolarization-stimulated uptake of ^{45}Ca nor the release of transmitter in synaptosomes;[18,60-62] therefore, the function of the nitrendipine binding sites in synaptosomes is uncertain.)

D. EFFECTS OF HEAVY METALS ON CURRENT FLOW THROUGH CALCIUM CHANNELS

This section will be divided into two parts: effects on invertebrate neurons (mostly in gastropods) and effects on vertebrate neurons (mostly cultured dorsal root ganglion and sympathetic neurons).

1. Invertebrate Neurons

Heavy metals, particularly cobalt and cadmium, have long been known to block calcium channels in invertebrate neurons.[1-3] Voltage clamp measurements in gastropod neurons show that heavy metals strongly inhibit calcium current flow through voltage-dependent calcium channels (Figure 2). Depending on the preparation and the experimental technique, heavy metals also suppress outward currents in gastropod neurons (Figure 2), somewhat complicating the voltage clamp picture for large depolarizations. Nevertheless, it is clear that current flow can be essentially completely abolished by sufficient concentrations of heavy metals.

Invertebrate neurobiologists have studied the effects of heavy metals on calcium channels almost entirely for the purpose of identifying membrane currents as flowing through calcium channels and as tools for understanding channel structure and function. Therefore, there has been very little screening of heavy metals of toxicological interest, particularly for concentration dependence. Among the more commonly studied metals, there is a wide disparity in the sequence of effectiveness and in the concentrations needed to block calcium channels, depending on the animal (and neuron type?) studied. In barnacle muscle fibers, metal cations slow the rate of rise of the calcium-dependent action potential, with the following IC_{50}s: La^{3+} (approximately 0.5 mM), Co^{2+} (5 to 10 mM), Mn^{2+} (10 to 20 mM), Ni^{2+} (10 to 20 mM), Mg^{2+} (>100 mM) (interpolation from Figure 2 in Hagiwara and Byerly[1]). Kostyuk and Krishtal[66] report an IC_{50} of approximately 50 μM for Cd^{2+} block of inward calcium currents in internally perfused *Helix pomatia* neurons. Akaike et al.[21] report the IC_{50}s for inhibition of calcium currents in internally perfused neurons of *Helix aspersa* as Co^{2+}, 4.2 mM; Cd^{2+}, 2.2 mM; La^{3+}, 0.2 mM; and Ni^{2+}, 0.03 mM. Byerly et al.[34] found that Cd^{2+} blocks calcium currents in internally perfused neurons of *Lymnaea stagnalis* at concentrations about 200 times lower than those required for

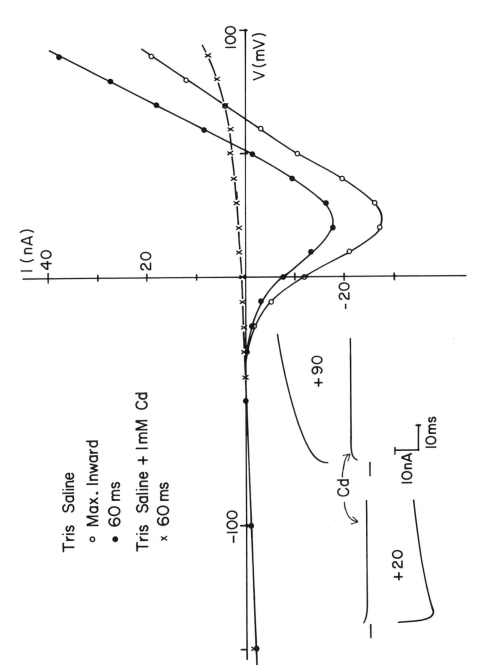

FIGURE 2. Effects of Cd^{2+} on calcium currents in internally perfused, voltage-clamped, neurons of *Lymnaea stagnalis*. I-V curve shows calcium currents before (open and closed circles) and after (crosses) addition of 1 mM Cd^{2+} to the external bathing medium. Note reduction in both inward and outward currents. Inset shows currents elicited by voltage steps to +20 and +90 mV from a holding potential of −50 mV, before and after addition of Cd^{2+}. (Reprinted from Byerly, L. and Hagiwara, S., *J. Physiol.*, 322, 503, 1982. With permission.)

equivalent blocking by Ni^{2+} and Co^{2+} (approximate IC_{50}s, interpolated from their Figure 9: Cd^{2+}, 5 μM ; Ni^{2+} and Co^{2+}, 1 mM). They found that Cd^{2+} is also much more potent in *Helix aspersa*, although the effective concentrations were about fourfold greater (approximate IC_{50}s: Cd^{2+}, 20 μM ; Ni^{2+} and Co^{2+}, 3 mM). Although some of the disparities among these results can be explained by differences in experimental conditions, particularly the concentration of calcium in the external medium, there nevertheless appear to be substantial differences in both the sequence of blocking potency and the effective concentrations.

Gastropod nervous systems contain uniquely identifiable neurons that differ in behavioral functions and electrophysiological properties. Although this has been of great importance in gastropod neuroethology, little attention has been paid in recent years to potential variability of calcium channels in different identifiable gastropod neurons. This appears to be due to the desire to take advantage of the internal perfusion technique for voltage clamping. This technique requires dissociated neurons with "clean", glia-free membranes to which the suction pipettes are attached. Since neuronal dissociation is a low-yield process, entire ganglia are usually dissociated and whatever large neurons remain viable are used for voltage clamping. Neuron identity is usually lost during dissociation. Potentially, at least, the loss of neuron identity may cause true differences among neuron types to be interpreted as statistical variability of unknown cause. For example, Byerly and Hagiwara,[22] studying dissociated neurons of *Lymnaea stagnalis*, found that the amount of outward current inhibited by Cd^{2+} and Co^{2+} varies from cell to cell. I have found the effects of heavy metals on inward currents through calcium channels differ markedly between two identifiable cell types in *Lymnaea*.

Using the single-electrode voltage clamp technique, I investigated the effects of *in vitro* Pb^{2+} and Co^{2+} exposure on B neurons and the RPeD1 neuron in *Lymnaea stagnalis* (nomenclature from Winlow and Benjamin[67]). With Ba^{2+} as the charge carrier for voltage-dependent calcium channels, micromolar concentrations of Pb^{2+} *in vitro* irreversibly inhibit current flow through calcium channels of B neurons, with roughly equal inhibition at concentrations ranging from 0.25 to 14 μM (Figure 3a). Outward currents are not affected. In the RPeD1 neuron, similar degrees of inhibition occur only with Pb^{2+} concentrations at least tenfold greater (Figure 3b). Further, the magnitude of inhibition is concentration dependent and the inhibition is readily reversible. Outward currents are not affected. Co^{2+} at 5 to 10 mM reduces inward and outward currents in both neuronal types. These data suggest that gastropod neurons, like vertebrate neurons, must have at least two types of calcium channels (and probably multiple types of channels, as yet unidentified, that carry the outward currents seen in these voltage clamps).

2. Vertebrate Neurons

As with electrophysiological studies of invertebrate neurons, few voltage- or patch-clamp studies of vertebrate calcium channels have been carried out for the purpose of investigating the effects of heavy metals. However, attempts to elucidate the structure of the calcium channels and to distinguish one channel type from another have resulted in a great deal of information about the interactions between certain metal ions, especially Cd^{2+}, with calcium channels.

In patch clamp studies of L-type channels in guinea pig ventricular cells, Lansman et al.[6] found that the potency of channel blocking by metal cations is $La^{3+} > Cd^{2+} > Co^{2+} > Mg^{2+}$ (IC_{50}s not determined, but Cd^{2+} shows at least some blocking activity at concentrations as low as 2 μM; Figure 4). Narahashi et al.,[69] determined blocking potencies for a series of metal ions in both a transient (perhaps T- or N-like) and a sustained (L-like) calcium current in mouse neuroblastoma cells. For the transient current, the sequence of effectiveness (with apparent K_d) was La^{3+} (1.5 μM) > Ni^{2+} (47 μM) > Cd^{2+} (160 μM) = Co^{2+} (160 μM). For the sustained current, the sequence was La^{3+} (0.9 μM) > Cd^{2+} (7.0 μM) > Ni^{2+} (280 μM) > Co^{2+} (560 μM).

Several studies have found that the different types of vertebrate channels have different sensitivities to block by divalent cations, especially Cd^{2+} and Ni^{2+}. In general, Cd^{2+} is much more

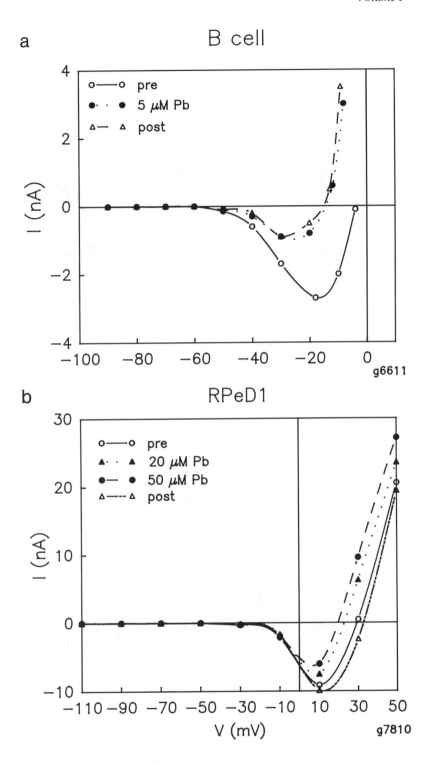

FIGURE 3. Effects of Pb²⁺ on current flow through calcium channels of B (a) and RPeD1 (b) neurons of *Lymnaea stagnalis*, under single-electrode voltage clamp. Ba²⁺ is the charge carrier. I-V curves show barium currents before and after the addition of Pb²⁺ to the external bathing medium. (a) Effects of 5 μM Pb²⁺ on B neuron. Note that inward barium currents are inhibited, while outward currents are almost unaffected. Compare with Figure 2. (b) Effects of 20 and 50 μM Pb²⁺ on RPeD1 neuron. Note dependence on Pb²⁺ concentration and recovery of barium currents after Pb²⁺ washout. (From Audesirk, unpublished results.)

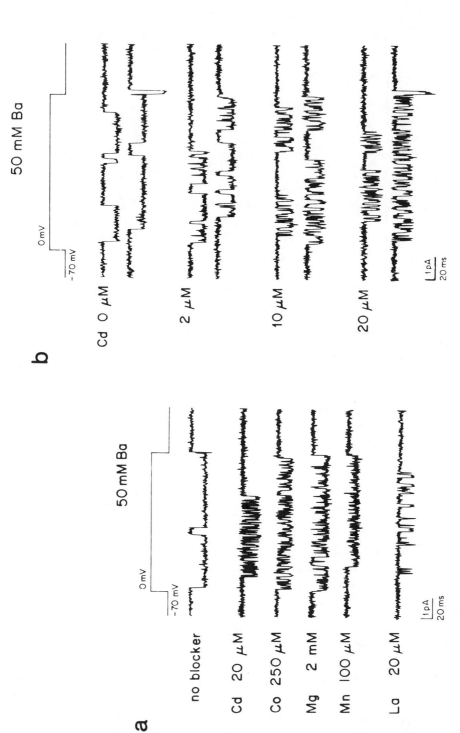

FIGURE 4. Effects of metal cations on L-type calcium channels in guinea pig ventricular cells, using the cell-attached patch clamp procedure. BAY K8644 added to the bathing medium to prolong channel openings. (a) Effects of several metal cations on channel opening. Note the different concentrations used. (b) Effects of 0 to 20 μM Cd^{2+} on channel opening. Note that even 2 μM Cd^{2+} has a significant inhibitory effect on the duration of channel opening. (Reproduced from Lansman et al., *J. Gen. Physiol.*, 88, 321, 1986. With permission.)

effective in blocking the sustained calcium channels (L-type or at least L-like) than the transient channels (T-type or T-like). In chick dorsal root ganglion neurons, 50 μM Cd^{2+} almost completely eliminates calcium currents in N-type and L-type channels, while about 55% of the T-type current remains.[11,15] Nilius et al.[70] found that guinea pig ventricular cells have both L- and T-type channels; L-type channels show significant block by 10 μM Cd^{2+}, while T-types are unaffected. Similarly, Cd^{2+} is much more potent in blocking the sustained calcium current in mouse neuroblastoma cells (see listing above from Narahashi et al.[69]). Bossu et al.[8] found that Cd^{2+} is much more effective in blocking the sustained calcium current in rat sensory neurons than in blocking the transient current (although the L, N, T classification scheme was not used by these authors, the time course and activation voltages suggest that these are L- and T-type channels, respectively). Extracellular recordings from mouse motor nerve terminal bundles under conditions that should isolate calcium currents also found a marked difference in Cd^{2+} sensitivity between a slow calcium current and a fast one.[17] The slow current was almost completely blocked by 10 μM Cd^{2+}, while the fast current required 15 mM Cd^{2+} for a similar degree of block. At least some blocking of the slow current appeared to occur with $10^{-10} M$ Cd^{2+}. In this experimental setup, it is not possible to identify channel types unambiguously; however, neither fast nor slow currents were sensitive to dihydropyridines, suggesting that neither is a "classical" L-type channel.

Ni^{2+} also discriminates among the vertebrate calcium channel types. Ni^{2+} at 100 μM virtually eliminates T-type current in chick dorsal root ganglion cells while leaving the L and N currents almost unaffected.[11] Similar sensitivity of sustained vs. transient currents were found in rabbit sinoatrial node cells[71] and mouse neuroblastoma cells.[69]

IV. EFFECTS OF CHRONIC *IN VIVO* HEAVY METAL EXPOSURE ON CALCIUM CHANNELS

Although there have been many studies of the effects of *in vivo* exposure to heavy metals on behavior and neurotransmitter metabolism, there have been few that examined calcium channel functioning. Carroll et al.[72] found that exposing mice to Pb^{2+} *in vivo* suppresses the release of acetylcholine from minces of cerebral cortex depolarized by high potassium *in vitro*. They attributed at least some of the decreased release to competition of Pb^{2+} with Ca^{2+}, presumably in calcium channel functioning, but could not rule out interference by Pb^{2+} with other aspects of the release process. Chronic Pb^{2+} exposure also inhibits the high potassium-stimulated release of GABA by rat brain synaptosomes.[73,74] Interestingly, *in vitro* exposure of synaptosomes to Pb^{2+}, even in concentrations as high as 100 μM, does not inhibit GABA release,[74] which makes it appear unlikely that chronic *in vivo* exposure suppresses release by effects on calcium channels.

Synaptosomes prepared from rats exposed to Pb^{2+} *in vivo* show enhanced binding of [³H]nitrendipine,[75] as do synaptosomes exposed to Pb^{2+} *in vitro*.[65] This effect can be reversed by washing the synaptosomes with an EDTA-EGTA mixture, which presumably removes Pb^{2+} otherwise permanently bound to the outside of nitrendipine-sensitive (L-type?) calcium channels. These data suggest that *in vivo* Pb^{2+} exposure may alter the structure of some calcium channels in an irreversible, or at least only slowly reversible, way by directly binding to the channels.

Audesirk[68] studied current flow through calcium channels in B neurons of *Lymnaea stagnalis* chronically exposed to Pb^{2+}. When B neurons were voltage clamped in a lead-free saline, neurons from lead-exposed animals showed currents about twice as large as neurons from control animals. There are several interpretations of these results; one interesting possibility is that chronic Pb^{2+} exposure may lead to synthesis of new calcium channels in quantities that overcompensate for the blockade by Pb^{2+}.

V. TOXICOLOGICAL IMPLICATIONS OF HEAVY METAL EFFECTS ON CALCIUM CHANNELS

What do the experimental data on the relationship between calcium channels and heavy metal exposure in animal systems, both *in vivo* and *in vitro*, tell us about the interactions between heavy metals and calcium channels in humans? With the exception of a few studies, nearly all the investigations of heavy metal effects on calcium channel functioning have involved acute *in vitro* exposure, usually for only a few minutes. Therefore, it is difficult to draw firm conclusions regarding chronic *in vivo* calcium channel functioning in humans exposed to small, asymptomatic levels of heavy metals. However, a few tentative observations might be offered. First, there are many similarities in calcium channel functioning among animals as distantly related as barnacles, gastropods, and mice. Therefore, there is every reason to expect that calcium channels in humans are very similar in structure and function to the channels of non-human animals. Data that show common mechanisms of action of heavy metals in animals almost certainly apply to humans as well, although the concentrations that induce effects may vary among species. Second, although there are essentially no data on the effects of combinations of heavy metals, it seems fairly certain that simultaneous exposure to several heavy metals will have at least additive effects in blocking calcium channels. Since humans in industrial societies are exposed to lead, cadmium, copper and other metals, their combined effects on calcium channels may be significant even if the concentration of each metal is very low. Third, some of the data indicate blocking effects at extremely low concentrations of heavy metals. For example, Nachshen[55] found that Pb^{2+} and La^{3+} inhibit calcium uptake in synaptosomes with IC_{50}s below 1 μM. In B cells of *Lymnaea stagnalis*,[68] Pb^{2+} causes approximately 50% inhibition of calcium channel currents at concentrations below 1 μM. Mouse nerve terminal bundles show about 25% inhibition of calcium currents when exposed to Cd^{2+} concentrations as low as 10^{-10} M.[17] These data suggest that heavy metals might alter calcium channel functioning at levels found in the fluids of asymptomatic humans. Fourth, the *in vivo* data suggest that certain heavy metals, notably Pb^{2+}, may have chronic effects on calcium channels that outlast the period of exposure, at least long enough for isolation of nervous tissue and experimental measurements to be performed. Finally, some heavy metals, including Mn^{2+} [6] and Pb^{2+},[41,53] appear to permeate through calcium channels into the neuronal cytoplasm. Once inside the cell, heavy metals may activate or inhibit many calcium-dependent and calcium-independent processes, including Ca-ATPase, Na-K-ATPase, calmodulin, and numerous cellular enzymes. These effects may be more prolonged and even more damaging than interference with calcium channel functioning.

One heavy metal cation seems to stand out as being quite different from the others: Pb^{2+}. All studies agree that Pb^{2+} is one of the most potent inorganic blockers of Ca^{2+} or Ba^{2+} flow through calcium channels.[49,50,68] However, at least some calcium channels also appear to be highly permeable to Pb^{2+},[53] whereas most blockers are either very slightly permeable (Mn^{2+}) or essentially impermeable. Further, Pb^{2+} stimulates nitrendipine binding to synaptosomes, whereas other potent calcium channel blockers (La^{3+}, Co^{2+}), suppress nitrendipine binding.[64,65] Finally, the effects of *in vitro* Pb^{2+} exposure have been found to be either reversible[41,43,44,55] or irreversible (Figure 3), depending on the system or even the neuronal type studied. Further studies of Pb^{2+} effects on calcium channels would probably prove to be very rewarding, and might provide insights both into Pb^{2+} toxicity and into the structures of calcium channels.

From a toxicological perspective, it is unfortunate that so few studies have examined the effects of *in vivo* heavy metal exposure on calcium channels, since *in vivo* exposure most closely approximates the human condition. Clearly, studies of calcium channel functioning in neurons of animals chronically exposed to heavy metals, or perhaps in chronically exposed cultured neurons, will be important to our understanding of the impact of heavy metal exposure on humans.

REFERENCES

1. **Hagiwara, S. and Byerly, L.,** Membrane biophysics of calcium currents, *Fed. Proc.,* 40, 2220, 1981.
2. **Hagiwara, S. and Byerly, L.,** Calcium channels, *Ann. Rev. Neurosci.,* 4, 69, 1981.
3. **Edwards, C.,** The selectivity of ion channels in nerve and muscle, *Neuroscience,* 7, 1335, 1982.
4. **Stanfield, P. R.,** Voltage-dependent calcium channels of excitable membranes, *Br. Med. Bull.,* 42, 359, 1986.
5. **Hess, P., Lansman, J. B., and Tsien, R. W.,** Calcium channel selectivity for divalent and monovalent cations. Voltage and concentration dependence of single channel current in ventricular heart cells, *J. Gen. Physiol.,* 88, 293, 1986.
6. **Lansman, J. B., Hess, P., and Tsien, R. W.,** Blockade of current through single calcium channels by Cd^{2+}, Mg^{2+}, and Ca^{2+}. Voltage and concentration dependence of calcium entry into the pore, *J. Gen. Physiol.,* 88, 321, 1986.
7. **Tsien, R. W., Hess, P., McCleskey, E. W., and Rosenberg, R. L.,** Calcium channels: mechanisms of selectivity, permeation, and block, *Ann. Rev. Biophys. Biophys. Chem.,* 16, 265, 1987.
8. **Bossu, J. L., Feltz, A., and Thomann, J. M.,** Depolarization elicits two distinct calcium currents in vertebrate sensory neurones, *Pfluegers Arch.,* 403, 360, 1985.
9. **Carbone, E. and Lux, H. D.,** A low voltage-activated, fully inactivating Ca channel in vertebrate sensory neurones, *Nature,* 310, 501, 1984.
10. **Cruz, L. J., Johnson, D. S., and Olivera, B. M.,** Characterization of the ω-conotoxin target. Evidence for tissue-specific heterogeneity in calcium channel types, *Biochemistry,* 26, 820, 1987.
11. **Fox, A. P., Nowycky, M. C., and Tsien, R. W.,** Kinetic and pharmacological properties distinguishing three types of calcium currents in chick sensory neurones, *J. Physiol.,* 394, 149, 1987.
12. **Fox, A. P., Nowycky, M. C., and Tsien, R. W.,** Single-channel recordings of three types of calcium channels in chick sensory neurones, *J. Physiol.,* 394, 173, 1987.
13. **Hirning, L. D., Fox, A. P., McCleskey, E. W., Olivera, B. M., Thayer, S. A., Miller, R. W., and Tsien, R. W.,** Dominant role of N-type Ca^{2+} channels in evoked release of norepinephrine from sympathetic neurons, *Science,* 239, 57, 1988.
14. **McCleskey, E. W., Fox, A. P., Feldman, D. H., Cruz, L. J., Olivera, B. M., Tsien, R. W., and Yoshikami, D.,** ω-conotoxin: direct and persistent blockade of specific types of calcium channels in neurons but not muscle, *Proc. Natl. Acad. Sci. U.S.A.,* 84, 4327, 1987.
15. **Nowycky, M. C., Fox, A. P., and Tsien, R. W.,** Three types of neuronal calcium channel with different calcium agonist sensitivity, *Nature,* 316, 440, 1985.
16. **Ogura, A. and Takahashi, M.,** Differential effect of a dihydropyridine derivative to Ca^{2+} entry pathways in neuronal preparations, *Brain. Res.,* 301, 323, 1984.
17. **Penner, R. and Dreyer, F.,** Two different presynaptic calcium currents in mouse motor nerve terminals, *Pfluegers Arch.,* 406, 190, 1986.
18. **Rivier, J., Galyean, R., Gray, W. R., Azimi-Zonooz, A., McIntosh, J. M., Cruz, L. J., and Olivera, B. M.,** Neuronal calcium channel inhibitors. Synthesis of ω-conotoxin GVIA and effects on ^{45}Ca uptake by synaptosomes, *J. Biol. Chem.,* 262, 1194, 1987.
19. **McCleskey, E. W., Fox, A. P., Feldman, D., and Tsien, R. W.,** Different types of calcium channels, *J. Exp. Biol.,* 124, 177, 1986.
20. **Miller, R. J.,** Multiple calcium channels and neuronal function, *Science,* 235, 46, 1987.
21. **Akaike, N., Lee, K. S., and Brown, A. M.,** The calcium current of *Helix* neuron, J. Gen. Physiol., 71, 509, 1978.
22. **Byerly, L. and Hagiwara, S.,** Calcium currents in internally perfused nerve cell bodies of *Limnea stagnalis,* *J. Physiol.,* 322, 503, 1982.
23. **Eckert, R. and Lux, H. D.,** A voltage-sensitive persistent calcium conductance in neuronal somata of *Helix,* *J. Physiol.,* 254, 129, 1976.
24. **Kostyuk, P. G., Krishtal, O. A., and Doroshenko, P. A.,** Calcium currents in snail neurones. I. Identification of calcium current, *Pfluegers Arch.,* 335, 83, 1974.
25. **Kostyuk, P. G., Krishtal, O. A., and Doroshenko, P. A.,** Calcium currents in snail neurones. II. The effect of external calcium concentration on the calcium inward current, *Pfluegers Arch.,* 348, 95, 1974.
26. **Standen, N. B.,** Voltage-clamp studies of the calcium inward current in an identified snail neurone: comparison with the sodium inward current, *J. Physiol.,* 249, 253, 1975.
27. **Augustine, G. J., Charlton, M. P., and Smith, S. J.,** Calcium entry into voltage-clamped presynaptic terminals of squid, *J. Physiol.,* 367, 143, 1985.
28. **Charlton, M. P., Smith, S. J., and Zucker, R. S.,** Role of presynaptic calcium ions and channels in synaptic facilitation and depression at squid giant synapse, *J. Physiol.,* 323, 173, 1982.
29. **Llinas, R., Steinberg, I. Z., and Walton, K.,** Presynaptic calcium currents in squid giant synapse, *Biophys. J.,* 33, 289, 1981.
30. **Chesnoy-Marchais, D.,** Kinetic properties and selectivity of calcium-permeable single channels in *Aplysia* neurones, *J. Physiol.,* 367, 457, 1985.

31. **Chad, J. E. and Eckert, R.,** An enzymatic mechanism for calcium current inactivation in dialyzed *Helix* neurones, *J. Physiol.,* 378, 31, 1986.

32. **Nerbonne, J. M. and Gurney, A. M.,** Blockade of Ca^{2+} and K^+ currents in bag cell neurons of *Aplysia californica* by dihydropyridine Ca^{2+} antagonists, *J. Neurosci.,* 7, 882, 1987.

33. **Nishi, K., Akaike, N., Oyama, Y., and Ito, H.,** Actions of calcium antagonists on calcium currents in *Helix* neurons: specificity and potency, *Circ. Res. Suppl. 1,* 52, 53, 1983.

34. **Byerly, L., Chase, P. B., and Stimers, J. R.,** Permeation and interaction of divalent cations in calcium channels of snail neurons, *J. Gen. Physiol.,* 85, 491, 1985.

35. **Gola, M. and Ducreux, C.,** D600 as a direct blocker of Ca-dependent K currents in Helix neurons, *Eur. J. Pharmacol.,* 117, 311, 1985.

36. **Reuter, H.,** The dependence of slow inward current in purkinje fibres on the extracellular calcium concentration, *J. Physiol.,* 192, 479, 1967.

37. **Fenwick, E. M., Marty, A., and Neher, E.,** Sodium and calcium channels in bovine chromaffin cells, *J. Physiol.,* 331, 599, 1982.

38. **Llinas, R. and Yarom, Y.,** Electrophysiology of mammalian inferior olivary neurones *in vitro.* Different types of voltage dependent ionic conductances, *J. Physiol.,* 315, 549, 1981.

39. **Llinas, R. and Yarom, Y.,** Properties and distribution of ionic conductances generating electroresponsiveness of mammalian inferior olivary neurones *in vitro, J. Physiol.,* 315, 569, 1981.

40. **Atchison, W. D. and Narahashi, T.,** Mechanism of action of lead on neuromuscular junctions, *Neuro Toxicol.,* 5(3), 267, 1984.

41. **Cooper, G. P. and Manalis, R. S.,** Interactions of lead and cadmium on acetylcholine release at the frog neuromuscular junction, *Toxicol. Appl. Pharmacol.,* 74, 411, 1984.

42. **Kolton, L. and Yaari, Y.,** Sites of action of lead on spontaneous transmitter release from motor nerve terminals, *Israel J. Med. Sci.,* 18, 165, 1982.

43. **Manalis, R. S. and Cooper, G. P.,** Presynaptic and postsynaptic effects of lead at the frog neuromuscular junction, *Nature,* 243, 354, 1973.

44. **Manalis, R. S., Cooper, G. P., and Pomeroy, S. L.,** Effects of lead on neuromuscular transmission in the frog, *Brain Res.,* 294, 95, 1984.

45. **Byerly, L. and Yazejian, B.,** Intracellular factors for the maintenance of calcium currents in perfused neurones from the snail, *Lymnaea stagnalis, J. Physiol.,* 370, 631, 1986.

46. **Cooper, G. P., Suszkiw, J. B., and Manalis, R. S.,** Presynaptic effects of heavy metals, in *Cellular and Molecular Neurotoxicity,* Narahashi, T., Ed., Raven Press, New York, 1984, 1.

47. **Cooper, G. P. and Manalis, R. S.,** Influence of heavy metals on synaptic transmission: a review, *Neuro Toxicol.,* 4(4), 69, 1983.

48. **Spence, I., Drew, C., Johnson, G. A. R., and Lodge, D.,** Acute effects of lead at central synapses *in vitro, Brain Res.,* 333, 103, 1985.

49. **Pocock, G. and Simons, T. J. B.,** Effects of lead ions on events associated with exocytosis in isolated bovine adrenal medullary cells, *J. Neurochem.,* 48, 376, 1987.

50. **Suszkiw, J., Toth, G., Murawsky, M., and Cooper, G. P.,** Effects of Pb^{2+} and Cd^{2+} on acetylcholine release and Ca^{2+} movements in synaptosomes and subcellular fractions from rat brain and *Torpedo* electric organ, *Brain Res.,* 323, 31, 1984.

51. **Minnema, D. J. and Michaelson, I. A.,** Differential effects of inorganic lead and δ-aminolevulinic acid *in vitro* on synaptosomal γ-aminobutyric acid release, *Toxicol. Appl. Pharmacol.,* 86, 437, 1986.

52. **Minnema, D. J., Greenland, R. D., and Michaelson, I. A.,** Effect of *in vitro* inorganic lead on dopamine release from superfused rat striatal synaptosomes, *Toxicol. Appl. Pharmacol.,* 84, 400, 1986.

53. **Simons, T. J. B. and Pocock, G.,** Lead enters bovine adrenal medullary cells through calcium channels, *J. Neurochem.,* 48, 383, 1987.

54. **Koenig, M. L. and Jope, R. S.,** Aluminum inhibits the fast phase of voltage-dependent calcium influx into synaptosomes, *J. Neurochem.,* 49, 316, 1987.

55. **Nachshen, D. A.,** Selectivity of the Ca binding site in synaptosome Ca channels. Inhibition of Ca influx by multivalent metal cations, *J. Gen. Physiol.,* 83, 941, 1984.

56. **Atchison, W. D., Joshi, U., and Thornburg, J. E.,** Irreversible suppression of calcium entry into nerve terminals by methylmercury, *J. Pharmacol. Exp. Ther.,* 238, 618, 1986.

57. **Gandia, L., Lopez, M. G., Fonteriz, R. I., Artalejo, C. R., and Garcia, A. G.,** Relative sensitivities of chromaffin cell calcium channels to organic and inorganic calcium antagonists, *Neurosci. Lett.,* 77, 333, 1987.

58. **Ladona, M. G., Aunis, D., Gandia, L., and Garcia, A. G.,** Dihydropyridine modulation of the chromaffin cell secretory response, *J. Neurochem.,* 48, 483, 1987.

59. **Hoshi, T. and Smith, S. J.,** Large depolarization induces long openings of voltage-dependent calcium channels in adrenal chromaffin cells, *J. Neurosci.,* 7, 571, 1987.

60. **Daniell, L. C., Barr, E. M., and Leslie, S. W.,** $^{45}Ca^{2+}$ uptake into rat whole brain synaptosomes unaltered by dihydropyridine calcium antagonists, *J. Neurochem.,* 41, 1455, 1983.

61. **Carvalho, C. A. M., Coutinho, O. P., and Carvalho, A. P.,** Effects of Ca^{2+} channel blockers on Ca^{2+} translocation across synaptosomal membranes, *J. Neurochem.,* 47, 1774, 1986.
62. **Suszkiw, J. B., O'Leary, M. E., Murawsky, M. M., and Wang, T.,** Presynaptic calcium channels in rat cortical synaptosomes: fast-kinetics of phasic calcium influx, channel inactivation, and relationship to nitrendipine receptors, *J. Neurosci.,* 6, 1349, 1986.
63. **Fairhurst, A. S., Thayer, S. A., Colker, J. E., and Beatty, D. A.,** A calcium antagonist drug binding site in skeletal muscle sarcoplasmic reticulum: evidence for a calcium channel, *Life Sci.,* 32, 1331, 1983.
64. **Gould, R. J., Murphy, K. M. M., and Snyder, S. H.,** [^{3}H]Nitrendipine-labeled calcium channels discriminate inorganic calcium agonists and antagonists, *Proc. Natl. Acad. Sci. U.S.A.,* 79, 3656, 1982.
65. **Rius, R. A., Govoni, S., Battaini, F., and Trabucchi, M.,** Differential sensitivity of [^{3}H]nitrendipine binding to cations of toxicological interest in various rat brain areas, *Toxicol. Lett.,* 27, 103, 1985.
66. **Kostyuk, P. G. and Krishtal, O. A.,** Separation of sodium and calcium currents in the somatic membrane of mollusc neurones, *J. Physiol.,* 270, 545, 1977.
67. **Winlow, W. and Benjamin, P. R.,** Neuronal mapping in the brain of the pond-snail, *Lymnaea stagnalis* (L.), in *Neurobiology of Invertebrates, Gastropoda Brain,* Salanki, J., Ed., Hungarian Academy of Sciences, Budapest, 1976, 41.
68. **Audesirk, G.,** Effects of *in vitro* and *in vivo* lead exposure on voltage-dependent calcium channels in central neurons of *Lymnaea stagnalis, Neuro Toxicol.,* 8, 579, 1987.
69. **Narahashi, T., Tsunoo, A., and Yoshii, M.,** Characterization of two types of calcium channels in mouse neuroblastoma cells, *J. Physiol.,* 383, 231, 1987.
70. **Nilius, B., Hess, P., Lansman, J. B., and Tsien, R. W.,** A novel type of cardiac calcium channel in ventricular cells, *Nature,* 316, 443, 1985.
71. **Hagiwara, S., Irisawa, H., and Kameyama, M.,** Transient-type calcium current contributes to the pacemaker potential in isolated rabbit sino-atrial node cells, *J. Physiol.,* 382, 104P, 1986.
72. **Carroll, P. T., Silbergeld, E. K., and Goldberg, A. M.,** Alteration of central cholinergic function by chronic lead acetate exposure, *Biochem. Pharmacol.,* 26, 397, 1977.
73. **Silbergeld, E. K., Miller, L. P., Kennedy, S., and Eng, N.,** Lead, GABA, and seizures: effects of subencephalopathic lead exposure on seizure sensitivity and GABAergic function, *Environ. Res.,* 19, 371, 1979.
74. **Silbergeld, E. K., Hruska, R. E., Miller, L. P., and Eng, N.,** Effects of lead *in vivo* and *in vitro* on GABAergic neurochemistry, *J. Neurochem.,* 34, 1712, 1980.
75. **Rius, R. A., Lucchi, L., Govoni, S., and Trabucchi, M.,** *In vivo* chronic lead exposure alters [^{3}H]nitrendipine binding in rat striatum, *Brain Res.,* 322, 180, 1984.

Chapter 2

ASSESSMENT OF THE EFFECTS OF LEAD AND MERCURY *IN VITRO* ON NEUROTRANSMITTER RELEASE

Daniel J. Minnema and Gary P. Cooper

TABLE OF CONTENTS

I. INTRODUCTION

Many heavy metals, in both their inorganic and organic forms, are recognized neurotoxicants. While no conclusive evidence exists that attributes the neurotoxic effects resulting from chronic *in vivo* heavy metal exposure to alterations in presynaptic transmitter release, acute *in vitro* and *in situ* exposure studies have demonstrated that at relevant concentrations many heavy metals do indeed affect the transmitter release process. The relatively simple preparations and procedures used to assess transmitter release *in vitro* have contributed to a better understanding of the effects of certain heavy metals on synaptic transmission, and in some cases, the mechansims by which such effects are mediated. *In vitro* release studies allow for strict control of experimental conditions (composition of physiological medium, concentration of chemicals and/or toxicants, pH, temperature), as well as the elimination of many confounding factors (blood-brain barrier, hepatic metabolism). Therefore, it has been possible to compare the effects of heavy metals on the release of transmitters from a variety of neuronal tissue preparations using and integrating results obtained form both electrophysiological and neurochemical studies. Despite the marked variety of neuronal characteristics in terms of structure, size, anatomical location, species of origin, biochemistry, physiology, organization, regenerating capacity, supporting cells, integration, and function, the mechanisms mediating transmitter release appear to be quite similar at all presynaptic nerve terminals.

Unlike selective pharmacological agents, the similarities in the *in vitro* actions of heavy metals thus far observed in a variety of different types of synapses suggest that many of the effects of metals on transmitter release are exerted through those mechanisms common to all transmitter systems. However, the potential for heavy metals to act on processes uniquely associated with a specific neurotransmitter system cannot be ignored, as indicated by the selective action of manganese on dopaminergic neurons.[1] *In vitro* studies by design ignore those factors associated with *in vivo* toxicant exposure (e.g., CNS integrative complexity, toxicokinetics, hepatic metabolisms, blood-brain barrier), thereby making extrapolation to the chronic "whole animal" exposure scenario difficult. However, it is hoped that by understanding the acute *in vitro* effects of these heavy metals on neurotransmission, testable hypotheses regarding the mechanisms of chronic neurotoxicity can be developed. Although the toxic consequences of chronic heavy metal exposure may not include alterations in release, the biochemical processes/mechanisms producing acute alterations in transmitter release may be identical or similar to those underlying the toxic effects observed under conditions of chronic exposure. Ultimately, the examination of the *in vitro* effects of heavy metals on transmitter release, along with pertinent *in vivo* pharmacokinetic data, may serve as a means of predicting the neurotoxic potential of heavy metals. The utility of *in vitro* techniques in neurotoxicology has been the subject of several reviews.[2-8]

The purpose of this review is to: (1) describe potential presynaptic sites of action of heavy metals, emphasizing those processes that are directly involved in the transmitter release process, (2) discuss some of the methodological considerations that impact on the interpretation of transmitter release data, since much of the inconsistency in the reported effects of heavy metals on the release process apparently reflects inattention to the limitations of the assays employed, and (3) review and discuss the effects of two neurotoxic heavy metals, lead (Pb) and mercury (Hg), on transmitter release. To obtain a broader perspective on the subject, the reader is encouraged to consult other recent reviews by Audesirk,[9] Cooper and Manalis,[10] Cooper et al.,[11] Manalis and Suszkiw,[12] and Shellenberger.[13]

The release of neurotransmitters from nerve terminals is a dynamic, continuously occurring phenomenon. For the purpose of this review, transmitter release can be separated into two categories: (1) spontaneous (asynchronous) release and (2) depolarization-evoked (synchronous) release. Both of these release processes occur under normal physiological conditions, and are not necessarily totally exclusive of one another. However, it is important to distinguish between depolarization-evoked vs. spontaneous release in that certain heavy metals can differentially affect these two processes.

A. DEPOLARIZATION-EVOKED TRANSMITTER RELEASE

In most electrophysiological studies the release of transmitter evoked by depolarization is described by the amplitude or magnitude of endplate potentials (EPPs), excitatory synaptic potentials (EPSPs), or inhibitory synaptic potentials (IPSPs). In neurochemical studies depolarization-evoked release is usually expressed as the difference in the amount of transmitter released over a fixed time period in the presence vs. the absence of a membrane-depolarizing stimulus. Depolarization-evoked transmitter release occurs, under normal physiological conditions, as a result of the arrival of an action potential at a nerve terminal. Depolarization of the presynaptic terminal membrane, normally associated with sodium (Na^+) influx, causes voltage-dependent calcium (Ca^{2+}) channels in the terminal membrane to open. It is the depolarization of the membrane, not the movement of Na^+, that triggers the opening of voltage-sensitive Ca^{2+} channels. The opening of these presynaptic membrane Ca^{2+} channels permits extracellular Ca^{2+} to move into the cytoplasmic compartment of the nerve terminal. This influx of Ca^{2+} causes a rise in the normally low intraterminal Ca^{2+} concentration, which in turn is linked to a yet-unidentified mechanism that facilitates transmitter release.[14,15] Physiological solutions containing high concentrations of potassium (K^+) which depolarize the membrane without the accompanying Na^+ influx, are commonly used experimentally to evoke transmitter release. Depolarization-evoked transmitter release can be prevented by either removing extraneuronal Ca^{2+} or inhibiting the influx of Ca^{2+} into the nerve terminal by the use of membrane Ca^{2+} channel blockers. This dependency on extracellular Ca^{2+} in mediating depolarization-evoked release is a necessary requirement for all identified transmitter substances.[16,17] The two distinguishing events in this process of depolarization-evoked release are: (1) depolarization of the nerve terminal membrane, and (2) movement of extraneuronal Ca^{2+} into the nerve terminal. The depolarization-evoked increase in the intraterminal Ca^{2+} concentration, along with the accompanying release of transmitter, is transient. The voltage-dependent Ca^{2+} channels normally remain open briefly (with a half-life of hundreds of milliseconds) following depolarization, and the resulting increase in the concentration of intraterminal Ca^{2+} is rapidly lowered by both intraterminal Ca^{2+}-buffering and neuronal membrane Ca^{2+} efflux mechanisms.[18,19] As the intraterminal Ca^{2+} concentration rapidly returns to resting (predepolarization) levels, the rate of transmitter release also returns to resting (spontaneous) levels.

B. SPONTANEOUS TRANSMITTER RELEASE

In most electrophysiological studies, the amount of spontaneous quantal release is described by the frequency of miniature endplate potentials (MEPPs), excitatory synaptic potentials (EPSPs), or inhibitory synaptic potentials (IPSPs). In neurochemical release studies, spontaneous release refers to the baseline rate of transmitter release that occurs in the absence of any apparent stimulus. In such studies, spontaneous release reflects both the normal physiological release process, as well as nonspecific transmitter release resulting from experimentally produced nerve terminal impairment and degradation. Spontaneous release is a random, continual "leakage" of transmitter from nerve terminals. Unlike depolarization-evoked release, spontaneous transmitter release is not dependent on Ca^{2+} influx into the terminal; it continues even in the absence of extraneuronal Ca^{2+}. The physiological mechanisms mediating spontaneous release remain unidentified. Spontaneous release may be the passive nonquantal independent passage of individual transmitter molecules across the nerve terminal membrane or may be "quantal" in nature, wherein many thousands of transmitter molecules, associated with synaptic vesicles, are simultaneously released by exocytosis.[14] The nonpassive movement of transmitter out of the terminal may reflect either random interactions of intraterminal Ca^{2+} (at basal concentration) with the intraterminal Ca^{2+} receptor that mediates release, or transient increases (above a set threshold) of intraterminal Ca^{2+} concentrations. Various techniques and/or treatments that elevate intraterminal Ca^{2+} concentrations have been shown to increase MEPP frequency. The physiological role of spontaneous release is also unclear, but it is possible that this continuous release of transmitter at rates lower than evoked by depolarization may serve as a tonic, sub-threshold, homeostatic mechanism regulating the postsynaptic tissue's ability to respond to transmitter released by depolarization. Spontaneous release may play an important role during nervous system development and maturation, particularly with respect to neuronal integration and regulation of postsynaptic receptor sensitivity.

II. POTENTIAL PRESYNAPTIC SITES OF HEAVY METAL ACTION

The functions of a number of presynaptic nerve terminal structures and processes, if targeted by a heavy metal, could result in altered transmitter release. Some of these sites of action are at the membrane level where the heavy metal could exert its action without necessarily entering into the cytoplasmic compartment. Others are intraterminal, most of which involve Ca^{2+} buffering mechanisms and/or putative Ca^{2+}-mediated processes. The following descriptions of possible sites of action of heavy metals are by no means all inclusive; nor are these sites necessarily exclusive of one another. In reality the interaction of metals with such sites is complex; many of these potential sites may be affected simultaneously, as determined by a variety of factors such as the metal species, metal concentration, and length of exposure. Furthermore, the effect of metal at one site (where it may act as a Ca^{2+} agonist) may not be the same as its effect at another site (where it may act as a Ca^{2+} antagonist). Our present ability to define potential sites of action of neurotoxicants is, in part, limited by our understanding of the release processes in general. In fact, certain heavy metals may be instrumental in providing a further characterization of the transmitter release process.

A. THE ACTION POTENTIAL

Since depolarization of the presynaptic nerve terminal is the critical event in initiating Ca^{2+} influx, toxicants that affect the action potential at the terminal will also influence transmitter release. Under resting conditions the nerve terminal is hyperpolarized, maintaining a potential of approximately -70 to -80 mV across the membrane. In addition, a chemical gradient is also maintained, where the extraneuronal Na^+ concentration is high ($\sim$140 mM) and the extraneuronal K^+ concentration is low ($\sim$5 mM) relative to the intracellular concentration of these ions

$(Na^+:\sim 10$ to 20 m$M)$ $(K^+:\sim 120$ m$M)$. This electrochemical gradient is maintained by a membrane that is relatively impermeable to Na^+ and K^+, and by energy-utilizing membrane Na^+,K^+-ATPase pumps that move Na^+ out of and K^+ into the cell. An action potential (membrane depolarization) is produced by changes in the membrane permeability to Na^+ ions, and subsequently K^+ ions. This increase in Na^+ permeability is controlled by specific membrane ion channels. The rapid opening and closing of these membrane channels permit the influx of Na^+ and the subsequent efflux of K^+, which are responsible for the changes in membrane potential from -75 mV to approximately +25 mV, and its eventual return to -75 mV. The important factors with respect to the effects of toxicants on the action potential are (1) membrane integrity and permeability, (2) membrane ion channels, and (3) membrane Na^+,K^+-ATPase pumps. These three factors will be dicussed only briefly since the action of neurotoxicants on these processes are associated more with the axon than with the nerve terminal. Still, it is important to recognize the potential roles these factors can play in altering transmitter release, as many of the *in vitro* techniques used to assess release will reflect alterations of these processes.

1. Membrane Permeability

The most obvious consequences of alterations in membrane permeability on transmitter release involve perturbations in action potential generation resulting from changes in the electrochemical gradient across the membrane. For convenience, changes in membrane permeability can be considered as either specific or nonspecific. Nonspecific membrane permeability changes refer to alterations of the protein-lipid bilayer, and would be expected to occur in all membranes. Toxicants which affect the organization of membrane lipids, structural proteins, and/or fluidity would alter the movement of many substances across the membrane (including ions), thus influencing transmitter release.[20] Ca^{2+} is known to play a role in stabilizing the membrane.[21] The potential interaction of heavy metals with membrane-bound Ca^{2+} could result in an increase in membrane permeability due to membrane destabilization. Changes in membrane permeability can also result from the interaction of toxicants with specific membrane proteins, particularly membrane-bound enzymes, receptors, and/or ion-channels. Some of these proteins may be specific for particular transmitter systems, such as receptor proteins and transmitter uptake carriers. Other proteins such as membrane ion pumps and channels are common to all transmitter systems.[20] Changes in the Na^+ electrochemical gradient, for example, could affect the Na^+-dependent high-affinity reuptake process associated with specific transmitter systems, which in turn could alter not transmitter flux. Another potential consequence of altered membrane permeability is the possible increase in susceptibility of a neuron to the intraterminal effects of another chemical which is normally excluded from the intracellular compartment.[22]

2. Membrane Ion Channels

The passive movement of Na^+, K^+, Ca^{2+}, and Cl^- across the relatively impermeable nerve terminal membrane is controlled by protein macromolecules, or ion channels, within the fluid lipid layer of the membrane.[23] Four generalized categories of ion channels have been described: (1) "nongated" or "leakage" channels which are always open, (2) "voltage-gated" channels which open or close in response to the potential across the membrane, (3) "receptor-gated" channels which open or close in response to the occupancy of a receptor on the membrane surface, and (4) "messenger-gated" channels that respond to shifts in intracellular ion concentrations or 2nd messengers (e.g., Ca^{2+}, inositol-triphosphate). However, there is significant overlap in this ion channel classification scheme. For example, some "receptor-gated" channels are also voltage sensitive. In addition, considerable variability can exist in the specificity of the various ion channels for the various ions. Neuronal ion channel variants are continually being characterized. The ability of the nerve terminal to depolarize following the arrival of an action potential is normally dependent on the movement of Na^+ and K^+ through specific nerve terminal membrane ion channels. Many neurotoxicants are known to act at the level of these channels. Tetrodotoxin acts at membrane Na^+ channels to antagonize the movement of Na^+ into the nerve. Other toxicants, such as the organochlorine and pyrethroid insecticides, act to open Na^+ channels or to prevent their closing.[24] Blocking the membrane Na^+ channels prevents the depolarization of the membrane, whereas continued membrane permeability to Na^+ prevents membrane repolarization.

The ultimate consequence of the inability of the neuronal membrane to regulate Na^+ movement is a loss of action potentials. While the loss of the ability of a neuron to propagate an action potential will undoubtedly alter transmitter release, such changes are usually associated with the axon rather than nerve terminal. The existence of multiple Na^+ channels offers the possibility that there exists a population of Na^+ channel subtypes specifically associated with the presynaptic nerve terminal, which may be selectively sensitive to a particular neurotoxicological insult. Similarly, there exist specific membrane channels for K^+, and selective agents that will block these channels. The inhibition of neuronal ^{36}Cl uptake at the GABA receptor-dependent Cl^- channel by certain cyclodiene insecticides demonstrates the importance of this ion channel in neurotoxicology. Cyclodienes act as $GABA_A$ receptor antagonists and as such they decrease Cl^- conductance and decrease the potential across the neuronal membrane (depolarization). The consequence of a less-negative membrane potential is hyperexcitability.[25] $GABA_A$ receptors are associated with postsynaptic rather than presynaptic processes. Although the processes that regulate presynaptic terminal Cl^- influx are not totally characterized, it is apparent from the cyclodiene example that effects on Cl^- movement can influence membrane function.

3. Na⁺,K⁺-ATPase

The neuronal Na^+ and K^+ ionic gradients are maintained by an intrinsic membrane-associated, energy-dependent, enzymatic ion pump, Na^+,K^+-adenosine triphosphatase (Na^+,K^+-ATPase). The energy dependency of this enzyme makes it vulnerable to compromised energy metabolism and/or ATP availability. The activity of this pump is dependent on the intraterminal Na^+ concentration; as the internal Na^+ concentration increases, the activity of the pump increases. This enzyme is reported to transport three Na^+ outward and two K^+ inward per terminal phosphate bond of cytoplasmic ATP hydrolyzed, although some evidence indicates that this Na^+/K^+ coupling ratio may range from 1.5 to 3, increasing in response to elevations in either internal Na^+ or K^+ concentrations.[26] The Na^+,K^+-ATPase pump is found in all mammalian cells, although there are species and tissue differences. For example, three distinct isoforms of Na^+,K^+-ATPase have been identified in rat brain.[27] The sensitivity of the pump to inhibition by cardiac glycosides, as reflected by differences in the rate of drug dissociation, will depend on the Na^+,K^+-ATPase isoform.[28] The type of isoform present, combined with the functional sensitivity of the release process to Na^+,K^+-ATPase inhibition, will ultimately determine the selectivity of Na^+,K^+-ATPase inhibitors on transmitter release. Na^+,K^+-ATPase activity is decreased by sulfhydryl blocking agents.[29] The toxic effects of certain heavy metals, because of their interaction with enzyme sulfhydryl groups, may involve Na^+,K^+-ATPase inhibition.

Inhibition of Na^+,K^+-ATPase by ouabain results in an increase in the intraterminal Na^+ concentration and a subsequent increase in both spontaneous and depolarization-evoked transmitter release.[30,31] It is believed that the sustained increase in intraterminal Na^+ increases Ca^{2+} efflux from intraterminal mitochondria via Na^+/Ca^{2+} exchange, which in turn increases the cytosolic Ca^{2+} concentration within the terminal, thus increasing transmitter release. Ouabain has been shown to produce a gradual increase in the intrasynaptosomal Ca^{2+} concentration, the temporal characteristics of which reflect the gradual ouabain-induced increase in spontaneous transmitter release.[31,32] We have also observed this gradual increase in transmitter release following Na^+,K^+-ATPase inhibition by either ouabain or by elimination of extrasynaptosomal K^+ (unpublished observations). If the increase in intraterminal Na^+ resulting from ouabain-induced Na^+,K^+-ATPase inhibition is prevented by either: (1) blocking the movement of Na^+ into the terminal using the Na^+-channel blocker tetrodotoxin,[33] or (2) removing the Na^+ gradient across the terminal membrane by lowering the external Na^+ concentration,[34] then ouabain fails to increase transmitter release. Other agents have effects on release similar to ouabain. Such agents include uncouplers of oxidative phosphorylation, intracellular-acting sulfhydryl reagents, and Ca^{2+}-ionophores such as A23187.[35,36] Although each of these agents act via different mechanisms, the common link is apparently an increase in intraterminal Ca^{2+} which, in turn, induces transmitter release.

B. CALCIUM (Ca^{2+}) CHANNELS

Neurons contain more than one type of Ca^{2+} channel.[37-42] Since these different Ca^{2+} channels apparently perform other functions in addition to mediating depolarization-dependent Ca^{2+} influx, the interaction of heavy metals with these channels could affect a variety of neuronal processes other than release. Nowycky et al. identified 3 basic subtypes of Ca^{2+} channels in sensory neurons using electrophysiological techniques.[37] The different voltage-dependences of these Ca^{2+} channel subtypes suggest that changes in the resting membrane potential will determine the relative involvement of each channel subtype in neuronal function, including transmitter release. However, the exact roles these Ca^{2+} channel subtypes serve in neurons are not clear, in part due to the different functions ascribed to these channel subtypes both within and between cell types. These channel subtypes can be classified, in a generalized fashion, as T (transient), L (long), and N (negative). The most common channel subtype is the L-Ca^{2+} channel, which opens for relatively long time periods (hundreds of milliseconds) following membrane depolarization, and appears to be the channel most important in allowing Ca^{2+} to enter the nerve terminal to facilitate transmitter release. In addition, when the resting membrane potential is very negative, a strong depolarization can also open N-type channels, which may also be involved in transmitter release. The T-type channels open briefly following small depolarizations if the membrane is first hyperpolarized. These T-channels may be involved in controlling action potential firing bursts, but appear to have little role in transmitter release since the Ca^{2+}-current produced by T-channels is one tenth the amount produced by N- or L-channels. These various subtypes of Ca^{2+} channels may be differentially controlled by various intraterminal 2nd messengers. A more thorough discussion on neuronal Ca^{2+} channels can be found in the accompanying review by Audesirk.

C. INTRATERMINAL Ca^{2+} REGULATION

Although the mechanisms are not yet understood, the critical event mediating transmitter release is the increase in the intraterminal Ca^{2+} concentration.[43] Insult to any of the steps that regulate intraterminal Ca^{2+} levels could be expected to alter transmitter release. When the nerve is at rest, the intracellular Ca^{2+} concentration is tightly maintained at about 0.1 to 0.3 μM.[19] Under resting conditions the rate of influx of Ca^{2+} into the terminal is balanced by the rate of Ca^{2+} efflux. The influx of Ca^{2+} across the relatively Ca^{2+}-impermeable nerve terminal plasma membrane under resting conditions involves mainly leakage (due to the large Ca^{2+} electrochemical gradient across the plasma membrane), although receptor and 2nd-messenger activated Ca^{2+} channels can contribute to this influx (see section below on Ca^{2+} cycling).

There are two basic means by which Ca^{2+} is extruded from the nerve terminal: by Na^+/Ca^{2+} exchange and by Ca^{2+}-ATPase pumps. Under resting conditions the extrusion of Ca^{2+} by Ca^{2+}-ATPase pumps is the dominant means by which the intraterminal Ca^{2+} concentration is regulated.[44] Neither of the two major intraterminal Ca^{2+} buffering organelles, the endoplasmic reticulum and mitochondria, play a role in regulating or defining the resting (steady state) intraterminal Ca^{2+} concentration. At steady state, Ca^{2+} is taken up by the mitochondrial uniporter at a rate which balances the efflux of Ca^{2+} from the mitochondrial matrix.[45]

When the nerve terminal is depolarized, voltage-sensitive Ca^{2+} channels open and there is a rapid influx of extracellular Ca^{2+} into the terminal. This transient inward flux of Ca^{2+} can elevate intraterminal free Ca^{2+} to 5 to 10 μM, thus activating the transmitter release mechanism.[19] Under these conditions both the membrane Ca^{2+}-ATPase pumps and the mitochondria will immediately begin to remove the increased cytosolic free Ca^{2+} from the terminal. The endoplasmic reticulum, already loaded with Ca^{2+}, does not act as a Ca^{2+} buffer under these conditions. Although a major role has been proposed for endoplasmic reticulum in terminal Ca^{2+} regulation, it now appears that this proposal was based on data obtained using an inappropriate tissue preparation.[46,19] The endoplasmic reticulum appears to serve rather as an intracellular Ca^{2+} store releasable by 2nd messengers to mediate receptor-activated signals. Ca^{2+} uptake by the intraterminal mitochondria serves initially as the predominate means by which elevated levels of Ca^{2+} are lowered. Since the entire respiratory capacity of the mitochondria can be employed for Ca^{2+} uptake, and since the maximum velocity of the membrane Ca^{2+}-ATPase pump is limited, the initial rate of cytoplasmic Ca^{2+} removal by the mitochondria can be 50 times greater than by the Ca^{2+}-ATPase pump.[19] The mitochondria continue to accumulate Ca^{2+} until the cytoplasmic level is decreased to about 1 μM. The plasma membrane Ca^{2+}-ATPase pump continues to remove cytoplasmic Ca^{2+} until a level of 0.1 to 0.3 μM is achieved. The elevated mitochondrial Ca^{2+} concentration at this point results in a net efflux of Ca^{2+} from the mitochondria into the cytoplasm, and subsequently this Ca^{2+} is removed from the terminal by the Ca^{2+}-ATPase pump.

D. SUSTAINED NEURONAL RESPONSES

Recent studies have demonstrated that transient increases in intracellular Ca^{2+} can induce a sustained cellular response that persists despite a return of the Ca^{2+} to basal (resting) concentrations.[47] In some cases this long-term change in cellular secretion is associated with increased Ca^{2+} cycling (e.g., increased cellular Ca^{2+} influx and efflux with no net increase in intracellular Ca^{2+}). Sustained neuronal responses, produced by either repeated depolarizations within a "short" time frame (increased neuronal activity) and/or activation of presynaptic receptors, may be characterized by changes in either the duration and/or magnitude of transmitter release, and/or the responsiveness of the nerve to depolarization-inducing stimuli. Regardless of whether produced by successive increases in Ca^{2+} concentration or by presynaptic receptor activation, the resulting sustained changes in transmitter release will be related to either: (1) altered intraterminal Ca^{2+} regulation, (2) altered ionic (e.g., K^+) current, and/or (3) altered responsiveness of the transmitter release machinery. The precise roles of various biochemicals (such as protein-kinase C and the various inositol-phosphates) and the sequence of cellular events that combine to produce sustained neuronal responses have yet to be elucidated. Nevertheless, the potential interaction of toxicants with those processes modulating long-term neuronal regulation of transmitter release represent an area of future concern with respect to subtle neurotoxic insult.

III. METHODOLOGICAL CONSIDERATIONS IN THE ASSESSMENT OF TRANSMITTER RELEASE

There are numerous tissue preparations and techniques useful for assessing neurotransmitter release, each possessing unique advantages and disadvantages. There are essentially two major approaches for examining transmitter release; electrophysiology and neurochemistry. The best strategy, however, for firmly establishing the actions of heavy metals on transmitter release is to employ complementary approaches. In contrast to earlier experimental efforts, recent studies have consistently demonstrated identical or similar effects of a particular heavy metal on transmitter release regardless of the preparation or technique employed. The inconsistency of earlier studies may reflect inherent limitations of some of the assays employed, or may be the result of inadequate attention to factors such as: buffer composition, solubility of the metal in the buffer, and/or relevant positive controls. An understanding of the advantages and limitations associated with the various tissue preparations and approaches used to examine transmitter release is an essential framework within which to discuss the reported effects of the heavy metals.

A. *IN VIVO* ASSESSMENT OF TRANSMITTER RELEASE

The ideal system in which to assess the effects of heavy metals on transmitter release is the intact *in vivo* nervous system. Although the thrust of this review is concerned with the acute *in vitro* effects of heavy metals, it is feasible to examine transmitter release *in vivo*. The ease of access coupled with the relatively simple organization of the PNS has facilitated the interpretation of release studies performed *in vivo* using preparations such as perfused sympathetic ganglia. Although more technically demanding and prone to interpretative difficulties, CNS transmitter release can be assessed *in vivo* using techniques such as push-pull perfusion, brain dialysis, and/or *in vivo* voltammetry.[48,49] Both push-pull perfusions and brain dialysis involve perfusing isolated brain regions or brain ventricles with a physiological buffer following the stereotaxic placement of either the perfusion cannula or dialysis tubing.

Ideally, perfusing a brain region allows small molecules (including transmitter substances) released from the surrounding tissue to diffuse, either directly or through the dialysis tubing, into the perfusate, which in turn can be collected and examined. There are, unfortunately, many inherent limitations to brain perfusion techniques. Since only 2 to 3% of the total CNS tissue mass represents presynaptic nerve terminals, most of the tissue being perfused is not the tissue of interest. This non-neuronal tissue may distort the diffusion of the transmitter from the nerve terminal to the perfusate by either accumulating, binding, or metabolizing the transmitter. This compromised diffusion is also of concern when chemicals (e.g., high-K^+, ^3H-transmitters, or toxicants) are applied directly to the CNS through the perfusate; percise control of the extraterminal environment is not possible. The tissue damage produced by implanting the cannula and/or dialysis tubing may obscure the effects of a toxicant, and may also result in a number of tissue reactions such as hypoxia, gliosis, proliferation of monocytes, and the encapsulation of the cannula/dialysis tubing by endothelial cells.[50] The perfusion process may also produce tissue damage, the extent of which can be monitored by measuring certain intracellular markers in the perfusate.[51-53] The use of a dialysis membrane, while offering the advantage of preventing the migration of proteins and cells into the perfusion medium,[54] also imposes additional tortuosity and constraint to the diffusion characteristics of the ions or molecules under study. Since the concentration of transmitter in the perfusate is usually very low, perfusate samples are often collected over relatively long periods of time (minutes) to obtain sufficient amounts of transmitter for detection, thus obscuring rapidly occurring events. All these factors indicate that the appearance of transmitter in the perfusion fluid may not reflect presynaptic events.

In contrast to perfusion techniques, *in vivo* voltammetry is able to detect changes in extracellular transmitter concentrations in seconds, or even milliseconds, and thus comes close to the real time scale of the neuronal release process. *In vivo* voltammetry, which involves the implantation of microvoltammetric electrodes into specific brain areas, can provide both quantitative and qualitative information regarding compositional changes in the neuronal extracellular fluid. The basic principle of *in vivo* voltammetry is that certain compounds will oxidize at the electrode surface when a sufficient potential, which varies from compound to compound, is applied. The electrons emitted by the oxidized compound are detected as a change in current, the magnitude of which is proportional to the concentration of the compound. Since a number of oxidizable compounds are present *in vivo,* it is necessary to rely upon the ability of voltammetric techniques to distinguish between these various compounds on the basis of the different potentials at which these compounds oxidize. Modification of the electrode material and surface has aided in the discrimination between oxidizable compounds,[55] but concern still remains as to whether observed current changes solely reflect released transmitter. The possibility that continuous application of voltage (necessary to oxidize the transmitter) to the implanted electrodes affects neuronal function cannot be dismissed. The application of *in vivo* voltammetry is limited to only a few, readily oxidizable transmitter substances, specifically dopamine (DA) and norepinephrine (NE). Similar concerns of tissue damage and tissue reactions, as recognized with the perfusion techniques, also apply to *in vivo* voltammetry. The complex, interactive nature of the intact CNS makes it difficult to directly relate any observed transmitter changes to effects on the presynaptic terminal, and even more difficult to speculate about subcellular mechanisms. Despite the limitations of these *in vivo* techniques, continual advances in the area of *in vivo* release assessment may make such approaches advantageous in the future, particularily in respect to determining the relevance of *in vitro* observations.

B. TISSUE PREPARATIONS USED TO ASSESS TRANSMITTER RELEASE *IN VITRO*

Ideally, the tissue preparation employed to assess transmitter release *in vitro* should possess all the properties of presynaptic terminals in the intact *in vivo* nervous system while eliminating any confounding factors that might affect neurotransmitter release. The isolated neuromuscular junction preparation has been used extensively in electrophysiological release studies. The utility of this preparation is discussed in the section detailing the electrophysiological assessment of release in PNS preparations.

In contrast to the relatively simple synaptic arrangement of the PNS neuromuscular junction, the intact CNS is characterized by limited accessibility, integrative complexity, and neuronal diversity that confound transmitter release assessments. However, there are distinct advantages of examining transmitter release in CNS tissue. In addition to the high ratio of nerve terminals to other tissue structures (relative to PNS tissues), a number of different transmitter systems are present in the CNS. Only cholinergic and noradrenergic transmission can be assessed in the mammalian PNS (although in certain invertebrate preparations other transmitter substances can be examined). While current evidence suggests many similarities of action among the heavy metals on transmitter release across a number of transmitter systems, this might not always be the case.[1] There do appear to be some major differences between the release processes of amino acid vs. monoamine transmitter systems.[56] To minimize the problems associated with CNS tissue preparations, "simpler" or "reduced" preparations derived from intact CNS tissue can be used to examine transmitter release. Such preparations include: cultured neurons, brain slices, brain minces, and synaptosomal preparations of varying purity. Unfortunately, none of these tissue preparations are ideal. The advantages, disadvantages, and limitations of each of neuronal preparations for examining the effects of heavy metals on transmitter release are discussed below.

1. Neuronal Cell Cultures

Neurons in culture offer the possibility of working with a homogenous intact population of cells in which the extracellular environment can be easily manipulated. Ideally, the viability of neurons in culture can be maintained over long periods of times (days, weeks, or even months). While cultured neuronal lines are available, there is some question as to how closely they resemble intact neurons.[57] The neurons used to establish these lines are abnormal, usually tumorigenic. That such cell lines are usually initiated with carcinogen-induced neurons is not surprising considering that normal mature neurons do not divide. Neuronal cell lines are much less differentiated than normal neurons, and exhibit abnormal characteristics. Primary neuronal cultures, in which cells are taken directly from the animal and maintained *in vitro,* are an advantage over cell lines since they consist of morphologically, bioelectrically, and biochemically differentiated diploid cells.[57-59] Since the starting material for primary CNS neuronal cultures can be obtained only from immature (early postmitotic differentiating) tissue, there is still the question as to whether these cells differentiate in the same manner in culture as *in vivo*. Although techniques are available for preparing primary cultures free of nonneuronal cell types (glia), such preparations usually contain several types of neurons. In contrast to primary CNS neuronal cultures, primary PNS neuronal cultures can be obtained from adult animal PNS tissue.[60] Neurons in primary cultures do not proliferate; their longevity is dependent on many factors, including the type of neuron culture, the composition of the culture media, and the scheduling of nutrient additions to the culture.

Establishing and maintaining neuronal cultures is often difficult, and many of the factors involved in their successful maintenance are not understood.[59] Although electrophysiological and neurochemical transmitter release studies using neuronal cultures have been very limited, the potential advantages of such preparations, particularily for chronic studies of *in vitro* toxicant exposure, will encourage the use of neuronal culture preparations in future neurotoxicological studies.

2. Brain Slices

Most studies evaluating the *in vitro* effects of heavy metals on CNS transmitter release have employed either mammalian brain slices or synaptosomes. Both preparations are relatively easy to obtain and the extra-neuronal medium can be readily controlled. The major drawback of brain slice and synaptosomal preparations, relative to neuronaı cultures, is their limited time of viability; they must be used within a few hours after the animal is sacrificed. Basal respiratory rates and energy metabolism of both preparations are similarly and noticeably compromised with respect to the intact brain, although in some aspects energy metabolism is reduced to a greater degree in synaptosomes than in slices.[61] Brain slices, which can be employed in either electrophysiological or neurochemical release studies, have the advantage of "intactness", which makes it easier to generalize any observed heavy metal effects to the more relevant *in vivo* situation. Since many of the neurons in slices are intact, it is possible to examine depolarization-evoked release electrophysiologically by stimulation of the nerve and recording from postsynaptic site. Although neurons within ~75 μm of the slice surfaces are damaged, many of the "deeper" neurons are intact and exhibit a tonic rate of action potential activity. The integrative nature of the relatively "intact" slice preparation, particularly the presence of neuronal feedback loops, interactions with other transmitter systems, and other compensatory processes, presents a confounding factor when relating the effects of a toxicant directly to the presynaptic release process. There are considerable non-neuronal cells and multiple diffusion barriers present in slices, including the nonviable slice surface neurons, which can limit exposure of the presynaptic terminals within the slice to various components of the physiological buffer solutions, including essential nutrients as well as experimental treatments (e.g., brief exposures to high-K^+ or heavy metals).[62] Also, slices may be disadvantageous in neurochemical studies in that the diffusion of transmitter out of the tissue may not directly reflect events at the presynaptic terminals.

3. Synaptosomal Preparations

When CNS tissue is homogenized under certain well-defined conditions, nerve terminals are "pinched" off their axons to form vesicles that are referred to as synaptosomes. Synaptosomes represent functional presynaptic entities as demonstrated by their ability to transport, synthesize, metabolize, store, and release transmitter.[63-65] Other aspects of presynaptic integrity are intact, including presynaptic receptors, intermediary metabolism, membrane permeability, membrane calcium gradients, and Ca^{2+} regulation.[66-68] However, with respect to the *in vivo* presynaptic terminal, synaptosomal preparations are compromised with respect to various aspects of cellular energetics, although the functional consequences of decreased respiration on transmitter release are not apparent.[61] Another potentially troubling aspect is that the magnitude of the Na^+ and K^+

gradients across the synaptosomal membrane is apparently less than in intact neurons, which affects membrane depolarization and the movement of ions across the membrane.[69] It is essentially impossible to obtain electrophysiological recordings from synaptosomes; these preparations are used almost exclusively in neurochemical studies. These isolated nerve terminals, unlike neurons in brain slice preparations, do not exhibit a basal rate of action potential activity. The major advantage of synaptosomal preparations, particularily purified synaptosomal preparations, is that the absence of other neuronal structures (axons, cell bodies, dendrites, etc.) facilitates the attribution of treatment effects on transmitter release directly to presynaptic processes. The lack of neuronal integration and accompanying supporting structures inherent to synaptosomal preparations eliminates the confounding influences of neuronal feedback loops and multiple diffussion barriers.[70] Synaptosomal preparations vary in purity, ranging from crude homogenates to relatively pure synaptosomal preparations.[71] Even these relatively pure mammalian synaptosomal preparations contain more than one type of transmitter system. This heterogeneity can be reduced by isolating specific brain regions having a high representation of synaptic terminals of the transmitter system of interest. A lack of homogeneity is of greater concern when preparations are loaded with a radiolabeled transmitter substance that can be accumulated in significant quantity by nerve terminals of other transmitter systems. Centrifugation on a sucrose gradient, in addition to separating synaptosomes from mitochondria and myelin, can be used to separate different types of synaptosomes.[72] However, even the purest synaptosomal preparations are contaminated by cellular fragments, which could confound release studies. When examining heavy metals, the presence of extrasynaptosomal mitochondria in the preparation should be avoided, as the mitochondria may act as source or sink for either the heavy metal and/or Ca^{2+}.[73] The potential for nonspecific uptake and release of transmitter substances, heavy metals, and/or essential ions increases as the proportion of contaminating non-synaptosomal to synaptosomal tissue in the preparation increases. A major drawback of synaptosomal preparations is their limited duration of viability. Preparations must be prepared and maintained (until used) at ~4°C, and used within 4 to 5 h following animal sacrifice.

C. *IN VITRO* AND *IN SITU* ELECTROPHYSIOLOGICAL ASSESSMENT OF RELEASE

Electrophysiological techniques do not directly measure transmitter release, but rather the postsynaptic consequences of release. An electrode placed in or near a postsynaptic cell is used to record the activity of that cell. Neuronal activity is characterized by frequency and/or magnitude changes in postsynaptic voltage or current. Electrophysiological assessment of transmitter release can be performed in either PNS or CNS tissue preparations, although data obtained from PNS preparations are usually easier to interpret. The electrophysiological approach, unlike the neurochemical approach, allows for the assessment of release within a physiological time frame (e.g., milliseconds).

Another advantage of the electrophysiological approach is the relative ease with which a number of important neurophysiological parameters can be measured, including: (1) nerve action potential conduction velocity, (2) nerve terminal depolarization, (3) spontaneous quantal transmitter release (MEPP frequency at the neuromuscular junction), (4) depolarization-evoked transmitter release (EPP amplitude at the neuromuscular junction), and (5) many parameters of postsynaptic receptor and cell membrane function. By appropriate adjustment of electrical stimulation parameters, pharmacological manipulations, and statistical evaluation, additional information can be acquired, including: (1) estimates of the probability of transmitter release, (2) readily available stores of transmitter, (3) rate of replenishment of transmitter, (4) fatigue or "run down", and (5) involvement of specific ions and ion channels.[74] Several important disadvantages or limitations of the electrophysiological approach should be noted. Since only indirect measurements of transmitter release are made, it is always necessary to control for possible postsynaptic effects. In the neuromuscular junction preparation possible postsynaptic effects of a neurotoxic agent can be determined by iontophoretically applying exogenous transmitter substance directly to the postsynaptic site. It is also important to distinguish whether a toxin affects depolarization-evoked release by altering generation and/or conduction of the action potential vs. direct effects on the exocytotic process.

1. Assessment of Release in PNS Preparations

Synapses in the peripheral nervous system, particularly the vertebrate motor nerve-skeletal muscle junction, represent a relatively simple and convenient preparation for the electrophysiological assessment of release. Almost all electrophysiological studies of the effects of heavy metals on neurotransmission have employed such preparations. The activation of the postsynaptic site (muscle endplate) is produced by the release of a single transmitter substance (ACh). This one-to-one relationship between transmitter release and changes in postsynaptic potential helps to simplify the interpretation of data. This relative simplicity, combined with the various measurements of postsynaptic function, make such preparations quite useful for screening potential neurotoxic agents, particularly since heavy metals and many other environmentally significant neurotoxins are highly nonspecific in action. However, due to the scarcity of nerve terminals in a large volume of mainly muscle tissue, the neuromuscular junction is not convenient to obtain measurements of nerve terminal transmembrane ion fluxes. Additionally, a note of caution is needed concerning the differences, or possible differences, between nominal ionic concentrations of solutions used to superfuse preparations and the concentrations actually in contact with the nerve terminals. Due to the large amount of muscle and connective tissue present in the neuromuscular preparation, heavy metals may be bound to an extent sufficient to lower the concentration present at the nerve terminal. Conversely, the tissue itself can be a source of troublesome or unwanted ions. For example, perfusion of a preparation with a Ca^{2+}-free solution does not guarantee the absence of Ca^{2+} near the nerve terminal. Though it may be too difficult or expensive to monitor the chemical microenvironment of the nerve terminal, several procedures are helpful in achieving better control and consistency; (1) remove as much of the fine connective tissue as possible from the surface of the muscle, (2) use only superficial endplates, (3) continuously superfuse preparations, and (4) use appropriate chelators when it is necessary to ensure that extremely low concentration of Ca^{2+} are present.

2. Assessment of Release in CNS Preparations

Although this approach has not yet been employed to our knowledge in the study of heavy metals, electrophysiological assessment of transmitter release can also be performed in mammalian CNS tissue. Most commonly hippocampal tissue slices are employed in such studies.[75] Intracellular recordings are obtained from large neuronal perikarya located in the CA3 and CA1 areas of the hippocamus. These intracellular recordings, obtained by using current-clamp and single electrode voltage-clamp procedures, can be used to determine the effects of neurotoxicants on the resting neuronal membrane conductance and firing threshold, spontaneous synaptic potentials, and synaptic potentials produced by stimulation of neuronal input pathways. Thus, intracellular recordings from CNS neurons offer essentially the same advantages and disadvantages as described above for the neuromuscular junction (PNS) preparation. The major difference between the PNS and CNS preparations is that CNS neurons receive multiple presynaptic inputs from a variety of neurons utilizing many different transmitter substances which can have different postsynaptic actions. Whereas increased activity at one presynaptic site may release a transmitter substance that facilitates postsynaptic neuronal activity, it is just as likely that increased activity at another presynaptic terminal releases a transmitter substance that inhibits postsynaptic neuronal activity. It is virtually impossible to determine the extent of excitatory vs. inhibitory synapses associated with a specific neuron, even assuming that the cell body of a particular neuron could be repeatedly identified and analyzed. In many cases the transmitter substances have not been identified, making it difficult if not impossible to differentiate presynaptic vs. postsynaptic toxicant effects by microiontophoretically applying exogenous transmitter.

An additional problem associated with CNS tissue slice preparations is the possible involvement of neuronal feedback loops which would further complicate the system. Thus, while this approach can be used to demonstrate differences in CNS neuronal activity resulting from heavy metal exposure, the complexity of the system makes it difficult to relate any observed changes in postsynaptic activity to effects on transmitter release. While the interpretational difficulty limits this approach as a sole means of assessing the effects of heavy metals on transmitter release, the combination of this approach with other approaches can be useful in confirming as well as assessing the potential relevance of the effects of toxicants observed in "simpler" systems.

D. NEUROCHEMICAL ASSESSMENT OF RELEASE IN CNS PREPARATIONS

Several neurochemical measures, such as transmitter to metabolite concentration ratios and transmitter turnover rates, have been proposed to reflect neuronal activity, which in turn could reflect transmitter release. Such measures are of limited value with respect to defining the sites of toxic metal action. While toxicant-induced changes in such measures may reflect alterations in transmitter release, effects on other neuronal processes, including transmitter reuptake, transmitter synthesis or metabolism, transmitter storage, or receptor activation, could just as likely cause changes in turnover rates or transmitter/metabolite ratios.

The best neurochemical approach for assessing transmitter release is to directly examine the efflux of transmitter from nerve tissue. Such a neurochemical approach, in contrast to the electrophysiological approach, provides a direct means of examining release since it is the concentration of transmitter emerging from the terminal that is quantified. A direct assessment of release provides a better indication as to the neuronal sites of toxicant action, as well as the mechanisms mediating the toxicant effect. Most neurochemical studies examining the effects of heavy metals on release have used CNS preparations. However, neurochemical techniques are by no means confined to CNS preparations. The techniques and concerns described below can also be applied to nerve tissues obtained from the mammalian PNS, as well as non-mammalian preparations such as synaptic vesicles from Tordepo electric organ.

1. Loading Tissue with Radiolabeled Transmitters

In most neurochemical release studies, the tissue is loaded with radiolabeled transmitter (or precursor); the radioactivity is used to quantify the amount of transmitter release. The radiolabeled transmitter is assumed to be selectively accumulated by, and uniformly distributed within, those nerve terminals that normally store and release that transmitter. These underlying assumptions are not completely valid. Transmitter substances can be accumulated, to varying degrees, at nerve terminals other than those normally associated with the transmitter, at neuronal sites other than the terminal, and at a variety of non-neuronal sites. Amino acid transmitters, because of their involvement in a variety of cellular processes, can be selectively accumulated at a number of cellular sites. Radiolabeled transmitter can also bind to specific membrane receptors as well as to a variety of non-specific neuronal and non-neuronal (glial, epithelial cells) membrane sites. All these sites can serve as a confounding source of radiolabel during transmitter release studies, the severity of which is lessened as the proportion of specific nerve endings in the tissue preparation increases. Even if the radiolabeled transmitter is assumed to be selectively accumulated into the appropriate nerve endings, it is not necessarily equilibrated or equally mixed within all the endogenous transmitter pools. There is considerable evidence that the newly taken-up and/or newly synthesized transmitter is preferentially released.[17] Certain pharmacological agents (such as amphetamine) have been proposed to release transmitter from selective intraterminal pools.[76] This problem can be reduced by (1) allowing enough time for the radiolabel to intermix with endogenous intraterminal transmitter stores, (2) minimizing transmitter reuptake during release to prevent reuptake of released transmitter and its preferential release, and (3) minimizing transmitter synthesis to prevent the dilution of radiolabeled transmitter pools and the preferential release of newly synthesized unlabeled transmitter.

Another concern when loading tissue for release studies is the chemical concentration of radiolabeled transmitter employed. If large concentrations of transmitter are used, then there is a possibility that the added radiolabeled transmitter can alter intraterminal transmitter pools and/ or interact at presynaptic autoreceptors to alter transmitter release. In most cases high specific-activity radiolabeled transmitters are available which allow for the loading of the tissue at concentrations far below those that would confound data interpretation. Since the high specific-activity radiolabels (particular ^{3}H-dopamine) may be subject to an increased rate of chemical degradation, it is necessary to periodically determine chemical purity.

2. Transmitter Metabolism

It must be demonstrated that the radiolabel subsequently released from the tissue reflects transmitter, and not a metabolic product. This concern does not necessarily apply if endogenous (unlabeled) transmitter is being quantified, or if the released material is subjected to chromatographic separation prior to quantification. In many cases, however, the radioactivity (corrected for efficiency and quench) is used as the sole means of quantifying transmitter release. Therefore, metabolic inhibitors are usually added to the incubation and release buffers to ensure that the label represents transmitter. However, the addition of such inhibitors may compromise the integrity of the preparations. For example, aminooxyacetic acid can be used to inhibit the metabolism of GABA. However, since GABA is both a transmitter substance and a cellular energy metabolite, transamination inhibition could potentially alter several cellular functions.[77] The purity of the radiolabel is also important when the tissue is loaded with transmitter precursor rather than the transmitter itself.

In the case of cholinergic transmission, it is necessary to load the tissue preparation with an ACh precursor, usually radiolabeled choline. Inside the nerve terminal the choline is incorporated into a variety of compounds, including ACh. Since the potential exists for the terminal to release the radiolabeled choline, as well as other radiolabeled choline-containing compounds, it is essential that the radioactivity released from cholinergic preparations be subjected to chromatographic analysis. Due to the unstable nature of ACh, care must be taken to prevent the breakdown of this transmitter subsequent to release.

3. Static vs. Dynamic Systems

There are basically two experimental techniques that are used to examine the release of transmitter from nerve tissue. The "static" approach involves examining release at only one point in time with any given preparation. In its simplest form this approach involves the loading of tissue with labeled transmitter, washing the tissue to remove any unaccumulated label, adding "release" buffer, incubating the tissue for a fixed period of time (usually minutes), separating the buffer from the tissue, and finally quantifying the radiolabel in the buffer and/or the tissue. When examining the effects of either heavy metal or membrane-depolarizing (high-K$^+$) treatments on release, parallel tissue preparations are employed by necessity. The major drawback of the "static" release approach is that all treatment comparisons are made *between* preparations with only one measure being obtained per tissue sample. The use of multiple tissue preparations to assess treatment effects increases the variability of the measurements, thus limiting the sensitivity of the technique. The other major limitation of the "static" approach is that release is assessed at only one time point; thus the temporal characteristics of a particular treatment are not available.[65]

The "dynamic" approach of assessing release, which involves synaptosomal superfusion, eliminates both of these drawbacks. This approach involves supporting the synaptosomes (usually on a filter membrane), passing physiological buffered solution at a constant rate across the layered synaptosomal bed, and then quantifying the transmitter in successively collected effluent fractions. The confounding problem of reuptake of release transmitter, as can occur in static systems, is alleviated to a great degree in superfusion systems by the continuous removal of released transmitter from the proximity of the synaptosomes. The composition of the superfusing buffer (e.g., high-K^+, heavy metal treatment) can be changed during the superfusion. Thus it is possible to get a baseline rate of transmitter release prior to heavy metal treatment within the same preparation. Temporal aspects (onset, duration, recovery) of toxicant effects on transmitter release can also be assessed using this approach.[78]

E. PHYSIOLOGICAL BUFFER SOLUTIONS

Of critical importance in any *in vitro* transmitter release study, but particularly in those studies involved with the heavy metals, is the composition and characteristics of the physiological medium to which the neuronal tissue is exposed. Osmolarity is a very important factor, particularly in electrophysiological studies employing tissue slices where little variation from 300 milliosmolar can be tolerated. The physiological buffer solutions should be well oxygenated.[79] Release experiments employing mammalian preparations are usually performed at 37°C, although lower temperatures (30°C) are sometimes employed.[80] Non-mammalian preparations, such as the frog neuromuscular junction, are usually studied at lower temperatures (15 to 25°C). The pH of the buffer should be maintained near 7.4 by employing a buffering compound having a pK_a near 7.4. Most commonly bicarbonate or phosphate-buffered solutions are employed. While these buffering compounds, particularly bicarbonate, are "more physiological" than other commonly employed buffers, certain heavy metals may interact with them and precipitate. For example, precipitation of $Pb_3(PO_4)_2$, $PbCO_3$ and/or $Pb(OH)_2$ can interfere with studies of the biological effects of Pb;[81] such problems may explain the apparently contradictory results of some Pb studies.

To eliminate the problem of precipitate formation, other compounds, such as N-2-hydroxyethylpiperazine-N'-2-ethanesulfonic acid (HEPES), tris (hydroxymethyl) aminomethane (TRIS), and/or 3-[N-Morpholine]propanesulfonic acid (MOPS), can be employed to buffer the physiological solution. Problems may also arise with these compounds, as suggested by the recent report that HEPES blocks chloride channels.[82] It should also be noted that the solubility product of $Pb(OH)_2$ limits the Pb concentration that can be obtained in physiological solutions to about 10 μM.[81] Another concern with heavy metals is the actual concentration of metal ions in solution. Pb can form complex ions with certain components of physiological solutions (e.g., $PbOH^+$, $PbCl^+$), so that the actual Pb^{2+} concentration is lower than the total Pb concentration.

Other components of the physiological solutions used in release studies include the ions Na^+, K^+, Mg^{2+}, Cl^-, and $SO_4^=$. The concentrations of these ions are based on either the plasma values[83] or cerebrospinal fluid values.[84] The concentration of Na^+, added mainly as NaCl, is usually adjusted between 130 and 140 mM. When release experiments are performed in which the Na^+ gradient across the nerve terminal membrane is removed (for example, to assess the role of Na^+,K^+-ATPase or Na^+ channels in the releasing effects of heavy metals), the Na^+ concentration is lowered to about 10 to 30 mM, and a "relatively inert" salt, such as choline chloride, is substituted for the remaining NaCl in order to maintain the ionic balance/osmolarity. Note that choline may not be an appropriate NaCl substitute if ACh release is being examined subsequent to loading with radiolabeled choline. The concentration of K^+, added as KCl, is usually around 5 mM. When high-K^+ is used to depolarize the preparation, the K^+ concentration is usually increased to 30 to 60 mM, with an equimolar decrease in the Na^+ concentration. The concentration of $MgSO_4$ is usually 1.2 mM. The total concentration of Ca^{2+} found in brain is normally about 1.2. mM (vs. 2.5 mM in plasma),[85] which is more than necessary to support optimal Ca^{2+}-

dependent transmitter release. Excessively high Ca^{2+} concentrations should be avoided as they suppress neuronal activity.[51]

In neuromuscular preparations, normal concentrations of Ca^{2+} will result in unwanted muscle contraction following nerve stimulation. Therefore, it is necessary to either curarize the preparations or use a superfusion solution containing a low ratio of Ca^{2+} to Mg^{2+}. Ca^{2+}-free solutions are often employed to assess potential heavy metal-Ca^{2+} interactions. While in some preparations (e.g., purified synaptosomes) a solution in which no Ca^{2+} is added is sufficient, other tissue preparations contain sufficiently high extracellular Ca^{2+} concentrations (>0.1 μM) that it is necessary to add a Ca^{2+}-chelator, such as EGTA, to the solution. Since Ca^{2+} is important in stabilizing the plasma membrane, the lack of extraneuronal Ca^{2+} (particularly if EGTA is used) may compromise the integrity of the terminal membrane. Glucose (~10 mM) is usually added as an energy source in order to maintain the respiratory capacity of the preparation. In some cases an antioxidant, such as 1 mM ascorbic acid, is also added, particularly if catecholamine release is being examined. In most neurochemical studies enzyme inhibitors are added to the buffers to prevent the breakdown of the neurotransmitter being examined. The presence of antioxidants and/or enzyme inhibitors could potentially confound transmitter release studies.

F. TISSUE DEPOLARIZATION TECHNIQUES

There are several means of depolarizing nerve tissue in order to evoke transmitter release. Ideally, the most appropriate (and most physiological) approach is to use direct nerve stimulation. This means of depolarization is routinely employed in PNS preparations such as the neuromuscular junction. Direct pathway stimulation (bipolar pulses of 200 μs duration, 20 to 80 V, 0.1 to 1 Hz) has been employed to measure the release of transmitters from CNS tissue slices.[86] However, in many CNS preparations, direct pathway stimulation is either difficult, impractical, or impossible.

1. Electrical Field Stimulation

Electrical field stimulation has been used to evoke transmitter release from CNS tissue preparations. The depolarization produced by subjecting the tissue to alternating voltage pulses (10 to 100 V, 10 to 100 Hz) is due to a large decrease in intracellular K^+.[87] In addition, electrical stimulation produces an increase in intraneuronal Na^+ that is equimolar to the decrease in K^+. The advantages of electrical field stimulation are (1) the parameters of stimulation can be accurately controlled, (2) the duration of electrical (depolarizing) pulses can be limited to very short (physiological) periods, (3) it is not necessary to alter the composition of the extraneuronal buffer, (4) unlike high-K^+ evoked release, electrical-field stimulated release involves membrane Na^+ channels and is therefore more physiological, and (5) there are minimal effects on glial swelling (unlike in high-K^+, see below). There are drawbacks to electrical field stimulation, however. It is sometimes technically difficult to produce trains of electrical pulses of significant voltage to evoke transmitter release without damaging the tissue. Although electrical stimulation produces increases in the release of a variety of transmitter substances that are qualitatively similar to high-K^+ evoked release, the dependence of release on extraneuronal Ca^{2+} is sometimes less than observed with pathway-evoked or high-K^+ evoked release.[88] For example, extraneuronal Ca^{2+} is a necessary requirement for pathway-evoked amino acid transmitter release.[89] However, electrical field-evoked amino acid transmitter release is independent of extraneuronal Ca^{2+},[90] indicating differences between release evoked by normal (physiological) vs. electrical field stimulation.

2. Na$^+$ Channel Ionophores

Another approach to producing depolarization is to use chemicals. Veratridine depolarizes the nerve terminal by opening membrane Na^+ channels. The release of transmitter evoked by veratridine can be blocked by the use of the Na^+ channel blocker, tetrodotoxin. Veratridine-evoked release of catecholamines and ACh is dependent on extraneuronal Ca^{2+}. However, as with electrical field stimulation, veratridine-evoked release of amino acid transmitter substances is Ca^{2+} independent.[88] This Ca^{2+}-independent release of amino acid transmitter substances may be related to the increase in intraneuronal Na^+ and the decrease in intraneuronal K^+ that result from both veratridine and electrical field stimulation. Since these Na^+ and K^+ changes apparently occur in all nerve terminals, the Ca^{2+} independent release of amino acid transmitters suggests a unique difference in the storage/release processes of amino acid transmitter containing nerve terminals.

3. High-K^+

The most common means of depolarizing CNS tissue is by increasing the extraneuronal K^+ concentration from about 5 mM to somewhere between 20 to 60 mM, with an equimolar reduction in Na^+ concentration to maintain ion balance and osmolarity. It is important to demonstrate significant transmitter release increases in the presence of high-K^+ (greater than 50% above normal K^+ baseline), and to demonstrate the Ca^{2+}-dependency of high-K^+ evoked release. Since the high-K^+ evoked release of all identified transmitter substances thus far examined, including the amino acid transmitters, is Ca^{2+} dependent, this technique of membrane depolarization may be more analogous to physiological action potentials than either veratridine or electrical field stimulation. Unlike action potential (pathway), electrical field, and veratridine-evoked release, the release of transmitter evoked by depolarization with high-K^+ cannot be antagonized by the Na^+ channel blocker, tetrodotoxin. Another problem with high-K^+ is that it produces significant tissue swelling.[91] Since most of the tissue swelling occurs in the glia,[92] the severity of the problem is markedly reduced in preparations having minimal glia representation.

It is important to limit the duration of the high-K^+ exposure to 1 s or less, since at these short time intervals there is a direct linear correlation of transmitter release with the uptake of extraneuronal Ca^{2+}.[93,94] There appears to be an inactivation of Ca^{2+} channels after short periods of depolarization; this inactivation appears to be mainly voltage-dependent, but may be also partially Ca^{2+}-dependent.[95] In addition, a significant Ca^{2+}-Na^+ exchange and/or Ca^{2+} efflux can occur at longer periods of depolarization.[95] Limiting the duration of high-K^+ depolarization is particularly important when examining the ability of heavy metals to inhibit depolarization-evoked release by blocking Ca^{2+} influx at the membrane Ca^{2+} channel. For example, low La^{3+} concentrations (<1 μM) block the influx of Ca^{2+} induced by a 1 s exposure to high-K^+, whereas high La^{3+} concentrations (>50 μM) are necessary to block the influx of Ca^{2+} associated with a 10 s high-K^+ exposure.[96] This differential sensitivity to La^{3+}, along with inability to completely abolish Ca^{2+} uptake after prolonged depolarization with high-K^+, suggest the existence of two populations of Ca^{2+} channels, an inactivating channel, and a non-inactivating channel.[95] Some of the heavy metals, including Pb^{2+}, act by competing with the uptake of Ca^{2+} selectively at the inactivating Ca^{2+} channels.[97]

IV. EFFECTS OF LEAD (Pb) ON TRANSMITTER RELEASE

Of all the neurotoxic metals, none has received more attention than Pb wtih respect to its effects on the nervous system. No physiological requirement has been established for this metal; therefore the body burden of Pb is solely an indicator of toxic potential. Inorganic Pb salts (Pb^{2+}), non-ionic organic Pb (tetraethyl-Pb, TEL), and monovalent organic Pb salts (triethyl-Pb^+, the toxic metabolite of tetraethyl-Pb),[98] are all capable of producing damage to the nervous system following ingestion, although the symptoms associated with organo-Pb poisoning differ from those of Pb^{2+}. The alkyl-Pb compounds, due to their high lipid solubility, preferentially accumulate in CNS tissue to produce nervous system impairment without accompanying effects on other systems.[99] With the marked reduction in gasoline TEL concentrations (and its eventual elimination), both the risk of accidental TEL exposure and the discharge of Pb^{2+} (formed by the combustion of TEL) from automobiles has been markedly reduced. Although the major environmental health concern is with exposure to Pb^{2+}, the methylation of Pb^{2+} by bacteria indicates the potential for future concern with organo-Pb exposure.[100] Due to the prevalence of Pb^{2+} in the environment, there is a large human pediatric and adult population chronically exposed to Pb^{2+}. The nature of the neurotoxic injury resulting from exposure to Pb^{2+} is dependent on a number of factors, including the level and duration of exposure, age at the time of exposure, and other factors such as diet. While higher Pb^{2+} exposure levels in children are associated with hydroencephalopathy, the major concern is with children who are clinically asymptomatic for Pb^{2+} poisoning but who exhibit elevated Pb body burdens. Although cognitive dysfunctions have been associated with exposures to low Pb levels, the precise nature of the insult to the nervous system, and the resulting funcitonal and/or behavioral consequences, remain unresolved. This uncertainty hinders the establishment of a "safe" Pb^{2+} body burden, if indeed one exists. In adults, the neurotoxic consequence of Pb^{2+} exposure are most often associated with the peripheral nervous system. The lack of CNS manifestations in Pb^{2+}-exposed adults may, in part, reflect difference in absorption, distribution, and retention of Pb^{2+} in adults vs. children.

A. EFFECTS OF INORGANIC LEAD (Pb^{2+}) ON TRANSMITTER RELEASE

The concern about the neuropsychological impact of pediatric exposure to low Pb^{2+} levels has resulted in a large number of animal studies of the CNS effects associated with *in vivo* (neonatal) Pb^{2+} exposure. Despite extensive research efforts, the loci of Pb^{2+} action remain elusive. The varied approaches of such studies, combined with confusing and often contradictory results, have made it difficult to determine whether altered neurotransmission is a consequence of *in vivo* Pb^{2+} exposure.[13,101] Establishing a role for altered transmitter release in Pb^{2+} toxicity is further hampered by the inconsistent results of acute *in vitro* Pb^{2+} exposure studies, particularly those studies employing CNS preparations. As previously discussed in the methodology section of this review, the apparent inconsistencies of such studies may be due to inappropriate attention to experimental/methodological factors. Instead of reporting the results of all *in vitro* Pb^{2+} studies examining transmitter release, which would result in the description of a number of unrelated, contrasting Pb^{2+} effects, the following discussion will focus on describing those effects of Pb^{2+} on transmitter release which are reproducible across a variety of preparations and techniques. This approach permits greater insight into the mechanistic processes mediating Pb^{2+}'s actions on transmitter release.

1. Effects of Pb^{2+} on Depolarization-Evoked Transmitter Release

Using the nictitating membrane of the cat, Kostial and Vouk noted that acute perfusion with low concentrations of Pb^{2+} (<5 μM) attenuated the force of membrane contraction produced by preganglionic nerve stimulation.[102] This Pb^{2+}-induced decrease in contraction was associated with reduced ACh liberation from the perfused superior cervical ganglion. The decrease in both membrane contraction and ACh release produced by Pb^{2+} could be reversed by increasing the Ca^{2+} concentration in the perfusing medium, suggesting that Pb^{2+} decreases depolarization-evoked ACh release by blocking the accompanying influx of Ca^{2+} into the terminal. This inhibition of depolarization-dependent Ca^{2+} influx by Pb^{2+} has been confirmed by a number of investigators employing a variety of preparations and techniques. Pb^{2+}'s ability to attenuate depolarization-evoked ACh release has been demonstrated at the frog neuromuscular junction,[11,103,104] the mammalian neuromuscular junction,[105,106] the bullfrog sympathetic ganglion,[107] and rat brain synaptosomes.[108,109] Similar Pb^{2+}-induced reductions in depolarization-dependent release have been observed with other transmitter systems: NE release from sympathetic nerves innervating the rabbit saphenous artery,[110] DA release from rat striatal synaptosomes,[111] and GABA release from rat cortical synaptosmes.[112]

As discussed previously, membrane depolarization opens nerve terminal membrane Ca^{2+} channels which allow extraneuronal Ca^{2+} to move down its electrochemical gradient into the terminal. Under physiological conditions (nerve stimulation) these channels remain open briefly, on the order of hundreds of milliseconds. The ability of certain heavy metals (e.g., lanthanum) to affect nerve terminal Ca^{2+} accumulation has been shown to be largely dependent on the length of depolarization.[96] Recent studies employing rat brain synaptosomes depolarized by brief exposure to high-K^+ have shown that Pb^{2+} inhibits depolarization-dependent accumulation of radioactive Ca^{2+}.[97,108] These results support the earlier observation that Pb^{2+} reduces radioactive Ca^{2+} uptake into the electrically stimulated frog sympathetic ganglion.[107] Pb^{2+} also blocks the inward Ca^{2+} currents into neurons of the snail.[9] Kinetic analysis of the interaction between Pb^{2+} and Ca^{2+} on neurotransmission at the neuromuscular junction indicates a competitive inhibition by Pb^{2+} ($K_i \sim 1.0 \ \mu M$) of depolarization-dependent Ca^{2+} influx.[97,104,106,107] A similar K_i ($<1.0 \ \mu M$) was determined for Pb^{2+} in synaptosomal ^{45}Ca uptake studies, indicating that this inhibition of Ca^{2+} influx by Pb^{2+} is the major means by which Pb^{2+} is blocking depolarization evoked release.[108] In most of the studies cited above care was taken to minimize Pb^{2+}-induced increases in spontaneous transmitter release which potentially could have confounded data interpretation. The competitive inhibition of Ca^{2+}-mediated transmitter release appears to be identical to that observed for many other divalent cations, including Mg^{2+}, Co^{2+}, Mn^{2+}, and Cd^{2+}. Pb^{2+}, however, is approximately a thousand times more potent than Mg^{2+}, Co^{2+}, or Mn^{2+} and about twice as effective as Cd^{2+}.[11,99,108] It is believed that Pb^{2+} acts by reversibly binding to a Ca^{2+}-receptor site located on the exterior of the presynaptic nerve terminal membrane since the blocking action of Pb^{2+} on the EPP can be immediately reversed by superfusing frog neuromuscular preparations with meso-2,3-dimercaptosuccinic acid, a heavy metal chelator.[113]

2. Effects of Pb^{2+} on Spontaneous Transmitter Release

Pb^{2+}, in addition to its effects on depolarization-evoked release, also produces an increase in spontaneous transmitter release. The Pb^{2+}-induced increase in spontaneous transmitter release, as reflected by an increase in MEPP frequency, has been demonstrated for ACh at the neuromuscular junction.[103-106,114] Similar Pb^{2+}-induced increases in spontaneous release have been demonstrated for several CNS transmitter substances using rat brain synaptosomes: DA,[111,115] ACh,[108,109] and GABA.[112,116] In contrast to the rapid (<1 s) onset of Pb^{2+}'s attenuation of depolarization-evoked release, the Pb^{2+}-induced increase in spontaneous release is delayed by about 15 to 60 s, depending on the preparation and/or the Pb^{2+} concentration employed. Thus, it is possible to separately examine the differential effects of Pb^{2+} on the two types of transmitter release by employing experimental protocols that take advantage of this temporal difference. The delay in the onset of spontaneous release suggests that Pb^{2+} may traverse the presynaptic membrane, and then act intraterminally to facilitate release. The slow reversal of Pb^{2+}-induced augmentation of spontaneous release by 2,3-dimercaptosuccinic acid, in contrast to the rapid reversal of Pb^{2+}-induced attenuation of depolarization-evoked release, also suggests that Pb^{2+} acts intraneuronally to increase spontaneous release.[113] Cooper and Manalis observed that this Pb^{2+}-induced increase in spontaneous release at the neuromuscular junction can be inhibited in the presence of Cd^{2+} (a potent Ca^{2+} channel blocker), suggesting that Pb^{2+} may enter the terminal through the voltage-sensitive membrane Ca^{2+} channels.[117] Attempts to block the Pb^{2+}-induced increase in spontaneous DA release from synaptosomes with either Cd^{2+}, Ni^{2+} (another Ca^{2+} channel blocker), or a combination of Ni^{2+}, tetrodotoxin (a Na^+ channel blocker) and tetraethylammonium (a K^+ channel blocker) were unsuccessful.[111,118] Therefore, the means by which Pb^{2+} moves into the nerve terminal, as well as the relationship between intraterminal Pb^{2+} accumulation and increases in spontaneous release, have yet to be resolved.

The mechanisms by which Pb^{2+} produces its effect on spontaneous transmitter release are still speculative. Pb^{2+} does not appear to induce transmitter release by depolarizing the nerve terminal membrane and thereby increasing Ca^{2+} influx, since the Pb^{2+}-induced increase in spontaneous release is still observed when extraneuronal Ca^{2+} is removed.[109,111,112] In fact, Pb^{2+} has been reported to actually decrease ^{45}Ca uptake by "resting" nerve terminals of sympathetic ganglia.[107] Na^+,K^+-ATPase inhibitors, such as ouabain, are known to increase transmitter release even in the absence of extraneuronal Ca^{2+}.[31,34,119] Release of transmitter induced by Na^+,K^+-ATPase inhibitors can be blocked when the Na^+ gradient across the terminal membrane is eliminated.[31] Pb^{2+} also has been shown to inhibit membrane Na^{2+}, K^+-ATPase.[120-123] However, the Pb^{2+}-induced increase in transmitter release was not altered when the extrasynaptosomal Na^+ concentration was lowered to a level approximating the intraterminal Na^+ concentration, indicating that the increase in spontaneous release is probably not due to inhibition of Na^+,K^+-ATPase by Pb^{2+}.[112,118]

The two major hypotheses advanced to account for the effects of Pb^{2+} on spontaneous release are (1) Pb^{2+} may act as a "Ca^{2+}-mimetic", substituting for Ca^{2+} at those intraterminal sites where increases in Ca^{2+} concentration induce transmitter release, or (2) Pb^{2+} may alter intraterminal Ca^{2+} homeostasis resulting in an increase in the intraterminal Ca^{2+} concentration. There is no direct evidence that Pb^{2+} acts as a Ca^{2+}-mimetic to increase transmitter release. One of the problems in testing this hypothesis is that the intraterminal site at which Ca^{2+} acts to induce transmitter release has not yet been identified. Several studies indicate that Pb^{2+} can substitute for Ca^{2+} to activate calmodulin-dependent phosphorylation.[124-127] Although activation of calmodulin has been suggested in asynchronous, spontaneous transmitter release,[128] there exists at this time a lack of conclusive evidence for a direct association of calmodulin with the transmitter release process.[15] Recently, picomolar concentrations of Pb^{2+} were observed to stimulate partially purified brain protein kinase C in a manner similar to micromolar concentrations of Ca^{2+}.[216] Thus, one can only speculate at present as to whether the Pb^{2+}-induced increase in spontaneous transmitter release reflects Pb^{2+}'s ability to act as a potent Ca^{2+} agonist directly at that unidentified intraterminal Ca^{2+} receptor mediating trasnsmitter release.

Almost all studies examining the effects of *in vitro* Pb^{2+} exposure on a variety of Ca^{2+} related processes have observed that Pb^{2+} interacts in some fashion at most sites at which Ca^{2+} is active. Yet, despite the important role of intracellular Ca^{2+} regulation in nerve terminals, surprisingly few studies have examined the effects of Pb^{2+} on these processes. Presumably, if Pb^{2+} increases spontaneous transmitter release by altering Ca^{2+} regulation, the net effect of such an alteration would be an increase in the intraterminal Ca^{2+} concentration. There are three major aspects of nerve terminal Ca^{2+} regulation which, if altered by Pb^{2+}, would result in an increase of the intraterminal Ca^{2+} concentration. Pb^{2+} could interact with terminal Ca^{2+} regulatory process to either (1) increase the net influx of Ca^{2+} into the terminal, (2) decrease the net efflux of Ca^{2+} out of the terminal, and/or (3) increase the efflux of Ca^{2+} from intraterminal stores or buffers (mitochondria, endoplasmic reticulum) into the cytoplasm.

Although it is well established that Pb^{2+} inhibits depolarization-dependent Ca^{2+} uptake into nerve terminals, the effect of Pb^{2+} on Ca^{2+} fluxes across the non-depolarized nerve terminal has not been thoroughly examined. It does not seem likely that Pb^{2+} would increase Ca^{2+} influx into the terminal, based on its effectiveness in inhibiting the movement of Ca^{2+} at the membrane Ca^{2+} channel. Pb^{2+} appears to actually decrease ^{45}Ca uptake by nondepolarized nerve terminals. The dominant means by which the intraterminal Ca^{2+} concentration is maintained is through membrane Ca^{2+}-ATPase-mediated Ca^{2+} extrusion. Although Pb^{2+} has been reported to be a weak inhibitor of Ca^{2+}-ATPase,[129] Pb^{2+} does not inhibit the efflux of Ca^{2+} from synaptosomes at concentrations that produce large increases in spontaneous transmitter release.[109] In fact, using hippocampal synaptosomes preloaded with ^{45}Ca, Pb^{2+} was found to increase ^{45}Ca efflux in a concentration-dependent manner at the same Pb^{2+} concentrations observed to increase spontaneous transmitter release.[109] The slight temporal differences in onset and peak effects (i.e., the effect of Pb^{2+} on transmitter release precedes its effect on ^{45}Ca efflux) are consistent with the view that Pb^{2+} increases the intraterminal Ca^{2+} concentration by increasing Ca^{2+} efflux from intraterminal Ca^{2+} stores. The elevated intraterminal Ca^{2+} concentrations would produce increased Ca^{2+} interactions at the intraterminal Ca^{2+}-receptor mediating transmitter release. Subsequently this Ca^{2+} would be extruded from the nerve terminal by the membrane Ca^{2+}-ATPase. Since the activity of the Ca^{2+} pump is stimulated by activated calmodulin,[130] it should be recognized that an alternative explanation for the Pb^{2+}-induced increase in ^{45}Ca efflux is that Pb^{2+} may act as a "Ca^{2+}-mimetic" to simultaneously activate both transmitter release and the presynaptic Ca^{2+} pump. However, this Pb^{2+}-induced increase in synaptosomal ^{45}Ca efflux was observed despite the assumed activation of calmodulin-dependent intrasynaptosomal Ca^{2+}-sequestration mechanisms.[131]

There are no direct studies on the effects of Pb^{2+} on intraneuronal Ca^{2+} concentration, although there have been several studies suggestive of possible intraneuronal sites which could serve as a source of increased Ca^{2+} following Pb^{2+} exposure. Studies using nonexcitable cells have shown that Pb^{2+} increases whole cell Ca concentration, most of which is associated with the mitochondrial compartment.[132,133] It is difficult to reconcile the Pb^{2+}-induced increase in synaptosomal ^{45}Ca efflux with the results of these studies, since the observed increase in ^{45}Ca efflux from the nerve terminal would apparently be expected to lower intraneuronal Ca^{2+}. One possibility is that the rate of "new" (unlabeled) Ca^{2+} entering the cell following Pb^{2+} exposure is greater than the efflux of Ca^{2+} (as reflected by ^{45}Ca). Alternatively, this apparent discrepancy could reflect a biphasic Pb^{2+} effect, either in a temporal and/or a concentration-related fashion. For example, low Pb^{2+} concentrations ($<10\ \mu M$) have been shown to inhibit microsomal and mitochondrial Ca^{2+} uptake, whereas higher Pb^{2+} concentrations ($>10\ \mu M$) increased Ca^{2+} uptake.[108,134-136] Decreased mitochondrial and/or microsomal Ca^{2+} uptake would be expected to increase cytosolic Ca^{2+}. Similarly, increases in mitochondrial Ca^{2+} efflux would also increase cytosolic Ca^{2+}. In kidney mitochondria preloaded with ^{45}Ca, Kapoor and van Rossum noted that Pb^{2+} stimulated the release of ^{45}Ca.[135] Using superfused rat brain mitochondria preloaded with ^{45}Ca, we have also observed a Pb^{2+}-induced increase in ^{45}Ca release.[137] The results of Chavez et al. confirm this effect of Pb^{2+} on mitochondrial Ca^{2+} efflux.[138] In addition, this latter study notes that Pb^{2+} uses the mitochondrial Ca^{2+} carrier system, supporting the X-ray microanalysis study in which Pb^{2+} was detected in association with Ca^{2+} in mitochondria from Pb^{2+} exposed synaptosomes.[139] Chavez et al. suggest that Pb^{2+} stimulates mitochondrial Ca^{2+} release by interacting with sulfhydryl groups located on the cytosolic side of the inner mitochondrial membrane.[138]

B. EFFECTS OF ORGANIC LEAD COMPOUNDS ON TRANSMITTER RELEASE

Surprisingly few studies have examined the effects of organo-Pb compounds on neurotransmission. Since organo-Pb is considerably more lipid soluble than Pb^{2+}, it is assumed that organo-Pb compounds readily accumulate inside the nerve terminal.[140] However, the actual intraterminal concentrations of either Pb species have never been measured following *in vitro* exposure. Furthermore, the molecular form(s) in which these Pb compounds exist inside the cell is unkown, making it difficult to compare the results of Pb^{2+} vs. organo-Pb exposure studies. Bondy et al.[141,142] examined the effect of tri-n-butyl-Pb^+ on the release of GABA, glutamate, ACh (choline), glycine, and DA from mouse brain homogenates. The conditions of release (high-K^+, Ca^{2+}-free) make interpretation difficult, but the results suggest that triorgano-Pb^+ at concentrations greater than $0.5\ \mu M$ increases transmitter release by either substituting for Ca^{2+} under depolarizing conditions or acting intraterminally to induce spontaneous transmitter release. Although the data were not presented, Bondy et al. note that similar increases in tri-n-butyl-Pb^+-induced transmitter release were obtained when nondepolarizing concentrations of K^+ were employed, indicating that this triorgano-Pb^+ compound increases spontaneous transmitter release.[142] Preliminary unpublished results from our studies indicate that triethyl-Pb^+ is a potent inducer of spontaneous ACh and GABA release from rat brain synaptosomes.[143] Although the effects of triethyl-Pb^+ on release were not directly assessed, Komulainen and Tuomisto attributed the inhibition of ^{3}H-DA and ^{3}H-5HT uptake into rat brain synaptosomes by this triorgano-Pb$^+$ compound to the release of these transmitters from synaptosomal storage granules.[144] Within the time frame of the transmitter uptake study, triethyl-Pb$^+$ facilitated the removal of newly incorporated ^{3}H-transmitters from the synaptosomes, suggesting that this triorgano-Pb$^+$ compound increases spontaneous transmitter release. In contrast, the nonionic organo-Pb compound, tetraethyl-Pb, did not alter synaptosomal transmitter uptake which by inference suggests that, within the time frame of the uptake study, this tetraorgano-Pb compound does not alter spontaneous transmitter release.[144] The proposed relationship of decreased transmitter uptake to increased transmitter release is only speculative; other explanations of decreased synaptosomal ^{3}H-transmitter uptake in the presence of triethyl-Pb$^+$ are possible.

The mechanism(s) by which triorgano-Pb$^+$ compounds apparently increase transmitter release are unresolved. In a broad sense, trialkyl-Pb$^+$ compounds are believed to alter cellular energy metabolism by their ability to stimulate inner mitochondrial matrix.[145,146] *In vivo* exposure to triethyl-Pb$^+$ produces morphological alterations in mitochondria.[147] Whether such mitochrondrial effects alter transmitter release is unknown. Low concentrations of triethyl-Pb$^+$ (5 μM) increase the intrasynaptosomal Ca^{2+} concentration,[32] which in turn would be expected to increase spontaneous transmitter release. These slowly developing triethyl-Pb$^+$-induced increases in intrasynaptosomal Ca^{2+} concentration were independent of organo-Pb$^+$ concentration (between 5 and 30 μM), reaching a plateau of ~30% above basal Ca^{2+} levels.[32] When extrasynaptosomal Ca^{2+} was removed, triethyl-Pb$^+$ did not increase intrasynaptosomal Ca^{2+} levels. Based on comparisons between the effects of veratridine (a membrane Na$^+$ channel ionophore), verapamil (a membrane Ca^{2+} channel blocker), and ouabain (a membrane Na$^+$,K$^+$-ATPase inhibitor) on intrasynaptosomal Ca^{2+} concentrations, Komulainen and Bondy suggest that triethyl-Pb$^+$ both stimulates synaptosomal Na$^+$ influx through membrane Na$^+$ channels and inhibits synaptosomal Na$^+$ extrusion by decreasing Na$^+$,K$^+$-ATPase activity.[32] These actions would increase the intrasynaptosomal Na$^+$ concentration. Elevated intrasynaptosomal Na$^+$ concentrations, in turn, could increase the intrasynaptosomal Ca^{2+} concentrations by either gradually depolarizing the plasma membrane, thus activating Ca^{2+} influx through voltage-dependent Ca^{2+} channels, or inducing Ca^{2+} efflux from intrasynaptosomal mitochondria. The resulting increase in intraterminal Ca^{2+} concentration would induce the release of transmitter. The inability to detect increases in intrasynaptosomal Ca^{2+} concentrations in the absence of extrasynaptosomal Ca^{2+} argues against membrane depolarization as the means by which triethyl-Pb$^+$ increases spontaneous release. However, the effects of triethyl-Pb$^+$ on intraterminal Ca^{2+} concentration do not correlate well with the effects of triethyl-Pb$^+$ on transmitter release, where the triorgano-Pb$^+$-induced increases in transmitter release were concentration dependent and occurred in the absence of extraneuronal Ca^{2+}.[142,143] Alternatively, triorgano-Pb$^+$ may induce transmitter release through mechanisms not associated with increases in intraterminal Ca^{2+} concentrations. Futher studies are necessary to establish a relationship between triorgano-Pb$^+$-induced increases in transmitter release and intraterminal Ca^{2+} regulation.

V. EFFECTS OF MERCURY (Hg) ON TRANSMITTER RELEASE

Long-term exposure to inorganic mercury (Hg^{2+}), in addition to its toxic effects on the kidneys and intestinal tract, can produce neurotoxic symptoms, including behavioral changes, tremor, and erethism.[148] Current neurotoxicological concern is mainly with organic Hg compounds, specifically methylmercury (MeHg$^+$). "Minamata Disease", which results from exposure to MeHg$^+$, is characterized by considerable neurological involvement, including paresthesia (impairment of vision, hearing, and speech), ataxia, and muscular weakness.[149] The symptoms of poisoning are similar for both acute and chronic MeHg$^+$ exposure. The relative ease with which MeHg$^+$ is absorbed, its ability to concentrate in the brain (15% of total body burden), its relatively long retention by the body, as well as the ability of microorganisms to methylate

Hg^{2+}, increase current concerns about the health effects associated with this organometal. Studies with $MeHg^+$ have shown a variety of neuronal insults, including alterations in protein, DNA, and RNA biosynthesis, changes in phospholipid/phosphoprotein metabolites, abnormalities in mitochondrial function, and perturbations in membrane permeability. $MeHg^+$ increases axonal transport and, at higher concentrations, damages microtubuli. The symptoms of ethylmercury poisoning are quite similar to those described for $MeHg^+$. The high lipid solubility of the short-chain alkylmercury compounds, their ability to readily penetrate cell membranes, combined with their high affinity for membrane protein sulfhydryl groups, all suggest that these compounds may alter membrane structural, transport, and/or receptor processes.

A. EFFECTS OF INORGANIC MERCURY (Hg^{2+}) ON TRANSMITTER RELEASE

1. Effects of Hg^{2+} on Depolarization-Evoked Transmitter Release

An early study in which the effects of 250 μM Hg^{2+} on the excitability threshold of the frog sciatic nerve gastrocnemius muscle preparation were measured found that Hg^{2+} rapidly (<2 min) produced an inhibition of nerve excitability. As this inhibition could be reversed by the addition of thiols (cysteine or glutathione) to the buffer, the investigators concluded that Hg^{2+} interacts with those plasma membrane -SH groups important in maintaining membrane excitability.[150] Since thiols can reverse the 1 mM Hg^{2+}-induced block of the action potential amplitude and the accompanying depolarization of the resting membrane potential in the squid axon, any apparent decrease in evoked release by Hg^{2+} may actually represent a decrease in axonal conduction.[151] However, experiments employing the voltage clamped giant axon suggest that Hg^{2+} does not affect the resting membrane potential but rather decreases early peak conductance.[152] In addition to effects on the axonal membrane, relatively high concentrations of Hg^{2+} microelectrophoretically applied to the innervated surface of the isolated electroplax membrane and frog neuromuscular junction produce depolarization of the postsynaptic membranes.[153,154] In addition to its effect on the postsynaptic membrane, Hg^{2+} can also affect specific postsynaptic receptors, as demonstrated by the marked decrease in the binding of ^{3}H-spiperone, a D_2 dopamine receptor ligand, to rat striatal homogenates in the presence of 1 mM Hg^{2+}.[155] This ability of Hg^{2+} to affect both pre- and postsynaptic processes, presumably as a result of the affinity of Hg^{2+} for -SH groups, can make it difficult to interpret the results of some transmitter release studies. For example, 40 μM Hg^{2+} produced a decrease in the stimulation-produced twitch height of the isolated frog skeletal muscle which was apparently due largely to decreases in the muscle resting potential and action potential amplitude, as well as decreases in effective muscle electrical resistance, rather than to effects on neurotransmitter release.[156,157]

In addition to the effects on axonal conduction and postsynaptic processes there is good evidence that Hg^{2+} also alters evoked transmitter release. Manalis and Cooper reported that micromolar concentrations of Hg^{2+} produced increases in endplate potential amplitude which were independent of any postsynaptic effect.[158] This and subsequent studies[156,159,160] have shown that the increased endplate potential persists for a variable period of time, depending on the Hg^{2+} concentration, and then falls precipitously along with a decrease in the muscle membrane potential. Using mouse brain homogenates under membrane depolarization conditions (55 mM K^+) in both the presence and absence of extranueronal Ca^{2+}, Bondy et al. noted that only high concentrations of Hg^{2+} (50 μM) induced the release of radiolabeled transmitters (glutamate, DA, GABA, glycine, and ACh).[141] The authors noted that the Hg^{2+}-induced release of transmitters was independent of and partially additive to Ca^{2+}-dependent depolarization-evoked release, indicating that Hg^{2+} is inducing transmitter release in some fashion other than substituting for Ca^{2+} under membrane depolarization conditions. However, Bondy et al. used an incubation buffer containing 1 mM ascorbic acid which has been shown to precipitate Hg^{2+},[161] making it difficult to interpret the data.

The ability of low concentrations of Hg^{2+} to increase evoked transmitter release stands in great contrast to other divalent metals such as Pb^{2+} and Cd^{2+} which reduce evoked release by competitively blocking Ca^{2+} uptake into nerve terminals.[97,104,107,108] However, at relatively high concentrations Hg^{2+} does reduce the Ca^{2+} flux into nerve terminals. Nachshen found that a Hg^{2+} concentration of 120 μM was required to produce a 50% reduction (IC_{50}) in the depolarization induced (1 s) uptake of Ca^{2+} into rat brain synaptosomes.[97] By comparison, the IC_{50} values for Pb^{2+} and Cd^{2+} were 0.4 and 1.0 μM, respectively.

The mechanism underlying Hg^{2+}-induced increases in transmitter release are not entirely clear. It is possible that the binding of Hg^{2+} to membrane sulfhydryl groups may enhance Ca^{2+} influx during depolarization or may itself directly affect release. A more probable explanation is based upon the idea that any substance capable of increasing the level of intraterminal Ca^{2+} will also increase evoked transmitter release. In support of this hypothesis, Binah et al. reported that Hg^{2+} concentrations greater than 0.1 μM decreased Ca^{2+} uptake by rat brain mitochondria.[159]

2. Effects of Hg^{2+} on Spontaneous Transmitter Release

Kostial and Landeka observed a large, transient increase in spontaneous ACh release from the perfused cat superior cervical ganglion following exposure to 100 μM Hg^{2+}.[162] Exposure to low Hg^{2+} concentrations (3 to 10 μM) produces an increase in spontaneous release of ACh at the frog neuromuscular junction.[156,158,159,163] This Hg^{2+}-induced increase in spontaneous release has also been demonstrated for other transmitter substances, specifically DA from rat brain striatal synaptosomes,[164,165] although other investigators have failed to observe this effect on synaptosomal release of DA, NE, and 5HT.[166] However, the absence of an effect in this case must be viewed with caution because the release buffer contained ascorbic acid which precipitates Hg^{2+}.[155,161] In studies carried out on the frog neuromuscular junction using Hg^{2+} concentrations of 1 μM or greater the MEPP frequency initially increases and then falls rapidly.[156,163] Since both the rise and fall in MEPP frequency essentially parallel the changes in EPP amplitude,[10,156] a common mechanism is probably responsible for the effects of Hg^{2+} on spontaneous and evoked ACh release. In the frog neuromuscular junction as well as rat brain synaptosomes the increased spontaneous transmitter release occurs in the absence of any Ca^{2+} added to the superfusion solutions.[163,164] Therefore, the effect of Hg^{2+} is not the result of either Ca^{2+} influx resulting from membrane depolarization or Na^+-Ca^{2+} exchange at the terminal membrane. If the Hg^{2+} effect on spontaneous release is ultimately due to an increase in intraterminal Ca^{2+}, then the Ca^{2+} must come from intraterminal Ca^{2+} stores. Binah et al. have suggested that Hg^{2+} affects intracellular Ca^{2+} regulatory mechanisms based on their observation that concentrations of Hg^{2+} greater than 0.5 μM inhibited ^{45}Ca uptake into rat brain mitochondria without affecting oxidative phosphorylation.[159] Similarly, an interaction of Hg^{2+} with Ca^{2+} at the mitochondrial level is indicated by the observation that the binding of Hg^{2+} to -SH groups on two specific mitochondrial membrane proteins (20 and 30 kDa) results in the release of intramitochondrial Ca^{2+} which is accompanied by the collapse of the transmembrane potential and diminution of the NAD(P)H/NAD(P) ratio.[167] Miyamoto reported that simultaneous treatment of frog neuromuscular preparations with a Na^+-channel blocker (tetrodotoxin) and a Ca^{2+}-channel blocker (Co^{2+}) delayed or prevented the increase in MEPP frequency as well as the decrease in muscle membrane potential normally produced by Hg^{2+}.[163] Neither of the blocking agents acting alone reduced the Hg^{2+} effect. It was concluded that Hg^{2+} gains access to the intracellular compartment via Na^+ and Ca^{2+} channels.

Hg^{2+} is also a potent inhibitor of the membrane-bound sulfhydryl-containing Na^+,K^+-ATPase,[168,169] with an observed IC_{50}=160 nM in synaptosomal membrane fragments.[165] Inhibition of Na^+,K^+-ATPase produces alterations in membrane cation permeability, cellular osmotic balance and eventual cell lysis.[168] Initial inhibition of Na^+,K^+-ATPase with cardiac glycosides such as ouabain produces an increase in spontaneous release.[170] There is some evidence that Hg^{2+} may produce its actions on spontaneous release by Na^+,K^+-ATPase inhibition. Both ouabain and

Hg^{2+} produce an increase in spontaneous transmitter release in the absence of extraneuronal Ca^{2+}, indicating that membrane depolarization per se does not mediate the effects of either of these agents. As with ouabain, the Hg^{2+}-induced increase in spontaneous release is attenuated when the Na^+ gradient across the neuronal membrane is markedly decreased. Using the frog neuromuscular preparation, Hare et al.[164] demonstrated that after complete or partial replacement of the extraneuronal Na^+ with equimolar amounts of Li^{3+}, the increase in the MEPP frequency normally induced by 3 μM Hg^{2+} was significantly reduced.[164] Similar studies employing rat striatal synaptosomes showed only a slight attenuation of the Hg^{2+}-induced increase in spontaneous DA release when the extrasynaptosomal Na^+ concentration was lowered from 140 to 32 mM.[165] It should be noted that complete replacement of Na^+ in the superfusion solutions does not prevent the increase in spontaneous neurotransmitter release. In experiments on the frog neuromuscular junction superfused with Ca^{2+}-free Ringer solutions, repeated 90 s periods of tetanic nerve stimulation (800 msec trains at 50 Hz repeated once per second) produced progressively greater increases in MEPP frequency following a brief exposure to 3 μM Hg^{2+}.[164] Ordinarily, tetanic nerve stimulation produces a small increase in MEPP frequency which quickly returns to control levels immediately after stimulation. These results closely resemble those obtained in similar experiments on neuromuscular preparations treated with ouabain but do not clearly implicate an inhibition of membrane Na^+,K^+-ATPase. It is doubtful that Na^+,K^+-ATPase inhibition is solely responsible for the effects of Hg^{2+}.[161] However, it does appear that Hg^{2+} alters membrane structure or function in such a manner as to prevent the maintenance of normal transmembrane ionic balances. Indeed, if the rapid fall in muscle membrane resting potential as well as the fall in effective muscle membrane ("input") resistance noted by Juang[157] also occur in nerve this would suggest that the effects of Hg^{2+} on spontaneous transmitter release cannot be attributed solely to inhibition of membrane Na^+,K^+-ATPase. The available data suggest that the effects of Hg^{2+} on synaptic transmission are probably a complex combination of actions on intracellular Ca^{2+} buffering, ATPase inhibition and, possibly, direct alterations of exocytosis and membrane permeability.

B. EFFECTS OF METHYLMERCURY (MeHg⁺) ON TRANSMITTER RELEASE

1. Effects of MeHg⁺ on Depolarizaition-Evoked Transmitter Release

$MeHg^+$, at relatively high concentrations (>20 μM), attenuates depolarization-evoked ACh release at the neuromuscular junction. This attenuation of the EPP continues despite the removal of $MeHg^+$ from the extracellular solution.[157,170-173] Since depolarization-evoked transmitter release is dependent on the influx of Ca^{2+} into the nerve terminal through voltage-dependent Ca^{2+} channels, some effort has been made to relate $MeHg^+$'s effect on depolarization-evoked release to Ca^{2+} influx. $MeHg^+$ reduces both phases of synaptosomal depolarization-dependent ^{45}Ca uptake;[174,175] the "fast phase" which occurs in less than 1 s, and the "slow phase" which occurs over longer periods of time (>1 s).[94,96] The inhibition of "slow phase" ^{45}Ca uptake by $MeHg^+$ could be in part reversed by either increasing the external Ca^{2+} or decreasing the external Na^+ concentrations, indicating that a portion of the blocking action of $MeHg^+$ on "slow phase" Ca^{2+} uptake involves Na^+-Ca^{2+} exchange.[175]

Failure to reverse the $MeHg^+$-induced block of the EPP with increased extracellular Ca^{2+} or 4-aminopyridine (which appears to block K^+ efflux and increase Ca^{2+} influx) suggests that $MeHg^+$ does not produce its effect by a competitive block of divalent cation influx.[176,177] This observation supports the observation that $MeHg^+$ produces a non-competitive block of depolarization-induced ^{45}Ca uptake into synaptosomes.[174] However, washing of the neuromuscular junction preparation with 1 mM D-penicillamine, a Hg chelator, did in some cases reverse the $MeHg^+$-induced block of EPPs.[177] Apparently the continued presence of $MeHg^+$ is necessary to maintain the block of synchronous evoked release. The block of synchronous release by $MeHg^+$ persisted when the intracellular free Ca^{2+} was increased by perfusing the preparations with Ca^{2+}-containing liposomes.[177] Alternatively, $MeHg^+$ may in part block depolarization-evoked transmitter release by blocking action potential conduction into the nerve terminal, since the MeHg$^+$-induced block of the EPP could be reversed by simply increasing the intensity and/or duration of the electrical stimulus applied to the phrenic nerve.[177] Blockage of impulse propagation along the axonal membrane or at the unmyelinated portion of the nerve terminal could result from MeHg$^+$-induced increases in membrane permeability,[178] increases in membrane leakage conductance,[179] and/or conformational changes in axonal or terminal membrane proteins resulting from MeHg$^+$'s interaction with protein sulfhydryl, carbonyl, amino, and/or hydroxyl groups.[180,181] This latter possibility is supported by the observation that sulfhydryl reagents can block nerve conduction in a fashion apparently similar to MeHg$^+$.[151,179,182-186] The Na$^+$ channel protein is one possible site where MeHg$^+$ could act to block conductance, as suggested by the decrease in membrane Na$^+$ conductance produced by organomercurial exposure.[179,183,187]

Subsequent to a second MeHg$^+$ exposure (following reversal of the EPP block by increase stimulus intensity), washout with either high extraneuronal Ca^{2+} or 4-aminopyridine was effective at partially reversing the MeHg$^+$ block, suggesting that in addition to conduction block, MeHg$^+$ is exerting direct effects at the nerve terminal on Ca^{2+}-dependent transmitter release.[177] This direct effect may be due to inhibition of depolarization-dependent Ca^{2+} influx, resulting from MeHg$^+$'s interaction with either the membrane Ca^{2+} channel (as discussed above) or processes that regulate intraterminal Ca^{2+}. This latter possibility is supported by observations that MeHg$^+$ decreases mitochondrial Ca^{2+} uptake (Binah et al., 1975) and increases mitochondrial Ca^{2+} efflux.[188] If MeHg$^+$ indeed alters intraterminal Ca^{2+} regulation, then changes in asynchronous (spontaneous) transmitter release would be expected. In fact, MeHg$^+$-induced increases in spontaneous transmitter release from rat brain synaptosomes are observed at concentrations below those that produce any change in the magnitude of depolarization-evoked release, as discussed below.

2. Effects of MeHg$^+$ on Spontaneous Transmitter Release

Initial exposure to MeHg$^+$ produces an increase in spontaneous transmitter release. This effect has been demonstrated both electrophysiologically in the neuromuscular junction[10,156,170,171,173,189-191] and neurochemically in CNS preparations.[141,142,144,188,192,193] The MeHg$^+$ concentrations necessary to induce transmitter release at the neuromuscular junction ($>10\,\mu M$) are somewhat greater than the effective MeHg$^+$ concentrations in the CNS preparations (>0.5 μM), perhaps reflecting the elimination of non-neuronal tissue barriers associated with tissue-disrupted CNS synaptosomes. A similar argument may explain the more rapid onset of MeHg$^+$'s effects observed in CNS preparations. The transmitter substances examined in CNS preparations include DA, GABA, ACh (choline), glutamate, glycine, 5HT, and NE.

Several transmitter release studies suggest that MeHg$^+$ is accumulated in the nerve terminal where it acts to increase the intraterminal free Ca^{2+} concentration which, in turn, increases spontaneous transmitter release. The means by which MeHg$^+$ enters the terminal is unknown. Since MeHg$^+$ is lipophilic, it can readily diffuse across the terminal membrane as demonstrated in model lipid membrane systems.[194-196] However, there have been no direct studies examining the transport of MeHg$^+$ into nerve terminals. Based on electrophysiological and pharmacological studies employing the neuromuscular junction, Atchison has suggested that movement of MeHg$^+$ through membrane voltage-dependent Ca^{2+} channels may serve as an additional pathway by which MeHg$^+$ can enter the nerve terminal.[190,191] Atchison proposes that MeHg$^+$ normally accumulates in the nerve terminal subsequent to passive diffusion. Since the rate of terminal MeHg$^+$ accumulation by this diffusion pathway is relatively slow, the time to the maximal effect of 100 μM MeHg$^+$ on MEPP frequency (spontaneous release) is also relatively slow (40 ± 5 min). However, if the nerve terminal membrane is depolarized with high-K$^+$ in a Ca^{2+}-free medium (which would open membrane voltage-dependent Ca^{2+} channels but not evoke transmitter release), the time to the MeHg$^+$-induced increase in spontaneous release is markedly shortened (10 ± 4 min). Similarly, treatment of the neuromuscular junction with the

Ca^{2+} channel activator, Bay-K-8644, in Ca^{2+}-deficient solutions also decreased the time to peak increase in MEPP frequency (13 ± 3 min). Based on these two observations, Atchison proposes that $MeHg^+$ can enter the nerve terminal through membrane Ca^{2+} channels when such channels are open. It is important to note that when the membrane Ca^{2+} channels are closed, $MeHg^+$ can still enter the terminal, although more slowly, via passive diffusion, thus explaining the inability of Co^{2+} (a Ca^{2+} channel blocker) to block the increase in spontaneous release induced by $MeHg^+$.[163]

$MeHg^+$ could induce transmitter release by either directly supporting transmitter release (i.e., acting as a Ca^{2+}-mimetic), displacing transmitter from intraterminal storage sites (in a fashion analogous to reserpine), and/or acting to increase the intraterminal Ca^{2+} concentration (either by increasing the net influx of Ca^{2+} into the terminal or by altering intraterminal Ca^{2+} regulation). $MeHg^+$ increases the efflux of ^{3}H-deoxyglucose phosphate from neuroblastoma cells[217] and synaptosomes.[188] The close correlation between $MeHg^+$-induced ^{3}H-deoxyglucose phosphate efflux and ^{3}H-transmitter release suggests that $MeHg^+$'s ability to increase transmitter release reflects, in part, leakage of transmitter that occurs subsequent to non-specific increases in nerve terminal membrane permeability.[188] Alternatively, there is considerable evidence that $MeHg^+$ increases the intraterminal Ca^{2+} concentration which, in turn, induces transitter release. A recent study using the fluorescent Ca^{2+} ion indicator dye, fura-2, has demonstrated that low $MeHg^+$ concentrations (<3 μM) produce marked concentration-dependent increases in the intrasynaptosomal Ca^{2+} concentration. However, the relationship of the $MeHg^+$-induced increases in intraterminal Ca^{2+} (as reported by Komulainen and Bondy)[32] to $MeHg^+$-induced increases in spontaneous transmitter release is unresolved because some of the data are contradictory.

One means by which the intraterminal Ca^{2+} concentration can be increased is by membrane depolarization. $MeHg^+$ has been shown to depolarize isolated squid axons,[179] and decrease the Na^+ current in neuroblastoma cells.[183] However, the extent of membrane depolarization alone does not appear to account for the large increase in spontaneous release induced by $MeHg^+$.[191] Since transmitter release resulting from membrane depolarization is dependent on the influx of extraneuronal Ca^{2+} into the nerve terminal via voltage-dependent membrane Ca^{2+} channels, the lack of any marked attenuation of $MeHg^+$-induced increases in transmitter release in the absence of extraterminal Ca^{2+} [188,190,191] argues against membrane depolarization as the mechanism of $MeHg^+$ action. In contrast, there is a marked reduction in the $MeHg^+$-induced increase in intrasynaptosomal Ca^{2+} concentration when extrasynaptosomal Ca^{2+} is removed.[32] That extrasynaptosomal Ca^{2+} serves as the major (but not sole) source of the $MeHg^+$-induced intrasynaptosomal Ca^{2+} is also supported by the observation that $MeHg^+$ (>30 μM) stimulates the non-depolarized accumulation of ^{45}Ca into synaptosomes from the extrasynaptosomal meduim.[32] However, it should be noted that under conditions of membrane depolarization, $MeHg^+$ (>25 μM) produces a concentration-dependent decrease in synaptosomal ^{45}Ca uptake.[174] This apparent discrepancy in ^{45}Ca uptake may involve selective effects of $MeHg^+$ on Ca^{2+} influx during resting vs. depolarizing conditions. The Ca^{2+} channel blocker, verapamil, while producing a 50% reduction in the depolarization-induced increase in intrasynaptosomal Ca^{2+} concentration,[32] did not alter the $MeHg^+$-induced increase in intrasynaptosomal Ca^{2+} concentration.[32] Despite the dependence on extrasynaptosomal Ca^{2+}, this increase in the intrasynaptosomal Ca^{2+} concentration produced by $MeHg^+$ is apparently not the result of membrane depolarization since the effect persists even when verapamil-sensitive Ca^{2+} channels are blocked. Thus $MeHg^+$ appears to increase net synaptosomal Ca^{2+} accumulation by either increasing Ca^{2+} influx via verapamil-insensitive membrane channels or decreasing Ca^{2+} efflux (i.e., Ca^{2+}-ATPase inhibition). This lack of an effect of verapamil on $MeHg^+$'s action is also consistent with the results and interpretation of Atchison, who proposed that while $MeHg^+$ can enter the terminal through open Ca^{2+} channels as well as by passive diffusion, under normal (nondepolarization) conditions $MeHg^+$ will enter mainly by passive diffusion.[190] Since the membrane Ca^{2+} channels are closed under resting conditions, Ca^{2+} channel blockers will not alter the rate at which $MeHg^+$ is accumulated by the terminal.

The roles of Na^+ and membrane Na^+-channels in the actions of $MeHg^+$ are not resolved. Since treatment of the terminal with either monensin (a Na^+ ionophore)[190] or tetrodotoxin (a Na^+-channel blocker)[163,170] did not alter the effect of $MeHg^+$ on spontaneous release, it appears that $MeHg^+$ is neither using the Na^+ channel to enter the terminal, nor increasing the influx of Na^+ into the terminal. Similarly, tetrodotoxin did not block the increase in intrasynaptosomal Ca^{2+} concentration produced by $MeHg^+$.[32] Thus, if $MeHg^+$ is increasing the intrasynaptosomal Na^+ concentration as reported by Cheung and Verity,[197] this effect is apparently not mediated by increases in Na^+ influx through membrane Na^+ (tetrodotoxin-sensitive) channels. Although prolonged times to peak $MeHg^+$-induced increase in spontaneous release were noted when the extraneuronal Na^+ was removed and replaced with methylamine, Atchison suggests these changes may reflect methylamine effects rather than changes resulting from Na^+ removal.[190] Our data with synaptosomes indicate that similar $MeHg^+$-induced increases in spontaneous DA release occur when the extrasynaptosomal Na^+ is replaced with choline.[193] These results are important with respect to $MeHg^+$'s reported ability to inhibit brain Na^+,K^+-ATPase at low concentrations.[198-200] Na^+,K^+-ATPase inhibition increases the intraterminal Na^+ concentration which, in turn, induces Ca^{2+} leakage from sequestered intracellular stores, thus increasing transmitter release. Since increases in spontaneous transmitter release produced by known inhibitors of Na^+,K^+-ATPase (cardiac glycosides) can be blocked by reducing the external Na^+ concentration, the observations that $MeHg^+$ still induces release in the presence of low extraneuronal Na^+ suggests that inhibition of Na^+,K^+-ATPase is not the means by which $MeHg^+$ increases spontaneous release. This conclusion is supported by the results of Komulainen and Bondy which note that in contrast to the rapid onset of $MeHg^+$'s effect, the increase in the intrasynaptosomal Ca^{2+} concentration produced by ouabain developed relatively slowly.[32] Furthermore, since the increases in the intrasynaptosomal Ca^{2+} concentrations produced by a combination of ouabain and $MeHg^+$ were additive, it was concluded that any elevation in intrasynaptosomal Na^+ produced by $MeHg^+$ contributed minimally to observed increases in Ca^{2+}.

Most studies of neurotransmitter release indicate that $MeHg^+$ increases spontaneous transmitter release by affecting intraterminal Ca^{2+} stores or buffers. The mitochondria contain the majority of the intraterminal sequestered Ca^{2+} and serve as the major means by which large increases in intraterminal Ca^{2+} resulting from depolarization-evoked Ca^{2+} influx are initially reduced.[19] $MeHg^+$ has been shown to affect several mitochondrial functions, including respiration and coupled oxidative phosphorylation.[197,201-205] $MeHg^+$ produces ultrastructural and pathological changes in mitochondria that are consistent with inhibition of both respiration and/or ADP phosphorylation.[206,207] Inhibition of these processes would collapse the membrane potential across the mitochondria which, in turn, could facilitate Ca^{2+} efflux. For example, the combination of rotenone (which inhibits respiration), oligomycin (which inhibits ATP synthesis), and iodoacetic acid (which inhibits glycolysis) produces an inhibition of net mitochondrial Ca^{2+} uptake and decreases synaptosomal ATP.[208,209] Although Komulainen and Bondy suggest that the source of most of the $MeHg^+$-induced increase in intrasynaptosomal free Ca^{2+} concentration is extraneuronal Ca^{2+}, they do indicate that some of this increase is due to effects of $MeHg^+$ on intrasynaptosomal mitochondria, possibly by inhibition of mitochondrial Ca^{2+} uptake.[32] This conclusion is based on the decreased effectiveness of $MeHg^+$ in elevating the intrasynaptosomal free Ca^{2+} concentration in the presence of rotenone and oligomycin. These investigators also indicate that $MeHg^+$-induced increases in intrasynaptosomal Ca^{2+} are not the result of the inhibition of mitochondrial glycolysis.

By examining the ability of a variety of pharmacological agents (reported to exert specific actions on mitochondria) to modify the $MeHg^+$-induced increase in spontaneous release at the neuromuscular junction, Atchison and coworkers concluded that the mitochondria are a major site of $MeHg^+$ action. These investigators employed several agents known to inhibit mitochondrial function, including dinitrophenol, dicoumarol, and valinomycin. None of these agents blocked the MeHg+-induced stimulation of spontaneous release, indicating that simple inhibition of mitochondrial function does not prevent the MeHg+ effect.[210] However, ruthenium red, a putative inhibitor of the mitochondrial Ca^{2+} uptake uniporter, was found to block the stimulatory effect of MeHg+ on MEPP frequency, suggesting that the increase in spontaneous transmitter release was due to an increase in cytosolic Ca^{2+} resulting from MeHg+'s ability to either block mitochondrial Ca^{2+} uptake or promote its release.[210] This finding supports the observation that MeHg+ decreases mitochondrial Ca^{2+} uptake.[159] A subsequent study of the effect of MeHg+ in the presence of Ba^{2+}, Sr^{2+}, and Ca^{2+} on asynchronous evoked transmitter release suggests that MeHg+ actually releases Ca^{2+} from intraterminal stores as opposed to merely preventing Ca^{2+} uptake by intraterminal Ca^{2+}-buffering systems.[176] Exposure to drugs which putatively either release Ca^{2+} from endoplasmic reticulum (caffeine) or mitochondria (ouabain), or block Ca^{2+} release from endoplasmic reticulum (N,N-dimethyl-amino-8-octyl-3,4,5-trimethoxybenzoate), did not prevent the action of MeHg+.[211] However, a putative inhibitor of mitochondrial Ca^{2+} uptake and release [N,N-bis-(3,4-dimethoxyphenethyl)-N-methylamine] was found to inhibit the MeHg+-induced increase in MEPP frequency; this further indicates that MeHg+ acts to increase spontaneous transmitter release by increasing the net mitochondrial Ca^{2+} efflux which leads to an increase in intraterminal free Ca^{2+}.[211]

Our results using both superfused rat brain synaptosomes and mitochondria also indicate that MeHg+ may act to increase Ca^{2+} efflux from mitochondria.[188] MeHg+, at concentrations similar to those that increase spontaneous transmitter release (>0.5 μM), produced a concentration-dependent increase in the efflux of ^{45}Ca from superfused isolated rat brain mitochondria preloaded with the radiolabel. Despite the large MeHg+-induced increases in both synaptosomal spontaneous transmitter release and mitochondrial Ca^{2+} efflux, only a very minimal increase in synaptosomal ^{45}Ca efflux was observed, and only at higher (>3 μM) MeHg+ concentrations. This observation contrasts with the close relationship between transmitter release and Ca^{2+} efflux observed with Pb^{2+}.[109] The efflux of ^{45}Ca from cells preloaded with ^{45}Ca has been employed as an indirect indicator of changes in cytosolic Ca^{2+} concentration.[212-214] Presumably, any treatment that increases the intracellular Ca^{2+} concentration would activate the various cellular mechanisms involved in maintaining the cytosolic Ca^{2+} concentration at approximately 0.2 μM, including the membrane Ca^{2+}-ATPase pump.[44,215] Therefore, if MeHg+ increases spontaneous transmitter release by increasing the intracellular free Ca^{2+} concentration, the MeHg+ should have produced an increase in synaptosomal ^{45}Ca efflux that was related, both temporally and quantitatively, to the increases in transmitter release. Since such a relationship between synaptosomal transmitter release and Ca^{2+} efflux was not observed, MeHg+ must have actions other than increasing mitochondrial Ca^{2+} efflux. If MeHg+'s mitochondrial action indeed increases the intraterminal Ca^{2+} concentration, then either this increased Ca^{2+} is readily absorbed by other (nonmitochondrial) neuronal Ca^{2+} buffers, or the membrane Ca^{2+}-ATPase pump is inhibited by MeHg+.

The MeHg+-induced increase in spontaneous transmitter release is eventually followed by a suppression of release.[156,163,170,171,189,191] This depression in both spontaneous and depolarization-evoked transmitter release by MeHg+ has only been observed in neuromuscular junction preparations. In synaptosomal preparations washed with a MeHg+-free buffer following a short (1.5 min) exposure to MeHg+, transmitter release decreased to baseline (pre-exposure) rates from which subsequent release evoked by depolarization was not impaired.[188] The inability to observe a suppression of release in the synaptosomal preparation may reflect the lower MeHg+ concentrations and/or shorter exposure times than employed in the neuromuscular junction preparation. Although simple washing of the neuromuscular junction with MeHg+-free buffer was ineffective, when a MeHg+-chelator (penicillamine) was added to the washing buffer the MeHg+-induced suppression of transmitter release was reversed.[191] The observation that the continued presence of MeHg+ is apparently necessary in order to maintain the latent suppression of release suggests that MeHg+ is acting directly in the terminal to prevent the efflux of neurotransmitter. In support of this "direct" action of $MeHg^+$ are the observations that the suppression of release is neither a preparation "artifact" reflecting post-synaptic ACh receptor inhibition, since the response to iontophoretic ACh application was not altered,[170] nor a consequence of depleted neurotransmitter stores, since treatment with La^{3+} was still effective at stimulating spontaneous release.[191] Whether the binding or interaction of $MeHg^+$ at intraterminal sites directly involved with the discharge of transmitter from the terminal is responsible for the suppression of release is unresolved. Similarly, it is uncertain what role this "direct" action of $MeHg^+$ may play in the $MeHg^+$-induced attenuation of depolarization-evoked release (in addition to the noncompetitive inhibition of Ca^{2+} influx as discussed previously).

VI. CONCLUDING REMARKS

Acute *in vitro* exposure to relatively low concentrations of several neurotoxic heavy metals has been shown to produce alterations in transmitter release from both PNS and CNS nerve terminals. Relatively consistent effects of various heavy metals on transmitter release obtained from PNS preparations such as the neuromuscular junction have been described. In early neurochemical studies employing various CNS preparations (particularily those studies examining the effects of Pb^{2+}), such consistency of heavy metal effects was not evident. This inconsistency was apparently a result of inattention to a number of confounding factors, such as limitations of the preparations and/or techniques employed, composition of the physiological buffers, duration of the depolarization, and solubility of the metal in the buffer. However, after these various factors had been recognized and addressed, there emerged a remarkable similarity, both quantitatively and qualitatively, between the description of heavy metal effects on transmitter release obtained from electrophysiological experiments employing PNS preparations and neurochemical experiments employing CNS preparations. Furthermore, a combination of electrophysiological and neurochemical approaches has facilitated the elucidation of the mechanisms by which the heavy metals exert their effects on transmitter release. While a variety of potential sites of heavy metal action are recognized, our current ability to fully characterize those processes affected by heavy metal exposure is limited by our incomplete knowledge of the mechanisms involved in transmitter release.

Decreases in depolarization-evoked transmitter release can result from reductions in voltage-dependent Ca^{2+} influx across the nerve terminal membrane, usually at the level of the membrane voltage-dependent Ca^{2+} channel. Pb^{2+} is a potent competitive inhibitor of Ca^{2+} influx at the membrane Ca^{2+} channel. The Pb^{2+}-induced attenuation of depolarization-evoked transmitter release can be reversed by increasing the extraneuronal Ca^{2+} concentration. In contrast, the reduction in depolarization-evoked release produced by exposure to $MeHg^+$ appears to involve several processes. These processes include a blockage of conduction, a "direct" interaction of $MeHg^+$ inside the terminal that prevents transmitter discharge, as well as a noncompetitive inhibition of voltage-dependent Ca^{2+} influx resulting from an unidentified intraterminal action of $MeHg^+$. Hg^{2+} produces an initial increase in depolarization-evoked transmitter release, which probably reflects an increase in the basal intraterminal Ca^{2+} concentration that, when combined with depolarization-dependent influx of Ca^{2+}, results in an increase in the Ca^{2+} concentration available to facilitate transmitter release.

The rate of spontaneous transmitter release is increased by several of the neurotoxic heavy metals, including Pb^{2+}, triorgano-Pb^+, Hg^{2+}, and $MeHg^+$. Pb^{2+} and $MeHg^+$ are examples of metals that differentially affect the two release processes. They decrease depolarization-evoked release but increase spontaneous transmitter release. With Pb^{2+} the mechanisms mediating these differential actions on release are not directly related, whereas the $MeHg^+$-induced decrease in EPP amplitude may be associated, in part, with those processes that mediate the $MeHg^+$-induced increases in MEPP frequency. Both Pb^{2+} and $MeHg^+$ appear to increase spontaneous release by increasing the intraterminal free Ca^{2+} concentration. The source of this Ca^{2+} is, in part, intraterminal mitochondria. Both Pb^{2+} and $MeHg^+$ increase the efflux of Ca^{2+} from mitochondria. However, other mechanisms likely contribute to the increases in spontaneous release induced by these metals. Pb^{2+} has been suggested to act as a Ca^{2+}-mimetic, and as such may act to directly trigger transmitter release. $MeHg^+$ may act to either increase Ca^{2+} influx into, and/or decrease Ca^{2+} efflux out of the terminal, thus increasing the intraterminal free Ca^{2+} concentration.

Ultimately it is hoped that an understanding of the acute, *in vitro* effects of the neurotoxic heavy metals on transmitter release will lead to the development of testable hypotheses regarding the chronic *in vivo* effects of these metals. Since considerable expense and effort is required for examining the *in vivo* effects of these heavy metals, a thorough understanding of their acute *in vitro* actions, as well as the mechanism by which these metals produce their actions, will facilitate chronic exposure studies. Since many of the neurotoxic metals appear to produce their actions by interacting with Ca^{2+}-mediated processes, it is necessary to understand the potential detrimental effects that chronic alterations in intraneuronal Ca^{2+} regulation may have with respect to nervous system function. The potential consequences of alterations in either Ca^{2+} regulation or transmitter release, particularly during development and maturation of the CNS, are not well understood. Perhaps acute metal-induced changes in neurotransmitter release during critical developmental periods will be ultimately expressed as chronic perturbations in neuronal activity at either the presynaptic, postsynaptic, and/or the integrative neuronal level. Potential changes in up/down regulation of neuronal activity may be a major factor mediating some of the neurotoxic effects, particularly subtle cognitive deficits, associated with chronic exposure to relatively low concentrations of many of the neurotoxic metals.

ACKNOWLEDGMENTS

The authors thank Dr. I. Arthur Michaelson for his helpful suggestions and discussions during the preparation of this manuscript, and Annette Townsley for typing the manuscript. The authors research was supported by NIH grants ES-03399 and ES-03992.

REFERENCES

1. **Donaldson, J. and Bardeau, A.,** Manganese neurotoxicity: possible clues to the etiology of human brain disorders, in *Metal Ions in Neurology and Psychiatry,* Gaabay, S., Harris, J., and Ho, B., Eds., Alan R. Liss, New York, 1985, 259.
2. **Rowan, M. J.,** Central nervous system toxicity evaluation *in vitro:* neurophysiological approach, in *Neurotoxicology,* Blum, K. and Manzo, L., Eds., Marcel Dekker, New York, 1985, chap. 28.
3. **Dewar, A. J.,** Neurotoxicity testing, in *Testing for Toxicity,* Gorrod, J. W., Ed., Taylor and Francis, London, 1981, 199.
4. **Fournier, E. and Roux, F.,** Limits of the use in neurotoxicology of cellular models developed in neurobiology, in *Advances in Neurotoxicology,* Manzo, L., Ed., Pergamon, Oxford, 1980, 307.
5. **Goldberg, A. M.,** Mechanisms of neurotoxicology as studied in tissue culture systems, *Toxicology,* 17, 201, 1980.
6. **Hayes, A. W., Ed.,** *Principles and Methods in Toxicology,* Raven Press, New York, 1982.
7. **Kolber, T. J., Wong, T. K., Grant, L. O., DeWaskin, R. S., and Hughes, T. J., Eds.,** *In Vitro Toxicity Testing of Environmental Agents: Current and Future Possibilities,* Elsevier, Amsterdam, 1982.
8. **Habermann, E.,** Advantages and pitfalls of particulate brain and membranes in neurotoxicological studies, *Neurobehav. Toxicol. Terat.,* 4, 613, 1982.
9. **Audesirk, G.,** Effects of lead exposure on the physiology of neurons, *Prog. Neurobiol.,* 24, 199, 1985.
10. **Cooper, G. P. and Manalis, R. S.,** Influence of heavy metals on synaptic transmission, *Neurotoxicology,* 4, 69, 1983.

11. **Cooper, G. P., Suszkiw, J. B., and Manalis, R. S.,** Heavy metals: effects on synaptic transmission, *Neurotoxicology,* 5, 247, 1984.

12. **Manalis, R. S. and Suszkiw, J. B.,** Effects of heavy metal ions on transmitter release, in *Presynaptic Regulation of Transmitter Release,* Feigenbaum, J. J. and Hanani, M., Eds., Freund Publishing, Tel Aviv, 1988.

13. **Shellenberger, M. K.,** Effects of early lead exposure on neuotransmitter systems in the brain: a review with commentary, *Neurotoxicology,* 5, 177, 1984.

14. **Katz, B. and Miledi, R.,** The release of acetylcholine from nerve endings by graded electrical pulses, *Proc. R. Soc. London, Ser. B.,* 167, 23, 1967.

15. **Augustine, G. J.,** Calcium action in synaptic transmitter release, *Ann. Rev. Neurosci.,* 10, 633, 1987.

16. **Siggins, G. R. and Gruol, D. L.,** Synaptic mechanisms in the vertebrate central nervous system, in *Handbook of Physiology, Intrinsic Regulatory Systems of the Brain,* Vol. 4, Bloom, F. E., Ed., American Physiological Society, Bethesda, MD, 1986.

17. **Cooper, J., Bloom, F., and Roth, R.,** *Biochemical Basis of Neuropharmacology,* 5th ed., Oxford University Press, New York, 1986.

18. **Llinas, R., Sugimori, M., and Simon, S. M.,** Transmission by presynaptic spike-like depolarization in the squid synapse, *Proc. Natl. Acad. Sci.,* 79, 2415, 1982.

19. **Nicholls, D. G.,** Intracellular calcium homeostasis, *Br. Med. Bull.,* 42, 353, 1986.

20. **Mailman, R. and Morell, P.,** Neurotoxicants and membrane-associated functions, *Rev. Biochem. Toxicol.,* 4, 213, 1982.

21. **Rubin, R. P.,** Calcium and Cellular Secretion, Plenum Press, New York, 1982.

22. **Toth, G. P., Cooper, G. P., and Suszkiw, J. B.,** Effects of divalent cations on acetylcholine release from digitonin-permeabilized rat cortical synaptosomes, *Neurotoxicology,* 8, 507, 1987.

23. **Hille, B.,** *Ionic channels of Excitable Membranes,* Sinauer Associates, Sunderland, MA, 1984.

24. **Lund, A. E. and Narahashi, T.,** Kinetics of sodium channel modification as the basis for the variation in the nerve membrane effects of pyrethroids and DDT analogs, *Pestic. Biochem. Physiol.,* 20, 208, 1983.

25. **Gant, D., Eldefrawi, M., and Eldefrawi, T.,** Cyclodiene insecticides inhibit $GABA_A$ receptor-regulated chloride transport, *Toxicol. Appl. Pharmacol.,* 88, 313, 1987.

26. **Mullins, L. and Brinley, F.,** Potassium fluxes in dialyzed squid axons, *J. Gen. Physiol.,* 53, 504, 1969.

27. **Shull, G. and Greeb, J.,** Molecular cloning of two isoforms of the plasma membrane Ca^{2+}-transporting ATPase from rat brain, *J. Biol. Chem.,* 263, 8646, 1988.

28. **Wallich, E.,** Personal communication, 1987.

29. **Skou, J. and Hilberg, C.,** The effect of sulphydryl-blocking reagents and of urea on the $(Na^+ + K^+)$-activated enzyme system, *Biochim. Biophys. Acta,* 110, 359, 1965.

30. **Elmquist, D. and Feldman, D. S.,** Effects of sodium pump inhibitors on spontaneous acetylcholine release at the neuromuscular junction, *J. Physiol.,* 181, 498, 1965.

31. **Baker, P. F. and Crawford, A. C.,** A note on the mechanisms by which inhibitors of the sodium pump accelerate spontaneous release of transmitter from motor nerve terminals, *J. Physiol. (London),* 247, 209, 1975.

32. **Komulainen, H. and Bondy, S.,** Increased free intrasynaptosomal Ca^{2+} by neurotoxic organometals: distinctive mechanisms, *Toxicol. Appl. Pharmacol.,* 88, 77, 1987.

33. **Meyer, E. M. and Cooper, J. R.,** Role of synaptosomal Na-accumulation in transmitter release, *Neurochem. Res.,* 9, 815, 1984.

34. **Sweadner, K.,** Ouabain-evoked norepinephrine release from intact rat sympathetic neurons: evidence for carrier-mediated release, *J. Neurosci.,* 5, 2397, 1985.

35. **Glagoleva, I. M., Liberman, E. A., and Khashaev, Z.,** Effect of uncouplers of oxidative phosphorylation on output of acetylcholine from nerve endings, *Biofizika,* 15, 76, 1970.

36. **Carmody, J. J.,** Enhancement of acetylcholine secretion by two sulfhydryl reagents, *Eur. J. Pharmacol.,* 47, 457, 1978.

37. **Nowycky, M. C., Fox, A. P., and Tsien, R. W.,** Three types of neuronal calcium channel with different calcium against sensitivity, *Nature,* 316, 440, 1985.

38. **Miller, R. J.,** Multiple calcium channels and neuronal function, *Science,* 235, 46, 1987.

39. **Hagiwara, S. and Byerly, L.,** Calcium channel, *Ann. Rev. Neurosci.,* 4, 69, 1981.

40. **Llinas, R., Steinberg, I. Z., and Walton, K.,** Relationship between presynaptic calcium current and postsynaptic potential in squid giant synapse, *Biophys. J.,* 33, 323, 1981.

41. **Tsien, R. W.,** Modulation of calcium currents in heart cells and neurons, in *Neuromodulation,* Levitan, L. and Kaezmarek, L., Eds., Osford University Press, Oxford, 1986.

42. **Stanfield, P. R.,** Voltage-dependent calcium channels of excitable membranes, *Br. Med. Bull.,* 42, 359, 1986.

43. **Katz, B.,** *The Release of Neural Transmitter Substances,* Liverpool University Press, Liverpool, 1969.

44. **Snelling, R. and Nicholls, D. G.,** Calcium efflux and cycling across the synaptosomal membrane, *Biochem. J.,* 226, 225, 1985.

45. **Nichols, D. G. and Scott, I. D.,** The regulation of brain mitochondrial calcium-ion transport, the role of ATP in the discrimination between kinetic and membrane potential dependent calcium-ion efflux mechanisms, *Biochem. J.,* 186, 833, 1980.

46. **Blaustein, M. P., Razlaff, R. W., and Kendrick, N. K.,** The regulation of intracellular calcium in presynaptic nerve terminals, *Ann. N.Y. Acad. Sci.,* 307, 195, 1978.
47. **Alkon, D. and Rasmussen, H.,** A spatial-temporal model of cell activation, *Science,* 239, 998, 1988.
48. **Joseph, H., Fillenz, M., MacDonald, I., and Marsden, C., Eds.,** *Monitoring Neurotransmitter Release During Behavior,* VCH, Deerfield, Beach, FL, and Horwood, Chichester, U.K., 1986.
49. **Myers, R. D. and Knatt, P. J.,** *Neurochemical Analysis of the Conscious Brain: Voltammetry and Push-Pull Perfusion,* New York Academy of Science, New York, 1986.
50. **Hamberger, A., Berthold, C., Karlsson, B., Lehmann, A., and Nystrom, B.,** *Glutamine, Glutamate, and GABA in the Central Nervous System,* Hertz, L., Kramme, G., McGeer, E., and Schousbrou, A., Eds., Alan R. Liss, New York, 1983, 473.
51. **Pittman, Q. J., Disturnal, J., Riphagen, C., Veale, W., and Bauce, L.,** Perfusion techniques for neuronal tissue, in *Neuromethods: 1. General Neurochemical Techniques,* Boulton, A. and Baker, G., Eds., Humana Press, Clifton, NJ, 1985, 279.
52. **Yaksh, T. L. and Yamamura, H. I.,** Factors affecting the performance of the push-pull cannula in brain, *J. Appl. Physiol.,* 37, 428, 1974.
53. **Hunchar, M. P., Hartman, B. K., and Sharpe, L. G.,** Evaluation of *in vivo* brain site perfusion with the push-pull cannula, *Am. J. Physiol.,* 236, R48, 1979.
54. **Westerink, B., Damsma, G., Rollema, H., DeVries, J., and Horn, A.,** Scope and limitations of *in vivo* brain dialysis: a comparison of its application to various neurotransmitter systems, *Life Sci.,* 41, 1763, 1987.
55. **Kuhr, W., Ewing, A., Caudill, W., and Wightman, R. M.,** Monitoring the stimulated release of dopamine with *in vivo* voltammetry. I. Characterization of the response observed in the caudate nucleus of the rat, *J. Neurochem.,* 43, 560, 1984.
56. **Levi, G., Gallo, V., and Raiteri, M.,** A reevaluation of veratridine as a tool for studying the depolarization-induced release of neurotransmitter from nerve endings, *Neurochem. Res.,* 5, 281, 1980.
57. **Hertz, L., Juurlink, B., Szuchet, S., and Walz, W.,** Cell and tissue cultures, in *Neuromethods 1: General Neurochemical Techniques,* Boulton, A. and Baker G., Eds., Humana Press, Clifton, NJ, 1985, 117.
58. **Icard, C., Licpkalns, V., Yates, H., Singh, N., Stephens, R., and Hart, R.,** Growth characteristics of human glioma-derived and fetal neural cells in culture, *J. Neuropath. Exp. Neurol.,* 40, 512, 1981.
59. **Laerum, O., Steinsvag, S., and Bjerkvig, R.,** Cell and tissue culture of the central nervous system: recent developments and current applications, *Acta Neurol. Scand.,* 72, 529, 1985.
60. **Scott, B. S.,** Adult mouse dorsal root ganglia neurons in cell culture, *J. Neurobiol.,* 8, 417, 1977.
61. **Lipton, P.,** Brain slices: uses and abuses, in *Neuromethods. 1: General Neurochemical Techniques,* Boulton, A. and Baker, G., Eds., Humana Press, Clifton, NJ, 1985, 69.
62. **Dunwiddie, T.,** The use of *in vitro* brain slices in neuropharmacology, in *Electrophysiological Techniques in Pharmacology,* Alan R. Liss, NY, 1986, 65.
63. **Dunkley, P., Rostas, J., Heath, J., and Powis, D.,** The preparation and use of synaptosomes for studying secretion of catecholamines, in *The Secretory Process: In Vitro Methods for Studying Secretion,* Poisner, A. and Trifaro, J., Eds., Elsevier, Press, NY, 1987.
64. **Bradford, H. F.,** Isolated nerve terminals as an *in vitro* preparation for the study of dynamic aspects of transmitter metabolism and release, in *Biochemical Principles and Techniques in Neuropharmacology,* Iversen, L., Iversen, S., and Snyder, S., Eds., Plenum Press, NY, 1975, 191.
65. **Raiteri, M. and Levi, G.,** Release mechanisms for catecholamines and serotonin in synaptosomes, *Rev. Neurosci.,* 3, 77, 1978.
66. **Bradford, H., Jones, D., Ward, H., and Booher, J.,** Biochemical and morphological studies of the short and long term survival of isolated nerve-endings, *Brain Res.,* 90, 245, 1975.
67. **Dunkley, P. and Robinson, P.,** Depolarization-dependent protein phosphorylation in synaptosomes: mechanisms and significance, *Prog. Brain Res.,* 69, 273, 1986.
68. **Chesselet, M. F.,** Presynaptic regulation of neurotransmitter release in the brain: facts and hypothesis, *Neuroscience,* 12, 347, 1984.
69. **Suszkiw, J., O'Leary, M., Murawsky, M., and Wang, T.,** Presynaptic calcium channels in rat cortical synaptosomes: fast-kinetics of phasic calcium influx channels inactivation, and relationship to nitrendipine receptors, *J. Neurosci.,* 6, 1349, 1986.
70. **deBellroche, J. and Bradford, H.,** The synaptosomes: an isolated, working, neuronal compartment, *Prog. Neurobiol.,* 1, 277, 1973.
71. **Gray, E. and Whittaker, V.,** The isolation of nerve endings from brain: an electron microscopic study of cell fragments divided by homogenization and centrifugation, *J. Anat. Lond.,* 96, 79, 1962.
72. **Rodriquez de Lores Arnaiz, G. and deIraldi, A.,** Subcellular fractionation, in *Neuromethods 1: General Neurochemical Techniques,* Boulton, A. and Baker, G., Eds., Humana Press, Clifton, NJ, 1985, 19.
73. **Silbergeld, E. and Adler, H.,** Subcellular mechanisms of lead neurotoxicity, *Brain Res.,* 148, 451, 1978.
74. **Hubbard, J. I., Llinas, R., and Quastel, D. M. J.,** *Electrophysiological Analysis of Synaptic Transmission,* Edward Arnold, London, 1969.

75. **Skrede, K. and Westgard, R.,** The transverse hippocampal slice: a well defined cortical structure maintained *in vitro, Brain Res.,* 35, 589, 1971.

76. **Kelly, R. B., Deutsch, J. W., Carloon, S. S., and Wagner, J. A.,** Biochemistry of neurotransmitter release, *Ann. Rev. Neurosci.,* 2, 399, 1979.

77. **Kuriyama, K., Weinstein, H., and Roberts, E.,** Uptake of gamma-aminobutyric acid by mitochondrial and synaptosomal fractions from mouse brain, *Brain Res.,* 16, 479, 1969.

78. **Minnema, D. and Michaelson, I. A.,** A superfusion apparatus for the examination of neurotransmitter release from synaptosomes, *J. Neurosci. Methods,* 14, 193, 1985.

79. **Llinas, R., Yarom, Y., and Sugimoni, M.,** The isolated mammalian brain *in vitro:* a new technique for the analysis of the electrical activity of neuronal circuit function, *Fed. Prod.,* 40, 2240, 1981.

80. **Dingledine, R., Dodd, J., and Kelly, J.,** The *in vitro* brain slice as a useful neurophysiological preparation for intracellular recording, *J. Neurosci. Methods,* 2, 323, 1980.

81. **Simons, T.,** Cellular interactions between lead and calcium, *Br. Med. Bull.,* 42, 431, 1986.

82. **Yamamoto, D. and Suzuki, N.,** Blockage of chloride channels by HEPES buffer, *Proc. R. Soc. Lond.,* B230, 93, 1987.

83. **Krebs, H.,** Body size and tissue respiration, *Biochim. Biophys. Acta,* 4, 249, 1950.

84. **Wood, J. H.,** Physiology, pharmacology and dynamics of cerebrospinal fluid, in *Neurobiology of Cerebrospinal Fluid,* Wood, J. H., Ed., Plenum Press, NY, 1980, 1.

85. **Heinmann, V., Lux, H., and Gutnick, M.,** Extracellular free calcium and potassium during paroxysmal activity in the cerebral cortex of the cat, *Exp. Brain Res.,* 27, 237, 1977.

86. **Bradford, H. and Richards, C.,** Specific release of endogenous glutamate from piriform cortex stimulated *in vitro, Brain Res.,* 105, 168, 1976.

87. **Keesey, J., Wallgreen, H., and McIlwain, H.,** The sodium, potassium, and chloride of cerebral tissues; maintenance, charge on stimulation, and subsequent recovery, *Biochem. J.,* 85, 289, 1965.

88. **Szerb, J. C.,** Relationship between Ca^{2+}-dependent and independent release of ^{3}H-GABA evoked by high K^+, veratridine, or electrical stimulation from rat cortical slices, *J. Neurochem.,* 32, 1565, 1979.

89. **Malthe-Sorenson, D., Skrede, K., and Fonnum, F.,** Calcium-dependent release of D-^{3}-H-aspartate evoked by selective electrical stimulation of excitatory afferent fibers to hippocampal pyrimidal cells *in vitro, Neuroscience,* 4, 1255, 1979.

90. **Cunningham, J. and Neal, M.,** On the mechanism by which veratridine causes a calcium-dependent release of gamma-aminobutyric acid from brain slices, *Br. J. Pharmacol.,* 73, 655, 1981.

91. **Okamoto, K. and Quastel, J.,** Water uptake and energy metabolism in brain slices from rat, *Biochem. J.,* 25, 1970.

92. **Bourke, R., Kimelberg, H., Daze, M., and Church, G.,** Swelling and ion uptake in cat cerebrocortical slices: control by neurotransmitters and ion transport mechanisms, *Neurochem. Res.,* 8, 5, 1983.

93. **Drapeau, P. and Blaustein, M.,** Initial release of ^{3}H-dopamine from rat striatal synaptosomes: correlation with calcium entry, *J. Neurosci.,* 3, 703, 1983.

94. **Blaustein, M., Nachshen, D., and Drapeau, P.,** Excitation-secretion couplings: the role of calcium, in *Chemical Neurotransmission: Seventy-five Years,* Stjarne, L., Hedqvist, A., Wennmalm, A., and Langerevantz, H., Eds., Academic Press, NY, 1981, 125.

95. **Nachshen, D. A.,** Regulation of cytosolic calcium concentration in presynaptic nerve endings isolated from rat brain, *J. Physiol.,* 363, 87, 1985.

96. **Nachshen, D. A. and Blaustein, M.,** Influx of calcium, strontium, and barium in presynaptic nerve endings, *J. Gen. Physiol.,* 79, 1065, 1982.

97. **Nachshen, D. A.,** Selectivity of the Ca binding site in synaptosome Ca influx by multivalent metal cations, *J. Gen. Physiol.,* 83, 941, 1984.

98. **Cremer, J. E.,** Toxicology and biochemistry of alkyl lead compounds, *Occup. Health Rev.,* 17, 14, 1965.

99. **Beattie, A. D., Moore, M. R., and Goldberg, A.,** Tetraethyl-lead poisoning, *Lancet,* 2, 12, 1972.

100. **Wong, P. T., Chau, Y. K., and Luxon, P.,** Methylation of lead in the environment, *Nature,* 253, 263, 1975.

101. **Winder, C. and Kitchen, I.,** Lead neurotoxicity: a review of the biochemical, neurochemical and drug-induced behavioural evidence, *Prog. Neurobiol.,* 22, 59, 1984.

102. **Kostial, K. and Vouk, V.,** Lead ions and synaptic transmission in the superior cervical ganglion of the cat, *Br. J. Pharmacol.,* 12, 219, 1957.

103. **Manalis, R. S. and Cooper, G. P.,** Presynaptic and postsynaptic effects of lead at the frog neuromuscular junction, *Nature (London),* 243, 354, 1973.

104. **Manalis, R. S., Cooper, G. P., and Pomeroy, S. L.,** Effects of lead on neuromuscular junction transmission in the frog, *Brain Res.,* 294, 95, 1984.

105. **Atchison, W. D. and Narahashi, T.,** Mechanism of action of lead on neuromuscular junctions, *Neurotoxicology,* 5, 267, 1984.

106. **Pickett, J. B. and Bornstein, J. C.,** Some effects of lead at mammalian neuromuscular junction, *Am. J. Physiol.,* 246, C271, 1984.

107. **Kober, T. E. and Cooper, G. P.,** Lead competitively inhibits calcium-dependent synaptic transmission in the bullfrog sympathetic ganglion, *Nature (London),* 262, 704, 1976.

108. **Suszkiw, J., Toth, G., Murawsky, M., and Cooper, G.,** Effects of Pb^{2+} and Cd^{2+} on acetylcholine release and Ca^{2+} movements in synaptosomes and subcellular fractions from rat brain and Torpedo electric organ, *Brain Res.,* 323, 31, 1984.

109. **Minnema, D., Michaelson, I., and Cooper, G.,** Calcium efflux and neurotransmitter release from rat hippocampal synaptosomes exposed to lead, *Toxicol. Appl. Pharmacol.,* 92, 351, 1988.

110. **Cooper, G. P. and Steinberg, D.,** Effects of cadmium and lead on adrenergic neuromuscular transmission in the rabbit, *Am. J. Physiol.,* 232, C128, 1977.

111. **Minnema, D., Greenland, R., and Michaelson, I. A.,** Effect of *in vitro* inorganic lead on dopamine release from superfused rat brain synaptosomes, *Toxicol. Appl. Pharmacol.,* 84, 400, 1986.

112. **Minnema, D. and Michaelson, I. A.,** Differential effects of inorganic lead and δ-aminolevulinic acid *in vitro* on synaptosomal gamma-aminobutyric acid release, *Toxicol. Appl. Pharmacol.,* 86, 437, 1986.

113. **Cooper, G. P. and Minnema, D. J.,** Unpublished observations, 1985.

114. **Kolton, L. and Yaari, Y.,** Sites of action of lead on spontaneous transmitter release from motor nerve terminals, *Isr. J. Med. Sci.,* 18, 165, 1982.

115. **Silbergeld, E.,** Interactions of lead and calcium on the synaptosomal uptake of dopamine and choline, *Life Sci.,* 20, 309, 1977.

116. **Spence, I., Drew, C., Johnston, G., and Lodge, D.,** Acute effects of lead at central synapses *in vitro, Brain Res.,* 333, 103, 1985.

117. **Cooper, G. P. and Manalis, R. S.,** Interactions of lead and cadmium on acetylcholine release at the frog neuromuscular junction, *Toxicol. Appl. Pharmacol.,* 74, 411, 1984.

118. **Minnema, D. J.,** Unpublished observations, 1987.

119. **Powis, D.,** Cardiac glycosides and autonomic neurotransmission, *J. Auton. Pharmacol.,* 3, 127, 1983.

120. **Siegel, G. and Fogt, S.,** Inhibition by lead ion of electrophorus electroplax (Na^+-K^+)-adenosine triphosphatase and K^+-p-nitrophenylphosphatase, *J. Biol. Chem.,* 252, 5201, 1977.

121. **Nechay, B. and Saunders, J.,** Inhibitory characteristics of lead chloride in sodium- and potassium-dependent adenosine-triphosphatase preparations derived from kidney, brain, and heart of several species, *J. Toxicol. Environ. Health,* 4, 147, 1978.

122. **Bertoni, J. M. and Sprenkle, P. M.,** Inhibition of brain cation pump enzyme by *in vitro* lead ion: effects of low level [Pb] and modulation by homogenate, *Toxicol. Appl. Pharmacol.,* 93, 101, 1988.

123. **Chandra, S. V., Murthy, R. C., Husain, T., and Bansal, S. K.,** Effect of interaction of heavy metals on (Na^+-K^+) ATPase and the uptake of ^3H-DA and ^3H-NA in rat brain synaptosomes, *Acta Pharmacol. Toxicol.,* 54, 210, 1984.

124. **Chao, S. H., Suzuki, Y., Zysk, J. R., and Cheung, W. Y.,** Activation of calmodulin by various metal cations as a function of ionic radius, *Mol. Pharmacol.,* 26, 75, 1984.

125. **Cheung, W.,** Calmodulin: its potential role in cell proliferation and heavy metal toxicity, *Fed. Proc.,* 43, 2995, 1984.

126. **Habermann, E., Crowell, K., and Janicki, P.,** Lead and other metals can substitute for Ca^{2+} in calmodulin, *Arch. Toxicol.,* 54, 61, 1983.

127. **Goldstein, G. and Ar, D.,** Lead activates calmodulin sensitive processes, *Life Sci.,* 33, 1001, 1983.

128. **Publicover, S.,** Calmodulin, synchronous and asynchronous release of neurotransmitter, *Comp. Biochem. Physiol.,* 82A, 7, 1985.

129. **Thompson, J. and Nechay, B.,** Inhibition by metals of canine renal calcium, magnesium-activated adenosinetriphosphatase, *J. Toxicol. Environ. Health,* 7, 901, 1981.

130. **Kuo, C., Ichida, S., Matsuda, T., Kakiuchi, S., and Yochisda, M.,** Regulation of ATP-dependent Ca-uptake of synaptic plasma membranes by Ca-dependent modulatory protein, *Life Sci.,* 125, 235, 1979.

131. **Ross, D. and Carderas, H.,** Calmodulin stimulation of Ca-dependent ATP hydrolysis and ATP-dependent Ca transport in synaptic membranes, *J. Neurochem.,* 41, 161, 1983.

132. **Pounds, J., Wright, R., Morrison, D., and Casciano, D.,** Effect of lead on calcium homeostasis in the isolated rat hepatocyte, *Toxicol. Appl. Pharmacol.,* 63, 389, 1982.

133. **Pounds, J. and Rosen, J.,** Cellular metabolism of lead: a kinetic analysis in cultured osteoclastic bone cells, *Toxicol. Appl. Pharmacol.,* 83, 531, 1986.

134. **Goldstein, G. W.,** Lead encephalopathy: the significance of lead inhibition of calcium uptake by brain mitochondria, *Brain Res.,* 136, 185, 1977.

135. **Kapoor, S. C. and Van Rossum, G. D.,** Effects of Pb^{2+} added *in vitro* on Ca^{2+} movements in isolated mitochondria and slices of rat kidney cortex, *Biochem. Pharmacol.,* 33, 1771, 1984.

136. **Parr, D. R. and Harris, E. J.,** The effect of lead on the calcium-handling capacity of rat heart mitochondria, *Biochem. J.,* 158, 289, 1976.

137. **Minnema, D. and Greenland, R.,** Unpublished observations, 1987.

138. **Chavez, E., Jay, D., and Bravo, C.,** The mechanism of lead-induced mitochondrial Ca^{2+} efflux, *J. Bioenerg. Biomemb.,* 19, 285, 1987.

139. **Silbergeld, E., Adler, H., and Costa, J.,** Subcellular localization of lead in synaptosomes, *Res. Commun. Chem. Pathol. Pharmacol.,* 17, 715, 1977.

140. **Wood, J. M., Cheh, A., Dizikes, L., Ridley, W., Rakow, S., and Lakowicz, J.,** Mechanisms for the biomethylation of metals and metalloids, *Fed. Proc.,* 37, 16, 1978.

141. **Bondy, S. C., Anderson, C. L., Harrington, M. E., and Prasad, K. N.,** Effects of organic and inorganic lead and mercury on neurotransmitter high-affinity transport and release mechanisms, *Environ. Res.,* 19, 102, 1979.

142. **Bondy, S. C., Harrington, M. E., Anderson, C. L., and Prasad, K. N.,** Effect of low concentrations of an organic lead compound on the transport and release of putative neurotransmitters, *Tox. Letters,* 3, 35, 1979.

143. **Minnema, D., Greenland, R., and Evans, J.,** Unpublished observations, 1988.

144. **Komulainen, H. and Tuomisto, J.,** Effects of heavy metals on monoamine uptake and release in brain synaptosomes and blood platelets, *Neurobehav. Toxicol. Terat.,* 4, 647, 1982.

145. **Aldridge, W., Street, B., and Skilleter, D.,** Oxidative phosphorylation: halide-dependent and halide-independent effects of triorganic tin and triorganic lead on mitochondrial functions, *Biochem. J.,* 168, 353, 1977.

146. **Aldridge, W. N.,** Effects on mitochondria and other enzyme systems, in *Biological Effects of Organolead Compounds,* Grandjean, P. and Grandjean, E., Eds., CRC Press, Boca Raton, FL, 1984, 137.

147. **Seawright, A., Brown, A., Ng, J., and Hrdlicka, J.,** Experimental pathology of short-chain alkyllead compounds, in *Biological Effects of Organolead Compounds,* Grandjean, P. and Grandjean, E., Eds., CRC Press, Boca Raton, FL, 1984, 177.

148. **Van Natta, F. C. and Whitehouse, M. W.,** Heavy metal poisoning: mercury and lead, *Ann. Internal Med.,* 76, 779, 1972.

149. **Berlin, M.,** Mercury, in *Handbook on the Toxicology of Metals,* 2nd ed., Friberg, L., Nordberg, G., and Vouk, V., Eds., Elsevier, NY, 1986, 387.

150. **del Castillo-Nicolau, J. and Hufschmidt, H.,** Reversible poisoning of nerve fibers by heavy-metal ions, *Nature,* 167, 146, 1951.

151. **Huneeus-Cox, F., Fernandez, H. L., and Smith, B. H.,** Effects of redox and sulfhydryl reagents on the bioelectric properties of the giant axon of the squid, *Biophys. J.,* 6, 675, 1966.

152. **Pennock, B. and Goldman, D.,** The action of lead and mercury on lobster axon, *Fed. Proc.,* 31, 319, 1972.

153. **Del Castillo, J., Escobar, I., and Gijon, E.,** Effects of the electrophoretic application of sulfhydryl reagents to the endplate receptors, *Int. J. Neurosci.,* 1, 199, 1971.

154. **Del Castillo, J., Bartels, E., and Sabrino, J.,** Microelectrophoretic application of cholinergic compounds, protein oxidizing agents, and mercurials to the chemically excitable membrane of the electroplax, *Proc. Natl. Acad. Sci.,* 69, 2081, 1972.

155. **Scheuhammer, A. M., and Cherian, M. G.,** Effects of heavy metal cations, sulfhydryl reagents and other chemical agents on striatal D$_2$ dopamine receptors, *Biochem. Pharmacol.,* 34, 3405, 1985.

156. **Juang, M. S.,** An electrophysiological study of the action of methylmercury chloride and mercuric chloride on the sciatic nerve-sartorius muscle preparation of the frog, *Toxicol. Appl. Pharmacol.,* 37, 339, 1976.

157. **Juang, M. S.,** Depression of frog muscle contraction by methylmercuric chloride and mercuric chloride, *Toxicol. Appl. Pharmacol.,* 35, 183, 1976.

158. **Manalis, R. S. and Cooper, G. P.,** Evoked transmitter release increased by inorganic mercury at frog neuromuscular junction, *Nature (London),* 257, 690, 1975.

159. **Binah, O., Meiri, U., and Rahamimoff, H.,** The effects of HgCl$_2$ and mersalyl on mechanisms regulating intracellular calcium and transmitter release, *Eur. J. Pharmacol.,* 51, 453, 1978.

160. **Cooper, G. P., Suszkiw, J. B., and Manalis, R. S.,** Presynaptic effects of heavy metals, in *Cellular and Molecular Neurotoxicology,* Narahashi, T., Ed., Raven Press, NY, 1984, 1.

161. **Hare, M. F., Minnema, D. J., Cooper, G. P., and Michaelson, I. A.,** Effects of mercuric chloride on dopamine release from rat brain striatal synaptosomes, *Toxicol. Appl. Pharmacol.,* 99, 266, 1989.

162. **Kostial, K. and Landeka, M.,** The action of mercury ions on the release of acetylcholine from presynaptic nerve endings, *Experientia,* 15, 834, 1975.

163. **Miyamoto, M. D.,** Hg^{2+} causes neurotoxicity at an intracellular site following entry through Na and Ca channels, *Brain Res.,* 267, 375, 1983.

164. **Hare M., Cooper, G., and Minnema, D.,** Effects of mercury on neurotransmitter release in rat striatal synaptosomes and frog skeletal neuromuscular junction, *Toxicologist,* 6, 195, 1986.

165. **Hare, M. F. and Minnema, D. J.,** Effects of mercuric chloride (Hg) on spontaneous transmitter release and Na$^+$, K$^+$-ATPase in synaptosomes, *Toxicologist,* 8, 45, 1988.

166. **Komulainen, H. and Tuomisto, J.,** Effects of heavy metals on dopamine, noradrenaline, and serotonin uptake and release in rat brain synaptosomes, *Acta Pharmacol. Toxicol.,* 48, 199, 1981.

167. **Chavez, E. and Holguin, J.,** Mitochondrial calcium release as induced by Hg²⁺, *J. Biol. Chem.,* 263, 3582, 1988.

168. **Rothstein, A.,** Sulfhydryl groups in membrane structure and function, in *Current Topics in Membranes and Transport,* Bronnen, F. and Kleinzeller, A., Eds., Academic Press, NY, 1970, 135.

169. **Kinter, W. B. and Pritchard, J. B.,** Altered permeability of cell membranes, in *Handbook of Physiology,* Lee, D. K., Falk, H. L., and Murphy, S. D., Eds., American Physiology Society, 1977, 563.

170. **Atchison, W. and Narahashi, T.,** Methylmercury-induced depression of neuromuscular transmission in the rat, *Neurotoxicology,* 3, 37, 1982.

171. **Barrett, J., Botz, D., and Chang, D. B.,** Block of neuromuscular transmission by methylmercury, in *Behavioral Toxicology, Early Detection of Occupational Hazards,* Xintaras, C., Johnson, B. L., and de Groot, L., Eds., U.S. Department HEW, 1974, 177.

172. **Atchison, W. D., Clark, A. W., and Narahashi, T.,** Presynaptic effects of methylmercury at the mammalian neuromuscular junction, in *Cellular and Molecular Neurotoxicology,* Narahashi, T., Ed., Raven Press, NY, 1984, 23.

173. **Von Burg, R. and Landry, T.,** Methylmercury induced neuromuscular dysfunction in the rat, *Neurosci. Lett.,* 1, 169, 1975.

174. **Atchison, W., Joshi, U., and Thornburg, J.,** Irreversible suppression of calcium entry into nerve terminals by methylmercury, *J. Pharmacol. Exp. Therap.,* 230, 618, 1986.

175. **Shafer, T. J. and Atchison, W. D.,** Block of ⁴⁵Ca uptake by methylmercury into nerve terminals is Na-dependent and partially reversible, *Toxicologist,* 8, 45, 1988.

176. **Traxinger, D. L. and Atchison, W. D.,** Comparative effects of divalent cations on the methylmercury-induced alterations of acetylcholine release, *J. Pharmacol. Exp. Therap.,* 90, 23, 1987.

177. **Traxinger, D. L. and Atchison, W. D.,** Reversal of methymercury-induced block of nerve-evoked release of acetylcholine at the neuromuscular junction, *Toxicol. Appl. Pharmacol.,* 90, 23, 1987.

178. **Passow, H. and Rothstein, A.,** The binding of mercury by the yeast cell in relation to changes in permeability, *J. Gen. Physiol.,* 41, 621, 1960.

179. **Shrivastav, B. B., Brodwick, M. S., and Narahashi, T.,** Methylmercury: effects on electrical properties of squid axon membranes, *Life Sci.,* 18, 1077, 1976.

180. **Webb, J. L.,** *Enzyme and Metabolic Inhibitors,* Vol. II, Academic Press, NY, 1966, 742.

181. **Dales, L. G.,** The neurotoxicity of alkyl mercury compounds, *Am. J. Med.,* 53, 219, 1972.

182. **Heggli, D. E. and Roed, A.,** Diphenylhydatoin-induced block of the rat phrenic nerve-diaphragm preparation pretreated with p-hydroxymercuribenzoate, *Eur. J. Pharmacol.,* 70, 175, 1981.

183. **Quandt, F. N., Kato, E., and Narahashi, T.,** Effects of methylmercury on electrical responses of neuroblastoma cells, *Neurotoxicology,* 3, 205, 1982.

184. **Smith, H. M.,** Effects of sulfhydryl blockade on axonal function, *J. Cell. Comp. Physiol.,* 51, 161, 1958.

185. **Marquis, J. K. and Mautner, H. G.,** The effect of electrical stimulation on the action of sulfhydryl reagents in the giant axon of the squid; suggested mechanisms for the role of thiol and disulfide groups in electrically-induced confirmational changes, *J. Membrane Biol.,* 15, 249, 1974.

186. **Marquis, J. K. and Mautner, H. G.,** The binding of thiol reagents to axonal membranes: the effect of electrical stimulation, *Biochem. Biophys. Res. Comm.,* 57, 154, 1974.

187. **Shrager, P.,** Slow sodium channel inactivation in nerve after exposure to sulfhydryl blocking reagents, *J. Gen. Physiol.,* 69, 183, 1977.

188. **Minnema, D. J., Cooper, G. P., and Greenland, R. D.,** Effects of methylmercury on neurotransmitter release from rat brain synaptosomes, *Toxicol. Appl. Pharmacol.,* 99, 510, 1989.

189. **Juang, M. S. and Yonemura, K.,** Increased spontaneous transmitter release from presynaptic nerve terminal by methylmercuric chloride, *Nature (London),* 256, 211, 1975.

190. **Atchison, W. D.,** Effects of activation of sodium and calcium entry on spontaneous release of acetylcholine induced by methylmercury, *J. Pharmacol. Exp. Therap.,* 241, 131, 1987.

191. **Atchison, W. D.,** Extracellular calcium-dependent and -independent effects of methylmercury on spontaneous and potassium-evoked release of acetylcholine at the neuromuscular junction, *J. Pharmacol. Exp. Therap.,* 237, 672, 1986.

192. **Komulainen, H. and Tuomisto, J.,** Interference of methylmercury with monoamine uptake and release in rat brain synaptosomes, *Acta Pharmacol. Toxicol.,* 48, 214, 1981.

193. **Minnema, D. J.,** Methylmercury-induced dopamine release from superfused rat striatal synaptosomes, *Toxicologist,* 7, 156, 1987.

194. **Lakowicz, J. and Anderson, C.,** Permeability of lipid bilayers to methylmercuric chloride: quantification by fluorescence quenching of a carbazole-labeled phospholipid, *Chem. Biol. Interact.,* 30, 309, 1980.

195. **Paiement, J. and Joly, L.,** Effect of organic mercury on the electrical resistance of phosphatidylserine bilayers, *Biochim. Biophys. Acta,* 816, 179, 1985.

196. **Leblanc, R., Joly, L., and Paiement, J.,** pH-dependent interaction between methylmercury chloride and some membrane phospholipids, *Chem. Biol. Interact.,* 48, 237, 1984.

197. **Cheung, M. and Verity, M.,** Methylmercury inhibition of synaptosome protein synthesis: the role of mitochondrial dysfunction, *Environ. Res.,* 24, 286, 1981.

198. **Su, M. and Okita, G. T.,** Studies of CNS teratogenesis induced by methylmercury, *Fed. Proc.,* 34, 810, 1975.

199. **Henderson, G. R., Huang, W. H., and Askari, A.,** Transport ATPase: the different modes of inhibition of the enzyme by various mercury compounds, *Biochem. Pharmacol.,* 28, 429, 1979.

200. **Holmes, L. S. and Okita, G. T.,** The role of Na,K-ATPase in methymercury-induced teratogenesis, *Fed. Proc.,* 38, 680, 1979.

201. **Verity, M. A., Brown, W. J., and Cheung, M.,** Organic mercurial encephalopathy: *in vivo* and *in vitro* effects of methylmercury on synaptosomal respiration, *J. Neurochem.,* 25, 759, 1975.

202. **Varnbo, I., Peterson, A., and Walum, E.,** Effects of toxic chemicals on the respiratory activity of cultured mouse neuroblastoma cells, *Xenobiotica,* 15, 727, 1985.

203. **Fox, J. H., Patel-Mandlik, K., and Cohen, M. M.,** Comparative effects of organic and inorganic mercury on brain slice respiration and metabolism, *J. Neurochem.,* 24, 757, 1975.

204. **Sarafian, T., Cheung, M. K., and Verity, M. A.,** *In vitro* methylmercury inhibition of protein synthesis in neonatal cerebellar perikarya, *Neuropathol. Appl. Neurobiol.,* 10, 85, 1984.

205. **Sone, N., Larsstuvold, M., and Kagawa, Y.,** Effect of methylmercury on phosphorylation, transport, and oxidation in mammalian mitochondria, *J. Biochem.,* 82, 859, 1977.

206. **Desnoyers, P. A. and Chang, L. W.,** Ultrastructural changes in rat hepatocytes following acute methylmercury intoxication, *Environ. Res.,* 9, 224, 1975.

207. **O'Kusky, J.,** Methylmercury poisoning of the developing nervous system: morphological changes in neuronal mitochondria, *Acta Neuropathol.,* 61, 116, 1983.

208. **Akerman, K. E. and Nicholls, D. G.,** ATP depletion increases Ca^{2+} uptake by synaptosomes, *FEBS Lett.,* 135, 212, 1981.

209. **Akerman, K. E. and Nicholls, D. G.,** Intrasynaptosomal compartmentation of calcium during depolarization-induced calcium uptake across the plasma membrane, *Biochim. Biophys. Acta,* 645, 41, 1981.

210. **Levesque, P. and Atchison, W.,** Interactions of mitochondrial inhibitors with methylmercury on spontaneous quantal release of acetylcholine, *Toxicol. Appl. Toxicol.,* 87, 315, 1987.

211. **Levesque, P. and Atchison, W.,** Effect of alteration of nerve terminal Ca^{2+} regulation on increased spontaneous quantal release of acetylcholine by methylmercury, *Toxicol. Appl. Pharmacol.,* 94, 55, 1988.

212. **Reichardt, L. F. and Kelly, R. B.,** A molecular description of nerve terminal function, *Ann. Rev. Biochem.,* 52, 871, 1983.

213. **Pounds, J.,** Effect of lead intoxication on calcium homeostasis and calcium-mediated cell function: a review, *Neurotoxicology,* 5, 295, 1984.

214. **Gilman, S., Kumaroo, K., and Hallenbeck, J.,** Effects of pressure on uptake and release of calcium by brain synaptosomes, *J. Appl. Physiol.,* 60, 1446, 1986.

215. **Lin, S. and Way, E.,** Calcium transport in and out of brain nerve endings *in vitro*: the role of synaptosomal membrane Ca^{2+}-ATPase in Ca^{2+}-extrusion, *Brain Res.,* 298, 225, 1984.

216. **Markovae, J. and Goldstein, G. W.,** Picomolar concentrations of lead stimulate brain protein kinase C, *Nature,* 334, 71, 1988.

217. **Walum, E.,** Membrane lesions in cultured mouse neuroblastoma cells exposed to metal compounds, *Toxicology,* 25, 67, 1982.

Chapter 3

ACTION OF METALS ON CALMODULIN-REGULATED CALCIUM PUMP ACTIVITY IN BRAIN

Durisala Desaiah

TABLE OF CONTENTS

I. INTRODUCTION

Metals are among the oldest toxic substances known to man. They occur naturally and redistribute in the environment through biological and geological cycles. Added to this is the human contribution to the environment through industrial activity. Cadmium, lead, and mercury are good examples of environmental contamination through industrial activity. Metals, in general, when accumulated excessively become toxic to living cells. The toxicology of various metal ions is so vast that the allotted space in this chapter prevent even the outlining of the subject. Instead, the chapter will be restricted to the effects of certain neurotoxic metal ions on calmodulin-regulated calcium pump activity in the central nervous system. Whenever it is essential, some correlations will be made with other tissues. Since this topic is concerned with the interaction of metal ions with calcium pump and its regulation by calmodulin, it is necessary for the reader to get an overview of the calcium pump in the nerve cell. This subject is also touched on in Chapter 1 and 2 of this volume.

II. ROLE OF CALCIUM IN NEURONAL FUNCTION

The basic information on today's understanding of calcium as a second messenger in cellular functions came from the serendipitous discovery of its requirement for excitability of nerve and muscle cells.[1,2] Nearly a century later an explosion of information occurred substantiating the notion that calcium plays a second messenger role in all eukaryotic cells. Several reviews appeared on the calcium function in the nervous system.[3-8] In brief, calcium plays a major role in neurotransmitter release, exocytosis, protein phosphorylation, cytoskeleton integrity, and axonal transport in the nervous system. In other tissues, calcium is involved in muscle contractility, microtubule formation, blood clotting, fertilization, cell fusion and adhesion, etc. Such a diversified regulatory function of calcium requires systematic translocation to discrete loci in the cells. There are specialized active ion pumps, ion channels, and ion-ion exchangers involved in calcium transport. Additionally, there are calcium-binding proteins inside the neuron by which information carried by calcium is transduced into a number of metabolic pathways.[10,11] Let us examine the calcium flux mechanisms as the level of intracellular calcium $(Ca^{2+})_i$ is the key for the regulation of many cellular functions as outlined above.

A. CALCIUM ENTRY MECHANISMS

The physiological concentration of $(Ca^{2+})_i$ in the cytoplasm of many cells including the neurons is about $10^{-7}M$ at rest,[12,13] while the extracellular calcium is in the millimolar range,[14] creating a large electrochemical gradient across neuronal plasma membranes. Following an action potential, mainly through voltage-dependent calcium channels, calcium enters neurons down a concentration gradient raising the $(Ca^{2+})_i$ levels to 10^{-6} or $10^{-5}M$.[3,14,15] The increase in $(Ca^{2+})_i$ triggers the release of a transmitter.[16,17] The action potential-evoked release of transmitters declines rapidly in about 1 to 2 msec, paralleling an abrupt fall of $(Ca^{2+})_i$.[18] However, all the calcium that enters during depolarization cannot be extruded within 1 to 2 msec. There must be other calcium-sequestering mechanisms operative to ensue the original physiological levels of $(Ca^{2+})_i$; they will be briefly discussed below. There are modulators of calcium channels which alter calcium conductance. For example, norepinephrine treatment reduces calcium conductance in rat sympathetic and chick sensory neurons.[19] Metals, such as Cd^{2+}, Co^{2+}, and La^{3+}, and calcium antagonists such as verapamil, D600 and dihydroxypyridines inhibit voltage-dependent calcium channels.[20]

B. INTRACELLULAR SEQUESTRATION OF CALCIUM

Neuronal $(Ca^{2+})_i$ levels are maintained by mechanisms other than transport across the plasma membranes. These include high affinity binding to membranes, macromolecules, and accumulation by intracellular organelles.[21-23] Mitochondria from the brain accumulate calcium with half maximal concentrations between 1 to 5×10^{-5} M.[24] However, evidence also indicates that mitochondria are unable to sequester calcium when $(Ca^{2+})_i$ falls below 0.5 to 1 μM.[13,25] It was suggested that other organelles such as endoplasmic reticulum and synaptic vesicles accumulate calcium more efficiently when the synaptosomes are depolarized.[25-27] Purified synaptic vesicles were shown to possess an ATP-powered calcium uptake, a Ca^{2+}-Mg^{2+} ATPase, and a Na^+-Ca^{2+} exchange system.[28,29] Based on the extensive information available it was concluded[30] that $(Ca^{2+})_i$ in resting neurons is mainly under the control of the calcium pumps in the smooth endoplasmic reticulum. However, the calcium that enters during nerve activity is rapidly buffered by cytoplasmic proteins and then sequestered by smooth endoplasmic reticulum. Mitochondria, on the other hand, only participate in calcium buffering under pathological conditions of calcium loading.

C. CALCIUM EXTRUSION MECHANISMS

The ATP-dependent calcium pump and the Na^+-Ca^{2+} exchanger are the two main mechanisms operating to expel calcium up the 10,000-fold concentration gradient between the cytosol and the extracellular space. Since the focus of this chapter is on the metal ion interaction with calcium pump, this system will be discussed in greater detail than the Na^+-Ca^{2+} exchanger.

A large number of studies appeared demonstrating the Na^+-Ca^{2+} exchange mechanisms in many tissues including synaptosomes.[30] This carrier is purely plasma membrane in origin and the Na^+ electrochemical gradient provides the energy to drive the exchange. Depending upon the Na^+ gradient, calcium can be moved in either direction across the plasma membrane.[31] The stoichiometry of this exchanger is believed to be 3 Na^+-1 Ca^{2+} exchange.[31] ATP has been shown to alter the kinetic properties of the Na^+-Ca^{2+} exchanger in squid axons[32] but no such information is available in mammalian brain.

ATP-dependent calcium pumps to remove $(Ca^{2+})_i$ from cells of many types have gained recognition in the last decade. Erythrocyte plasma membrane Ca^{2+} ATPase as well as myocardial sacroplasmic reticulum Ca^{2+} ATPase received most of the attention as compared to other tissues. Several reviews have appeared on the general subject of calcium pumps.[33-35] Some of the properties of the plasma membrane calcium pump are: (1) the pump is of plasma membrane origin, (2) the pump has high Ca^{2+}-affinity Ca^{2+}-transport, and (3) it possesses a high affinity for Ca^{2+}. Several lines of evidence indicate that plasma membranes in a variety of tissues possess high Ca^{2+}-affinity ATPase and Ca^{2+}-transport.[35] The affinity of Ca^{2+} for ATPase was measured using ethylene glycol *bis* (β-amino-ethyl ether) N, N, N^1, N^1-tetracetic acid (EGTA) as a Ca^{2+} buffer and in the presence of calmodulin (endogenous or exogenous). Under these circumstances the Ca^{2+} affinity of Ca^{2+} ATPase is about 0.01 to 1 μM. High Ca^{2+} affinity Ca^{2+} transport was also demonstrated in purified plasma membrane vesicles. Demonstration of Ca^{2+} efflux in intact cells is somewhat difficult with the preloading with $^{45}Ca^{2+}$. The purified membrane vesicles which have inside-out formation serve well as models for Ca^{2+} transport studies. As the current topic is concerned with brain Ca^{2+} ATPase, the Ca^{2+} ATPase in neurons and its regulation by calmodulin will be discussed.

1. Neuronal Calcium ATPase

Neurotransmission is dependent on the Ca^{2+} stimulus for the neurotransmitter release and various other key reactions both in pre- and post-synaptic neurons. Most of these studies were carried out using the synaptosomes, which are nothing but the pinched off nerve endings. Synaptosomes are functionally "intact" in that they retain many of the properties of *in situ* presynaptic nerve terminals.[36] Hence, these preparations are highly useful in determining the Ca^{2+} pump activity. Mammalian brain synaptosomal plasma membranes are shown to contain a high affinity Ca^{2+} ATPase.[37] This enzyme was highly stimulated by calmodulin.[37-39] The purified Ca^{2+} ATPase from synaptosomal membranes has $M_r = 138,000$ and requires Mg^{2+} for its activity.[40] Antibodies raised against the human erythrocyte Ca^{2+} ATPase cross-reacted with the synaptosomal ATPase, suggesting its plasma membrane origin.[41] Synaptosomal membrane vesicles can exchange Na^+ for Ca^{2+} in the presence of Na^+ gradient with high Ca^{2+} affinity.[42] These data suggest a Ca^{2+} transport system with high affinity for its substrate in the same synaptosomal plasma membrane which also contains ATP-dependent Ca^{2+} ATPase. Ca^{2+} ATPase in synaptosomes, like in other tissues, is also under regulation by proteins, lipids, and ions. Information on the regulatory aspects has been extensively reviewed,[34,35] and will be briefly reviewed here for calmodulin.

2. Calmodulin and Its Regulation of Calcium ATPase

Calmodulin (CaM), an ubiquitous Ca^{2+}-binding protein, was originally discovered as a calcium-dependent activator of cyclic nucleotide phosphodiesterase.[43-45] Several workers since then have described the properties of CaM and several reviews have appeared on this subject.[7,11,46-48] CaM is a multifunctional regulatory protein present in all eukaryotic cells and shown to activate, in a calcium-dependent manner, a number of enzymes involved in many physiological processes.[46] CaM is distributed unevenly in various tissues, and brain and testis contains high levels of CaM. Within the brain certain neurons such as Purkinje and granular cell bodies contain more CaM than other neurons.[49,50] In the synaptic junctions, CaM was concentrated on the inner surface of the postsynaptic membranes.[51,52] Although CaM is mainly described as cytosolic protein, it has been shown that 30 to 40% of CaM in rat brain is associated with particulate fractions in the presence of calcium.[53] The primary structure with complete amino acid sequence of bovine brain CaM was determined.[54,55] These studies showed that brain CaM contains 4 calcium-binding sites. Each site is a helix-loop-helix consisting of 12 amino acid residues, six of which contribute oxygen ligands to calcium coordination. Physical and structural properties of CaM from vertebrate and invertebrates have been reviewed.[56] Calcium binds to brain CaM with high affinity with a K_d of 14 μM.[57] Binding of calcium induces conformational changes in CaM which then binds with its regulated target protein or enzyme.

Ca^{2+} ATPase isolated from many tissues with the exception of the liver has been shown to be activated by CaM. Activation and inactivation of this enzyme may be through phosphorylation and dephosphorylation. Ca^{2+} affinity of Ca^{2+} ATPase is altered by CaM by a factor of 10 to 100. In spite of the fact that CaM is known to regulate Ca^{2+} ATPase from many tissues, most of the studies were conducted with human erythrocyte Ca^{2+} ATPase.[34,35] Brain synaptosomal membrane Ca^{2+} ATPase, as stated before, was purified and shown to be under CaM regulation.[37,40] Recent studies with a combination of direct protein sequence analysis of the erythrocyte enzyme and molecular cloning from brain tissue demonstrate that the Ca^{2+} ATPase in both tissues contains the 138 kDa form which is under CaM regulation.[58]

The above information clearly demonstrates that calcium plays a pivotal role in neuronal functions, and that the intraneuronal calcium concentration is maintained by various plasma membrane ion pumps and ion-ion exchangers in addition to the intracellular organelles and calcium-binding proteins. CaM present in the brain regulates Ca^{2+} pump activity. Hereafter, I will discuss how metal ions alter the CaM activated Ca^{2+} pump activity and how these changes relate to the neurotoxicity of these metals.

III. NEUROTOXICITY OF METALS

It is difficult to determine the specific mechanism(s) of the neurotoxicity of metals as they are general toxicants affecting many subcellular systems. However, correlation of biochemical end points to observed symptoms (mostly behavioral), morphological and physiological changes will sometimes enable us to understand the neurotoxic insult. Neurochemical evaluation of a specific hypothesis will help in elucidating the neurotoxic action of certain metals. Many metals produce alterations in the nervous system at excessive doses but some of them are neurotoxic even at low doses. Mercury, lead, tin, aluminum, manganese, and to a lesser extent cadmium are examples of such neurotoxic metals. Organic derivatives of some of these metals such as methyl mercury, tri-substituted lead, and tin compounds are more hazardous to health than are inorganic metals. A review of the literature indicates that these neurotoxic metals cause behavioral, morphological, and biochemical changes in the nervous system.[59-69] An extensive description of the toxicity of each metal becomes a chapter by itself. Therefore the discussion of their effects on calcium-CaM activated Ca^{2+} pump activity in the brain is restricted.

The discussion presented in the earlier sections of this chapter clearly indicates that the Ca^{2+} pump is identified with high-affinity Ca^{2+} ATPase in plasma membranes of a variety of tissues including brain synaptosomal membranes. It is also established that this pump and Na^+-Ca^{2+} exchange are the two major mechanisms for removing excess calcium from the cell.[11] The literature on this topic has been extensively reviewed and also constituted part of the volume on "Transport ATPases".[70] However, relatively less information is available on the modulation of Ca^{2+} ATPase by metal ions.

A. METAL BINDING TO CALMODULIN

Calmodulin, an ubiquitous calcium-binding protein, is shown to confer calcium sensitivity to nearly 30 to 40 target proteins including plasma membrane Ca^{2+} ATPase. Binding of calcium to all four calcium-binding sites on CaM results in structural changes, increasing the α-helix and intrinsic tyrosine fluorescence. Binding of calcium also exposes hydrophobic sites on CaM which are active binding sites for drugs and proteins.[56,71,72] Besides calcium, there are several other metal ions which have been shown to bind to CaM. They are Al^{3+}, Cd^{2+}, Hg^{2+}, La^{3+}, Mn^{2+}, Pb^{2+}, Sm^{3+}, Sr^{2+}, Tb^{3+}, and Zn^{2+}.[73-81] These studies demonstrated that many of these metal ions bind with CaM effectively and increase the hydrophobic areas as well as the fluorescence. Some metals such as Cd^{2+} and Tb^{3+} were shown to be effectively substituted for Ca^{2+} in activating CaM and its regulatory functions.[77] Similarly, in flow dialysis experiments it was shown that other metals such as Pb^{2+}, Sr^{2+}, Mn^{2+}, and to a lesser extent Cd^{2+} activated CaM and its action on phosphodiesterase activity.[81] When bound to CaM they may modify CaM allosterically.[79] Al^{3+} increased the fluorescence by 400% when bound to bovine brain CaM, but in contrast to Ca^{2+} it decreased the α-helix content.[78] Whatever the end result, most of these studies revealed that a number of metals effectively bind to CaM and alter its conformation which in turn may modulate its target enzymes and proteins.

B. EFFECT OF METAL IONS ON CAM-ACTIVATED CALCIUM ATPASE

Plasma membrane Ca^{2+} ATPase along with Na^+-Ca^{2+} exchanger maintains the intracellular $(Ca^{2+})_i$. As discussed before, the $(Ca^{2+})_i$ is critical for proper neurotransmission. Alteration of the $(Ca^{2+})_i$ either directly or indirectly by modulation of Ca^{2+} pumps by exogenous toxicants such as metal ions result in nervous system dysfunction. Mercuric compounds, lead, aluminum, manganese, and to a lesser extent cadmium are neurotoxic in that they either alter one or many events associated with neurotransmission such as neurotransmitter synthesis, release, postsynaptic receptor binding, and reuptake.[63-69,82] The metals may be effecting these biochemical events directly[63-64,82] or indirectly by altering the Ca^{2+} transport mechanisms.[83-88] The effects of metal ions on neurotransmitter release are discussed elsewhere in this volume. It suffices to mention here that metals such as Pb^{2+}, Cd^{2+}, etc., may either act directly on transmitter release processes, or interact with CaM at Ca^{2+} sites, thus promoting transmitter release. Inhibition of CaM-activated Ca^{2+} ATPase in plasma membranes as mechanism for the increased transmitter release is an attractive hypothesis. Increase in intranueronal Ca^{2+} triggers the transmitter release and inhibition of Ca^{2+} ATPase may increase $(Ca^{2+})_i$ levels. We have shown both *in vitro* and *in vivo* that heavy metals decrease Ca^{2+} ATPase activity in synaptosomes.[87] This decrease was specific to CaM-activated Ca^{2+} ATPase at low concentrations of metals where the basal enzyme was not inhibited.[87,88] Other reports also showed such a relationship.[75,84] In contrast to these studies, others reported that CaM was shown to be activated by various metal ions.[74,77] These authors also showed that the metal-activated CaM stimulated its target enzymes such as phosphodiesterase. These studies were, however, not conducted *in vivo*. Chronic activation of CaM by metals may alter its properties which is detrimental to neuronal function.[77] On the other hand, if these metals reduce CaM activity or its levels in neurons, they may directly alter the neuronal activity. We believe that chronic exposure to metals may lead to the decreased levels of CaM as we have reported with other neurotoxic compounds.[38,39]

Future studies should be directed to determine the absolute levels of CaM in the nervous system in animals exposed chronically to sub-lethal doses of various neurotoxic metals. It is also necessary to know whether such an altered CaM would induce changes in its many targeted biochemical and neurochemical events associated with neurotransmission. The existing information on metal effects on CaM-activated Ca^{2+} ATPase may only partially explain the neurotoxicity of metals.

ACKNOWLEDGMENTS

I would like to thank Dr. Robert Currier, McCarty Professor and Chairman of the Neurology Department, for his constant encouragement in conducting my research. I also thank Ms. Margaret Collins for her expert typing of this manuscript.

REFERENCES

1. **Ringer, S. J.,** Regarding the action of hydrate of soda, hydrate of ammonia and hydrate of potash on the ventricle of the frog's heart, *J. Physiol. (London),* 3, 195, 1881.
2. **Locke, F. S.,** Notiz uberden Einfluss physiologischer Kochsalzlo sung auf die elektrische Erregbarkeit von Muskel and Nerv, *Centralbl. Physiol.,* 8, 166, 1884.
3. **Brinley, F. J., Jr.,** Calcium buffering in squid axon, *Ann. Rev. Biophys. Bioeng.,* 7, 363, 1978.
4. **Erulkar, S. D. and Fine, A.,** Calcium in nervous system, in *Reviews of Neuroscience,* Schneider, D. M., Ed., Raven Press, New York, 1979, 179.
5. **Llinas, R. R.,** Calcium in synaptic transmission, *Sci. Am.,* 247, 38, 1982.
6. **Blaustein, M. P. and Nelson, M. T.,** Na^+-Ca^{2+} exchange: its role in the regulation of cell calcium, in *Membrane Transport of Calcium,* Carafoli, E., Ed., Academic Press, New York, 1982, 217.
7. **Stefano, A.,** Calcium and brain proteins, in *Metal Ions in Biological Systems,* Vol. 17, Sigel, H., Ed., Marcel Dekker, New York, 1984, chap. 7.
8. **Hammerschlag, R.,** The role of calcium in the initiation of fast axonal transport, *Fed. Proc.,* 39, 2809, 1980.
9. **Ochs, S.,** Calcium requirement for axoplasmic transport and the role of the perineural sheath, in *Nerve Repair and Regeneration: Its Clinical and Experimental Basis,* Jewett, D. L. and McCarroll, H. R., Eds., C. V. Mosby, St. Louis, MO, 1980, 77.
10. **Kretsinger, R. H.,** Structure and evolution of calcium modulated proteins, *CRC Crit. Rev. Biochem.,* 8, 119, 1980.
11. **Cheung, W. Y.,** Calmodulin plays a pivotal role in cellular regulation, *Science,* 207, 19, 1980.
12. **DiPolo, R., Requena, J., Brinley, R. J., Jr., Mullins, L. J., Scarpa, A., and Tiffert, T.,** Ionized calcium concentrations in squid axons, *J. Gen. Physiol.,* 67, 433, 1976.
13. **Schweitzer, E. and Blaustein, M. P.,** Calcium buffering in presynaptic nerve terminals: free calcium levels measured with arsenaro III, *Biochim. Biophys. Acta,* 600, 912, 1980.
14. **Katz, B. and Miledi, R.,** Tetrodotoxin-resistant electric activity in presynaptic terminals, *J. Physiol. (London),* 203, 459, 1969.
15. **Nachshen, D. A. and Blaustein, M. P.,** Some properties of potassium-stimulated calcium influx in presynaptic nerve endings, *J. Gen. Physiol.,* 76, 709, 1980.
16. **Miledi, R.,** Transmitter release induced by the injection of calcium ions into nerve terminals, *Proc. R. Soc., London Ser. B,* 183, 421, 1973.
17. **Llinas, R. and Nicholson, C.,** Calcium role in depolarization secretion coupling: an aequorin study in squid giant synapse, *Proc. Natl. Acad. Sci. U.S.A.,* 72, 187, 1975.

18. **Katz, B. and Miledi, R.,** The role of calcium in neuromuscular facilitation, *J. Physiol. (London),* 195, 481, 1968.

19. **Dunlap, K. and Fishbach, G. D.,** Neurotransmitters decrease the calcium conductance activated by depolarization of embryonic chick sensory neurones, *J. Physiol. (London),* 317, 519, 1981.

20. **Fleckenstein, A.,** Specific pharmacology of calcium in myocardium, cardiac pacemakers, and vascular smooth muscle, *Ann. Rev. Pharmacol. Toxicol.,* 17, 149, 1977.

21. **Althaus-Salzmann, M., Carafoli, E., and Jokob, A.,** Ca^{2+}, K^+ redistributions and α-adrenergic activation of glycogenolysis in perfused rat livers, *Eur. J. Biochim.,* 106, 241, 1980.

22. **Baker, P. F. and Schlaepfer, W.,** Calcium uptake by axoplasm extruded from giant axons of Loligo, *J. Physiol. (London),* 249, 37, 1975.

23. **Glenney, J. R., Bretscher, A., and Weber, K.,** Calcium control of the intestinal microvillus cytoskeleton: its implications for the regulations of microfilament organizations, *Proc. Natl. Acad. Sci. U.S.A.,* 77, 6458, 1980.

24. **Nicholls, D. G. and Akerman, K. E. O.,** Biochemical approaches to the study of cytosolic calcium regulation in nerve endings, *Philos. Trans. R. Soc. London, Ser. B:,* 296, 115, 1981.

25. **Becker, G. L., Fiskum, G., and Lehninger, A. L.,** Regulation of free Ca^{2+} by liver mitochondria and endoplasmic reticulum, *J. Biol. Chem.,* 255, 9009, 1980.

26. **Blaustein, M. P., Ratzlaff, R. W., and Schweitzer, E. S.,** Control of intracellular calcium in presynaptic nerve terminals, *Fed. Proc.,* 39, 2790, 1980.

27. **Brinley, F. J., Jr.,** Regulation of intracellular calcium in squid axons, *Fed. Proc.,* 39, 2778, 1980.

28. **Blitz, A. L., Fine, R. E., and Toselli, P. A.,** Evidence that coated vesicles from brain are calcium-sequestering organelles resembling sarcoplasmic reticulum, *J. Cell. Biol.,* 75, 135, 1977.

29. **Rahaminoff, H. and Spanier, R.,** Sodium-dependent calcium uptake in membrane vesicles derived from rat brain synaptosomes, *FEBS Lett.,* 104, 111, 1979.

30. **McGraw, C. F., Nachshen, D. A., and Blaustein, M. P.,** Calcium movement and regulation in presynaptic nerve terminals, in *Calcium and Cell Function,* Vol. 2, Cheung, W. Y., Ed., Academic Press, New York, 1982, chap. 3.

31. **Blaustein, M. P. and Ector, A. C.,** Carrier-mediated sodium-dependent and calcium-dependent calcium efflux from pinched-off presynaptic nerve terminals (synaptosomes) *in vitro, Biochim. Biophys. Acta,* 419, 295, 1976.

32. **DiPolo, R.,** Effect of ATP on the calcium efflux in dialyzed squid giant axons, *J. Gen. Physiol.,* 54, 503, 1974.

33. **Carafoli, E. and Crompton, M.,** The regulation of intracellular calcium, *Curr. Top. Membr. Transp.,* 10, 151, 1978.

34. **Carafoli, E. and Zurini, M.,** The Ca^{2+} pumping ATPase of plasma membranes. Purification, reconstitution and properties, *Biochim. Biophys. Acta.,* 683, 288, 1982.

35. **Penniston, J. T.,** Plasma membrane Ca^{2+}-AtPases as active Ca^{2+} pumps, in *Calcium and Cell Function,* Vol. 4, Cheung, W. Y., Ed., Academic Press, New York, 1983, chap. 3.

36. **Bradford, H. F.,** Isolated nerve terminals as an *in vitro* preparation for the study of dynamic aspects of transmitter metabolism and release, in *Handbook of Psychopharmacology,* Vol. 1, Iverson, L. L., Iverson, S. D., and Snyder, S. H., Eds., Plenum, New York, 1975, 191.

37. **Sobue, K., Ichida, S., Yoshida, H., Yamazaki, R., and Kakiuchi, S.,** Occurrence of a Ca^{2+} and modulator protein-activatable ATPase in the synaptic plasma membranes of brain, *FEBS Lett.,* 99, 199, 1979.

38. **Desaiah, D., Chetty, C. S., and Prasada Rao, K. S.,** Chlordecone inhibition of calmodulin activated calcium ATPase in rat brain synaptosomes, *J. Toxicol. Env. Hlth.,* 16, 189, 1985.

39. **Prasada Rao, K. S., Trottman, C. H., Morrow, W., and Desaiah, D.,** Toxaphene inhibition of calmodulin dependent calcium ATPase activity in rat brain synaptosomes, *Fund. Appl. Toxicol.,* 6, 648, 1986.

40. **Hakim, G., Itano, T., Verma, A. K., and Penniston, J. T.,** Purification of the Ca^{2+} ATPase from rat brain synaptic plasma membrane, *Biochem. J.,* 207, 225, 1982.

41. **Verma, A. K., Gorski, J. P., and Penniston, J. T.,** Antibodies directed toward human erythrocyte Ca^{2+} ATPase: effect on enzyme function; immunoreactivity of Ca^{2+} ATPases from other sources, *Arch. Biochem. Biophys.,* 215, 345, 1982.

42. **Gill, D. L., Grollman, E. F., and Kohn, L. D.,** Calcium transport mechanisms in membrane vesicles from guinea pig brain synaptosomes, *J. Biol. Chem.,* 256, 184, 1981.

43. **Cheung, W. Y.,** Cyclic 3',5'-nucleotide phosphodiesterase: pronounced stimulation by snake venom, *Biochem. Biophys. Res. Commun.,* 29, 478, 1967.

44. **Cheung, W. Y.,** Cyclic 3',5'-nucleotide phosphodiesterase: demonstration of an activator, *Biochem. Biophys. Res. Commun.,* 38, 533, 1970.

45. **Kakiuchi, S. and Yamazaki, R.,** Calcium dependent phosphodiesterase activity and its activating factor from brain, *Biochem. Biophys. Res. Commun.,* 41, 1104, 1970.

46. **Wallace, R. W., Tallant, E. A., and Cheung, W. Y.,** Assay preparation and properties of calmodulin, in *Calcium and Cell Function,* Vol. 1., Cheung, W. Y., Ed., Academic Press, New York, 1980, chap. 2.

47. **Klee, C. B. and Vanaman, T. C.,** Calmodulin, *Adv. Protein Chem.,* 35, 213, 1982.

48. **Means, A. R., Tash, J. S., and Chafouleas, J. G.,** Physiological implications of the presence, distribution and regulation of calmodulin in eukaryotic cells, *Physiol. Rev.,* 62, 1, 1982.

49. **Wood, J. G., Wallace, R. W., Whitaker, J. N., and Cheung, W. Y.,** Immunocytochemical localization of calmodulin and a heat-labile calmodulin-binding protein (CaM-BP$_{80}$) in basal ganglia of mouse brain, *J. Cell. Biol.,* 84, 66, 1980.

50. **Lin, C. T., Dedman, J. R., Brinkley, B. R., and Means, A. R.,** Localization of calmodulin in rat cerebellum by immunoelectron microscopy, *J. Cell. Biol.,* 85, 473, 1980.

51. **Grab, D. J., Berzins, K., Cohen, R. S., and Siekevitz, P.,** Presence of calmodulin in postsynaptic densities isolated from canine cerebral cortex, *J. Biol. Chem.,* 254, 8690, 1979.

52. **Carlin, R. K., Bartlet, D., and Siekevitz, P.,** Identification of Fodrin as a major calmodulin-binding protein in post-synaptic density preparations, *J. Cell. Biol.,* 96, 443, 1983.

53. **Kakiuchi, S., Yamazaki, R., Yasuda, S., Sobue, K., Oshima, M., and Nakajima, T.,** Membrane-bound protein modulator and phosphodiesterase, *Adv. Cyclic Nucleotide Res.,* 9, 253, 1978.

54. **Vanaman, T. C., Sharief, F. S., and Watterson, D. M.,** Structural homolog between brain modulator protein and muscle TnC$_s$, in *Calcium Binding Proteins and Calcium Function,* Wasserman, R. H., Corradino, Z. A., Carafoli, E., Kretsinger, R. H., MacClennan, D. A., and Siegel, F. L., Eds., Elsevier, New York, 1977, 107.

55. **Watterson, D. M., Sharief, F. S., and Vanaman, T. C.,** The complete amino acid sequence of the Ca^{2+}-dependent modulator protein (calmodulin) of bovine brain, *J. Biol. Chem.,* 255, 962, 1980.

56. **Vanaman, T. C.,** Structure, function and evolution of calmodulin, in *Calcium and Cell Function,* Vol. 1, Cheung, W. Y., Ed., Academic Press, New York, 1980, chap. 3.

57. **Keller, C. H., Olwin, B. B., LaPorte, D. C., and Storm, D. R.,** Determination of the free-energy coupling for binding of calcium ions and Troponin I to calmodulin, *Biochemistry,* 21, 156, 1982.

58. **Brandt, P., Zurini, M., Sisken, B. F., Rhoads, R. E., and Vanaman, T. C.,** Structural studies, molecular cloning, and localization of the plasma membrane Ca^{2+} pumping ATPase, in *Calcium-Binding Proteins in the Health and Disease,* Norman, A. W., Vanaman, T. C., and Means, A. R., Eds., Academic Press, New York, 1987, 544.

59. **Tilson, H. A., Mactutus, C. F., Mclamb, R. L., and Burne, T. A.,** Triethyl and trimethyl lead: effects on rat behaviour, cerebral morphology and levels of lead in blood and brain, *Neurobehav. Toxicol. Teratol.,* 4, 671, 1982.

60. **Squib, R. E., Carmichael, N. S., and Tilson, H. A.,** Behavioural and neuromorphological effects of triethyl tin bromide in adult rats, *Toxicol. Appl. Pharmacol.,* 55, 188, 1980.

61. **Needleman, H. L. and LeViton, A.,** Lead and neurobehavioural deficit in children, *Lancet,* 2, 104, 1979.

62. **Shiraki, H. and Takenchi, T.,** Minamata Disease, in *Pathology of the Nervous System,* Minckler, J., Ed., McGraw-Hill, New York, 1971, 1651.

63. **Slotkin, T. A. and Bartolome, J.,** Biochemical mechanisms of developmental neurotoxicity of methyl mercury, *Neurotoxicology,* 8, 65, 1987.

64. **Mailman, R. B. and Lewis, M. H.,** Neurotoxicants and central catecholamine systems, *Neurotoxicology,* 8, 123, 1987.

65. **Hammond, P. B. and Foulkes, E. C.,** Metal ion toxicity in man and animals, in *Metal Ions in Biological Systems,* Vol. 20, Sigel, H., Ed., Marcel Dekker, New York, 1986, chap. 6.

66. **Barbeau, A.,** Manganese and extrapyramidal disorder, *Neurotoxicology,* 5, 13, 1984.

67. **Gabbiani, G. Baic, D., and Desiel, C.,** Toxicity of cadmium for the central nervous system, *Exp. Neurol.,* 18, 154, 1967.

68. **Wong, K. L. and Klassen, C. D.,** Neurotoxic effects of cadmium in young rats, *Toxicol. Appl. Pharmacol.,* 63, 330, 1982.

69. **Wisniewski, H. M., Korthals, J. K., Kopeloff, L. M., Ferszt, R., Chusid, J. C., and Terry, R. D.,** Neurotoxicity of aluminum, in *Neurotoxicology,* Vol. 1, Roizin, L., Shiraki, H., and Grcevis, N., Eds., Raven Press, New York, 1977, 313.

70. **Carafoli, E. and Scarpa, A., Eds.,** Transport ATPases, *Ann. N.Y. Acad. Sci.,* 402, 296, 1982.

71. **Tanaka, T. and Hidaka, H.,** Hydrophobic regions function in calmodulin-enzyme(s) interactions, *J. Biol. Chem.,* 255, 11078, 1980.

72. **Cox, J. A., Comte, M., Malone, A., Burger, D., and Stein, E. A.,** Mode of action of the regulatory protein calmodulin, in *Metal Ions in Biological Systems,* Vol. 17, Sigel, H., Ed., Marcel Dekker, New York, 1984, chap. 6.

73. **Wallace, R. W., Tallant, E. A., Dockter, M. E., and Cheung, W. Y.,** Calcium-binding domains of calmodulin, sequence of fill as determined with terbium luminescence, *J. Biol. Chem.,* 257, 1845, 1982.

74. **Chao, S. H., Suzuki, Y., Zysk, J. R., and Cheung, W. Y.,** Activation of calmodulin by various metal cations as a function of ionic radius, *Mol. Pharmacol.,* 26, 75, 1984.

75. **Cox, J. L. and Harrison, J. D., Jr.,** Correlation of metal toxicity with *in vitro* calmodulin inhibition, *Biochem. Biophys. Res. Commun.,* 115, 106, 1983.

76. **Baudier, J., Haglid, K., Haiech, J., and Gerar, D.,** Zinc ion binding to human brain calcium binding proteins, calmodulin and S100B protein, *Biochem. Biophys. Res. Commun.,* 114, 1138, 1983.
77. **Suzuki, Y., Chao, S. H., Zysk, J. R., and Cheung, W. Y.,** Stimulation of calmodulin by cadmium ion, *Arch. Toxicol.,* 57, 205, 1985.
78. **Siegel, N., Suhayda, C., and Haug, A.,** Aluminum changes the conformation of calmodulin, *Physiol. Chem. Phys.,* 14, 165, 1982.
79. **Mills, J. S. and Johnson, J. D.,** Metal ions as allosteric regulators of calmodulin, *J. Biol. Chem.,* 260, 15100, 1985.
80. **Martin, S. R. and Bayley, P. M.,** The effects of Ca^{2+} and Cd^{2+} on the secondary and tertiary structure of bovine testis calmodulin, *Biochem. J.,* 238, 485, 1986.
81. **Haberman, E., Crowell, K., and Janicki, P.,** Lead and other metals can substitute for Ca^{2+} in calmodulin, *Arch. Toxicol.,* 54, 61, 1983.
82. **Toth, G. P., Cooper, G. P., and Suszkiw, J. B.,** Effects of divalent cations on acetylocholine release from digitonin-permeabilized rat cortical synaptosomes, *Neurotoxicology,* 8, 507, 1987.
83. **DiPolo, R., Rojas, H. R., and Beague, L.,** Vanadate inhibits uncoupled Ca efflux but not Na-Ca exchange in squid axons, *Nature,* 281, 228, 1979.
84. **Robinson, J. D.,** Vanadate inhibition of brain (Ca^{2+} + Mg^{2+}) ATPase, *Neurochem. Res.,* 6, 225, 1981.
85. **Chiu, V. C., Mouring, D., and Haynes, D. H.,** Action of mercurials on the active and passive transport properties of sarcoplasmic reticulum, *J. Bioeng. Biomed.,* 15, 13, 1983.
86. **Prasada Rao, K. S., Chetty, C. S., Trottman, C. H., Uzodinma, J. E., and Desaiah, D.,** Effect of tricyclohexylhydroxytin on synaptosomal Ca^{2+}-dependent ATP hydrolysis and rat brain subcellular calmodulin, *Cell Biochem. Function,* 3, 267, 1985.
87. **Moorthy, K. S., Ahammad Sahib, K. I., Uzodinma, J. E., Trottman, C. H., and Desaiah, D.,** Effect of mercury, cadmium and lead on calmodulin regulated calcium pump activity in rat brain and heart, *Trends Life Sci. (India),* 1, 37, 1986.
88. **Nath, R., Vig, P. J. S., and Desaiah, D.,** Metal inhibition of calmodulin activity in monkey brain, *Toxicologist,* 8, 27, 1988.

Chapter 4

NEUROPATHOLOGY OF HEAVY METALS AND ITS MODULATION BY NUTRITIONAL INFLUENCES

Louis W. Chang and Casey S. Fu

TABLE OF CONTENTS

I. INTRODUCTION

Heavy metals, like many other elements in the environment, constantly interplay with human life environmentally, industrially, occupationally, and biologically. Many of these elements, such as lead, mercury, and cadmium, are nonessential to life. Continuous and excessive exposures to these elements may result in toxicity to the organism. Other elements, such as iron, calcium, selenium, copper, and zinc, are essential to human health. Imbalance of these elements in the biological system, either as a result of dietary intake or as a consequence of the presence of other metals or elements (e.g., cadmium), may also constitute an adverse condition for health.

The prime purpose of the present chapter is to summarize the toxicity and pathological impacts of some of the neurotoxic metals, including mercury, lead, cadmium, aluminum, and tin, and to present some current thoughts on the phenomena of "metal-element interaction", emphasizing the influences of mineral elements, such as calcium, zinc, copper, and selenium, on the toxicity of heavy metals. It becomes apparent that the complexity of "toxic mechanisms" for metals lies not only in their direct toxic impact on cells, but also in their secondary effects on the metabolism and the overall balances of certain essential metals and elements in the system. It is our hope that metal-metal or metal-element interactions will be considered whenever the "toxic effects" or "toxic mechanism" of a metal are investigated.

II. NEUROTOXICITY AND NEUROPATHOLOGICAL EFFECTS

The neurotoxicities of lead, mercury, and cadmium are well known and have been extensively reviewed in the past.[1-5] In our present treatise, we only intend to update, summarize, and emphasize the essence of pathology induced by these neurotoxic metals. Aluminum and tin (alkyltin) have also received increasing attention as potent neurotoxicants in recent years.[5-19] These two metals will also be included in our present review.

A. LEAD

Lead is an ubiquitous element and, because of its extensive usage by man, it has presented serious impact on human health.[20-24] Lead has no essential biological function. Its accumulation in tissues simply reflects exposures to contaminated environments.

There are two groups of lead compounds: organic and inorganic. Although both organic and inorganic leads are known to be neurotoxic, the neurological effects of inorganic lead compounds are by far more extensively investigated.

1. Neuropathological Changes Induced by Inorganic Lead

There is a reduction and retardation of brain growth when animals are exposed to lead during their developmental period.[25-30] Edematous and hemorrhagic changes, particularly in the cerebellum, are the most frequent observations.[27,31-36] Of all areas of the brain, cerebellum and hippocampus appear to be the most vulnerable to lead toxicity.[29,33] A reduction in the thickness of various cortical areas, in correlation with reduction in brain weight, has been demonstrated.[27,28,30,37]

The phenomena of cerebral and cerebellar edema and hemorrhagic changes are highly suggestive of vascular alterations in lead poisoning. Indeed, administration of radioactive lead to animals demonstrates a rapid accumulation of lead in the endothelial cells of brain capillaries.[38-41] Electron microscopic studies also revealed changes in the capillary endothelial cells after lead intixocation.[36,42] It is generally believed that lead has a toxic action on the cerebral capillary bed, inducing vascular permeability changes, edema, and hemorrhages. A breakdown in the blood-brain barrier as a result of lead poisoning has also been reported.[31,35,42,43]

Both dysmyelination (under-development of myelin) and demyelination (destruction of myelin) occur in lead intoxication. In developing animals, a decreased myelin sheath formation with a significant reduction in myelin basic proteins, phospholipids, galactolipids, sulfatides, and cholesterol is observed.[26,44] The reduction of myelination may be associated with reduction in axonal development in lead-poisoned animals. Despite a thinning of myelin sheath, there is no delay in the onset of myelin formation[44] and the myelin lamella itself is normal.[26] There is suggestion that some damage to the Schwann cells may occur[45] but demyelination, as observed in adult animals, is seldom found in the developing nervous system.

Segmental demyelination, noticeably in the peripheral nerves, is a prominent feature in lead poisoning.[46-49] Such form of demyelination involves only isolated segments (between internodes) of the nerve, suggesting random involvement of the Schwann cells.[50] Cavanagh[51] also suggested that lead intoxications produced selected injuries to Schwann cells leading to isolated segmental demyelination, and endoneurial edema may be followed by associative axonal degeneration. Myers and co-workers,[52] however, suggested that the initial injury in lead poisoning was the development of endoneurial edema resulting from blood-nerve barrier damage, and the increase in endoneurial pressure in turn induced secondary changes in the Schwann cells and axons. The precise pathological mechanism for lead-induced neuropathy still requires further investigation and clarification.

Abnormalities in certain neurons[26-28,54-58] and glial cells[27,57-59] are also observe, aside from the vascular and peripheral myelin changes. A reduction in cortical thickness and neuronal sizes was reported by Krigman and co-workers[27,28] in developing rats; the number of neurons was found to be unaffected. Other studies, however, showed a reduction in neuronal number[60] and size,[54] as well as stunted dendritic development and retarded maturation of the cerebellar Purkinje cells in lead intoxication.[54-58] Similarly, a reduction in dendritic development of neurons in the hippocampal dentate gyrus was also reported.[25] These investigators described a retarded dendritic development with alterations in dendritic branching and reduction in dendritic length of the hippocampal neurons. A reduced axonal development, at least in the hippocampal mossy fiber tract, was also noted after lead exposure.[25]

Although a reduction in synaptic density (number of synapses per unit area or per neuron) has been reported by various investigators,[27,30,62,63] there does not seem to be any significant alteration in the maturation or in morphological features of the synapses.[27,30,62]

2. Neuropathological Changes Induced by Organolead

The earliest comprehensive reports on alkyllead-induced neuropathological changes were made by Davis et al.[64] and Schepers[65] in 1963. By using tetraalkyllead compounds (R_4Pb), these investigators induced diffusely distributed lesions in the neocortex, brain stem, medulla, midbrain, cerebellum, thalamus, and spinal cord of dogs and rats. Later studies by Niklowitz[66] also described neuronal changes in the neocortex, Ammon's horn of the hippocampus and Purkinje cells of the cerebellum in rabbits exposed to tetraethyllead (Et_4Pb).

Et_4Pb-intoxicated rats exhibited signs and symptoms similar to those induced by trimethyltin, namely hyperexcitability, aggression, tremor, convulsion, and hyperactivity.[67,68] Biochemical studies *in vitro* revealed that Et_4Pb and Me_4Pb themselves were actually quite biologically inert.[65] However, after being dealkylated in the liver to the corresponding trialkyllead compounds (i.e., Et_3Pb and Me_3Pb), they became potent neurotoxic compounds.[69]

Et_3Pb induces rapid and marked degenerative changes in neurons, particularly those in the limbic system. Neuronal changes such as cytoplasmic eosinophilia and shrinkage, nuclear pyknosis and karyorrhexis, and severe cytoplasmic chromatolysis with or without nuclear alterations were observed.[70] These changes were first observed in the pyriform cortex followed by the entorhinal cortex and the hippocampus, including both the dentate granule neurons and Ammons' horn pyramidal cells. Recent studies by Chang and co-workers[71,72] also demonstrated neuronal changes in the hippocampus, brain stem, spinal cord, and dorsal root ganglia of rats exposed to Et_3Pb or Me_3Pb (Figures 1 and 2). Besides the non-specific cytoplasmic alterations, such as lysosomal accumulation, mitochondrial swelling, cytoplasmic membrane dilatation,

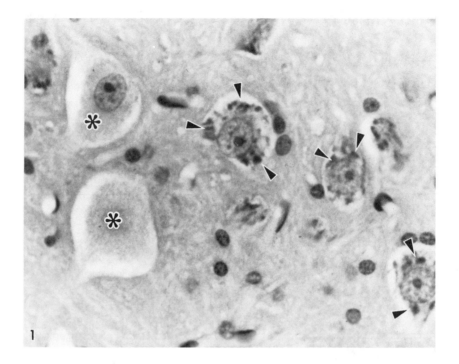

FIGURE 1. Brain stem neurons, rat. Trimethyllead, 22.0 mg/kg, 7 d. Extensive chromatolytic changes were observed in many brain stem neurons (*). Some neurons still retained their Nissl substance (arrowhead). H & E, magnification × 450.

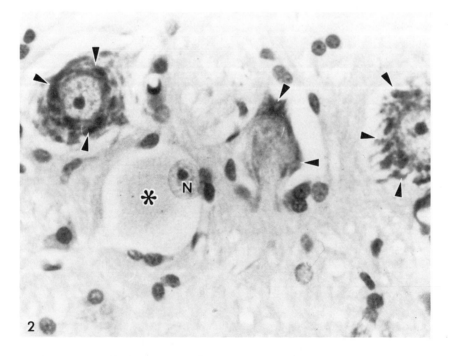

FIGURE 2. Spinal cord motoneurons, rat. Trimethyllead, 22.0 mg/kg, 7 d. A motoneuron showed chromatolysis (*) with eccentric nucleus (N). Other neurons still displayed normal Nissl pattern (arrowhead). H & E, magnification × 450.

and vacuolation, marked accumulation of neurofibrils was also observed in many affected neurons in rabbits.[66] The precise significance of such cytoplasmic alterations is still unclear at the present time.

B. ALKYLTINS

Although inorganic tin compounds are not known to induce any significant neurological changes, some organic tin compounds, particularly triethyltin (TET) and trimethyltin (TMT), are now recognized as potent neurotoxic metal compounds inducing rapid and prominent changes in the central nervous system.[10-17,73]

1. Neuropathological Changes Induced by TET

In 1954, over 200 individuals in France were intoxicated by an anti-infectious drug, Stalinon, which contained diethyltin iodine. Patients displayed prominent clinical signs of nausea, vertigo, visual disturbance, papilledema, paraplegia, and seizures suggesting increased intracranial pressures.[74,75] At autopsy cerebral edema, particularly in the white matter, was a prominent finding. Although 30% of the patients had complete clinical recovery, many patients displayed persistent symptoms of neurological damage.[74]

Experimental investigations confirmed that the principal pathological involvement in TET exposure was severe cerebral edema of the white matter[76-78] (Figures 3 and 4). Intramyelinic vacuoles, formed by accumulation of edematous fluid, can be demonstrated with electron microscopy.[79-82] Thus, in adult animals, TET induces cerebral edema with little or no damage to neurons.[18,83-85] Swelling of the glial cells, most noticeably the oligodendrocytes, together with vacuolization of the central myelin, resulting from splitting of the interperiod line[83,84] by the edematous fluid, were the most prominent pathological features.

Effects of TET on the developing nervous system were also investigated by several investigators.[86-90] Petechial hemorrhage of the cerebellum, diffused cerebral neuronal necrosis, and edematous changes of the myelinated fibers were observed in very young animals treated with TET sulfate.[86] With increasing age, the damage appeared to be confined only to the myelinated fibers.[73,87] It was also apparent that large myelinated fibers were more affected than smaller fibers. Hypomyelination[88] and poor dentritic development of some cortical neurons[89,90] were also observed in rats exposed to TET during early periods of their development.

2. Neuropathological Changes Induced by TMT

While TET is a potent myelinotoxicant inducing primarily changes in the central myelin, TMT is a neuronal toxicant inducing marked destruction of selected populations of neurons, particularly of those in the limbic system. Overt neurological and behavioral changes were induced in animals by TMT: aggression, hyperexcitability, tremor, seizures, hyperreactivity, learning deficits, and changes in schedule-controlled behavior.[91-93]

Pathological changes in the nervous system as a result of TMT intoxication have been investigated by various investigators. It is of interest to note that lesions produced by TMT in mice are different from those induced in rats.[13]

a. Neuropathology in Mice

Mice were found to be extremely susceptible to the toxicity of TMT. Severe tremors were observed in animals within 24 h of TMT exposure.[8,91] Histopathologic examination revealed extensive necrosis and destruction of the fascia dentata in the hippocampus, with only minimal damage in the hippocampal Ammon's horn[8] (Figure 5). Prominent chromatolytic and vacuolar changes in the brain stem neurons as well as in the spinal motorneurons were also observed[12,94] (Figure 6).

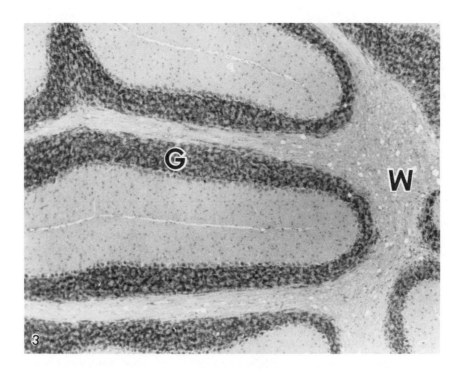

FIGURE 3. Cerebellum, rat. Triethyltin, 10.0 mg/kg, 4 h. Small microscopic vacuoles already developed in the white matter (W) at this early time of intoxication. G, granule cell layer. H & E, magnification × 250.

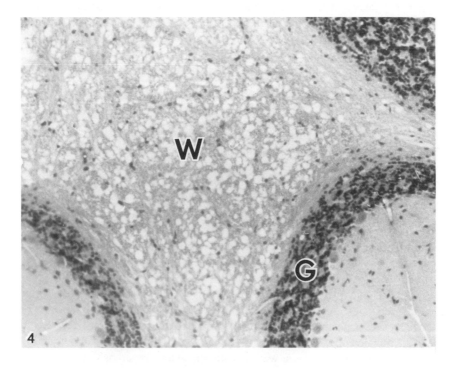

FIGURE 4. Cerebellum, rat. Triethyltin, 3.0 mg/kg, 17 d. Extensive vacuolar (edematous) change in the myelin sheaths of the white matter (W) was observed. No pathology was seen on the granule cells (G). H & E, magnification × 350.

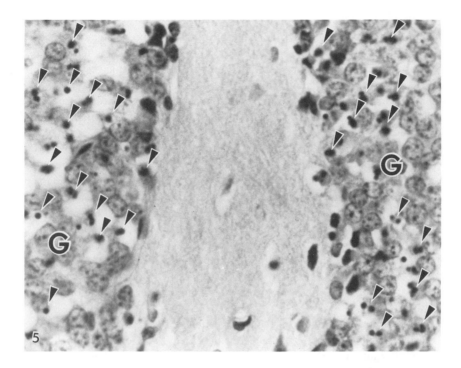

FIGURE 5.   Hippocampal fascia dentata, mouse.Triethyltin, 3.0 mg/kg, 48 h. Extensive pyknosis and necrosis (arrowhead) as well as vacuolar change in the fascia dentata granule cell layers (G) were observed. H & E, magnification × 450.

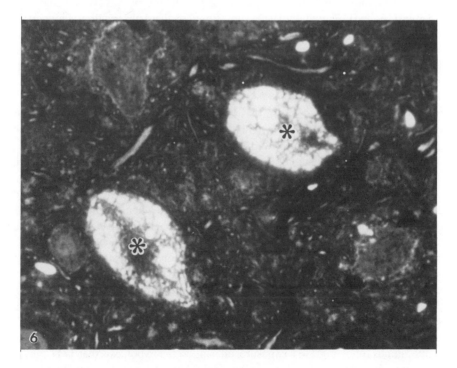

FIGURE 6.   Spinal motoneurons, mouse.Triethyltin, 3.0 mg/kg, 72 h. Two neurons (*) in the anterior horn showed extensive vacuolation. Toluidine blue stain, magnification × 450.

b. Neuropathology in Rats

The neuropathological effects of TMT in rats have a slower onset than those found in mice. The distribution of principal lesions is also different from that induced in mice. While fascia dentata, brain stem nuclei, and spinal cord are primarily affected in mice,[8,12,94] Ammon's horn (Figures 7, 8, 9), pyriform and entorhinal cortices (Figures 10, 11), and amygdaloid body are chiefly damaged in rats.[10,13-15,17,95,96] Besides neurons in the limbic system, other neurons such as those in the olfactory nuclei, retina, inner ear, and spinal ganglia, were also found to be affected by TMT.[11]

There was a discrepancy among several independent studies claiming susceptibility or resistance of CA_{3a} and CA_{3b} neurons in the Ammon's horn to TMT toxicity.[95-98] Careful reassessment of these phenomena has revealed that there is actually a differential septotemporal distribution of TMT-induced neuronal necrosis.[10,15] By means of step-serial sections in both coronal and sagittal planes, Chang and co-workers[15] had demonstrated that while the fascia dentata granule cells were most involved ventrally and temporally, the pyramidal cells of the Ammon's horn in the CA_{3a} and CA_{3b} sectors were most affected dorsomedially (septally). The CA_{3c} cells in the septal pole also showed slightly more involvement than those in the temporal pole. $CA_{1,2}$ cells, on the other hand, were most affected at the temporal pole. Thus, depending on the plane of sectioning (i.e., coronal or sagittal), the distribution of cellular involvement will appear differently.[15]

Another important observation on the interrelationship of lesion development between dentate granule cells and Ammon's pyramidal neurons in the hippocampal formation was also reported by Chang and co-workers.[15,17] These investigators demonstrated that those portions of hippocampus which had more severe loss of dentate granule cells (e.g., at temporal pole) showed only little damage to the CA_{3a} and CA_{3b} pyramidal neurons. Conversely, those areas displaying little granule cell damage (e.g., at septal pole) showed more severe involvement of the CA_{3a} and CA_{3b} pyramidal neurons. A similar reverse pathological relationship was also apparent between corresponding CA_3 and $CA_{1,2}$ pyramidal neurons. Based on these observations, these investigators postulated that induction of pathology in the Ammon's horn pyramidal neurons, particularly those in CA_3 subfield, required the presence and functional integrity of the dentate granule cells.[17] The preservation of Ammon's horn with rapid and extensive destruction of the fascia dentata in TMT-exposed mice[8,99] is also in good agreement with this hypothesis.

Studies with young animals demonstrated that rat hippocampi were most vulnerable to TMT toxicity between postnatal days 13 to 15.[14] Total and rapid destruction of the entire Ammon's horn was induced when rats were exposed to TMT at this age.[14,100] Exposure of younger animals to the same dose of TMT, however, showed a reduction in lesion development with decreasing age of the animals. In fact, no pathological lesion in the hippocampus was observed in animals less than 3 d of postnatal age.[14] Exposure of rats at different postnatal age periods (between day 1 and day 15) further revealed that specific neuronal destruction in various subfields of Ammon's horn (e.g., CA_{3a}, CA_{3b}, CA_{3c}, etc.) could be induced when the animals were exposed to TMT at specific neonatal ages (e.g., postnatal day 5, 7, 9, etc.).[14] The specific pathological pattern induced could also be correlated with the developmental and functional maturity of the hippocampal granule cells and their fibers.[14] That is, functional connections of dentate granule cells and their fibers (mossy fibers) with the Ammon's horn pyramidal neurons are believed to be essential in the induction of neuronal injury in Ammon's horn.[17,101]

Depletion of zinc in hippocampal mossy fibers was observed in TMT-treated animals, suggesting hyperexcitation in these cells.[102] Indeed, the reduction in recurrent inhibition in the dentate gyri, alterations in GABA metabolism,[104] and the early injury of dentate basket cells[16] in the hippocampus following TMT exposure, all support the hypothesis that TMT may act as a "hyperexcitotoxin", inducing cellular injury in the limbic system via hyperexcitation of these neurons.[17,101] Similar mechanisms in the induction of hippocampal damages[105] have also been postulated for alkyllead, kainic acid, bicuculline, and dipiperidinoethane.

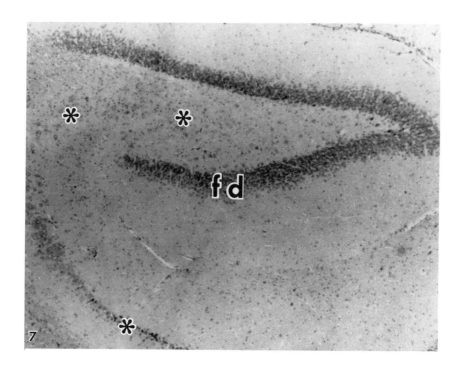

FIGURE 7. Hippocampus, rat. Trimethyltin, 6.0 mg/kg, 15 d. Significant cell loss in the Ammon's horn (*) was observed. Granule cells in the fascia dentata (fd) showed only minimal damage. H & E, magnification × 250.

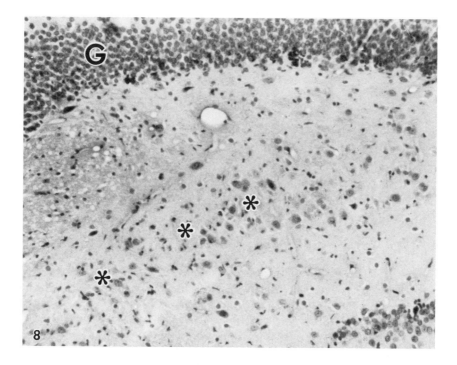

FIGURE 8. Hippocampus, rat. Trimethyltin, 6.0 mg/kg, 15 d. Total destruction of the CA_{3c} subfield of the Ammon's horn (*) was evident. No remarkable cell loss was seen in the granule layer (G) of the fascia dentata. H & E, magnification × 450.

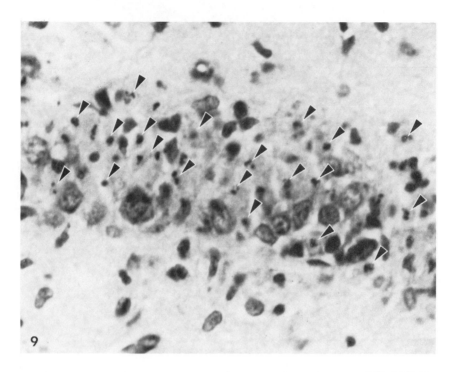

FIGURE 9. Ammon's horn, $CA_{1,2}$ subfield, rat. Trimethyltin, 6.0 mg/kg, 15 d. Extensive neuronal necrosis (arrowhead) was seen. H & E, magnification × 450.

C. METHYLMERCURY

Mercury, particularly organomercuric compounds, is known to have potent neurotoxic properties. Inorganic mercury, although reported to induce tremor and other neurological problems, does not cross the blood-brain barrier readily[106] and causes only minor changes in the central nervous system. Organomercury, on the other hand, crosses the blood-brain barrier rapidly[106] and induces extensive damage to the central nervous system, primarily the cerebellum and visual cortex. Prominent changes in the dorsal root ganglia and peripheral nerves have also been reported.[107,108] Since the subject of methylmercury intoxication has been extensively reviewed by this author in previous publications,[3,4,109,110] only the essence will be presented in this treatise.

1. Human Studies

Methylmercury is a heavy metal compound which has induced major episodes of poisoning in the human population. Since the outbreak of "Minamata disease" (methylmercury poisoning) in Japan in the early 1950's,[111-114] methylmercury has attracted much attention and is recognized as one of the most neurotoxic compounds. Among all clinical signs displayed by patients with Minamata disease, sensory disturbance (numness of the extremities), cerebellar ataxia, and constriction of visual field are the most prominent and consistent findings.[111,114] Autopsy of the patients revealed gross atrophy of the calcarine cortices and of the cerebellar cortex, particularly the folla at the lateral lobes and vermices. Histopathological examination[111-114] demonstrated extensive degeneration of the nerve cells and fibers in the calcarine cortex. The most marked cell loss was found in the second and upper-third lamina with accompanying astrogliosis. Pathological changes in the cerebellum mainly consisted of loss of granule cells, particularly those at the depth of the sulci. Purkinje neurons were found to be more resistant than the granule cells. Proliferation of Bergmann's glia and fiber (astrocytic proliferation) were also prominent. However, no remarkable change was found in the white matter.

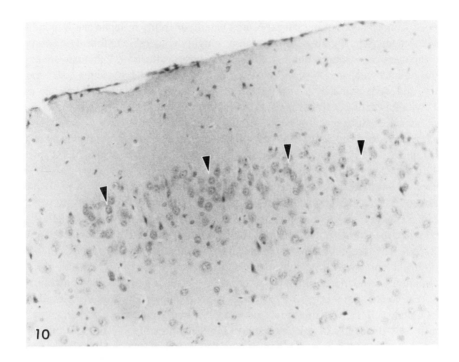

FIGURE 10. Pyriform cortex, rat. The normal pyriform cortex of rat showed a 4 to 5-cell thick layer of small granule neurons (arrowhead). H & E, magnification × 450.

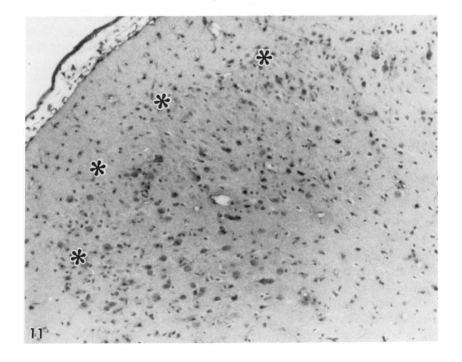

FIGURE 11. Pyriform cortex, rat. Trimethyltin, 6.0 mg/kg, 60 d. Total destruction and neuronal loss (*) was evident in the pyriform cortex (compare with Figure 10). H & E, magnification × 450.

Reports from episodes of human exposure to methylmercury in Japan and Iraq indicate that the development of the human nervous system is highly vulnerable to this toxic compound.[115-121] The Japanese cases represent a more chronic exposure situation. Patients were exposed at both pre- and neonatal ages. The Iraqi cases, on the other hand, represent a much more acute exposure and were primarily prenatal or in utero assaults. The Japanese episode presented patients with severe neurological signs, including mental retardation, impairment of movement and gait, ataxia, disturbances in speech and swallowing, and convulsions. Restriction of visual field was reported at a later stage in almost all patients. In the Iraqi case, early visual deficits and total blindness were reported in 30% of the patients.[115-117] The Iraqi patients displayed less mental disturbances than the Japanese cases.[116]

Autopsy examination[118-121] of patients who were exposed to MeHg in utero during the Japanese episode (fetal Minamata disease) revealed general atrophy and underdevelopment of the brain. Reduction in size and thickness of the cortex, corpus callosum, basal ganglia, and cerebral white matter was noted. Histopathology was less marked in anatomic selectivity than in adult cases. The cerebrum showed marked underdevelopment with hypoplastic and dysplastic neurons in the cortex, with resting matrix cells lining the periventricular region. Although demyelination was not noted, thinning of the white matter as a result of hypomyelination was evident. The cerebellum was severely affected with thinning in both granule cell and molecular layers. Granule cell loss was most prominent in the culmen and declive of monticulus and was not confined to the depths of sulci as displayed in the adult situation. Ectopic and disoriented Purkinje cells were found scattered in the molecular layer, indicating a disturbance of neuronal maturation, migration, and development.[109] Poor development of the Purkinje dendritic processes was also evident.[109,110]

Examination of brains from the Iraqi episode revealed abnormal gyral pattern and cerebral lobes.[122] Microscopically, marked disruption of cytoarchitecture with nests of heterotopic neurons in the white matters of the cerebrum and cerebellum were observed. Normal pattern of the cerebral cortex was disrupted with irregular columns of neuronal aggregates. Many cortical neuronal layers had an "undulating" pattern. Active neuronal necrosis, like that found in the Japanese cases, was not observed.

The variations in pathological entities between the Japanese and Iraqi cases are most likely attributed to the amount of methylmercury involved (acute high dose or chronic low dose) and the stages of development of the brain (prenatal or postnatal) when the exposures occurred. Nevertheless, brains from both episodes showed atrophy and under-development with heterotopic and ectopic neurons. These findings reflect disruptions of brain development, disturbance in maturation and migration of the nerve cells, and alterations in general cytoarchitecture of various brain structures. All these changes probably were related to the behavioral changes and neurological problems observed in these patients.

2. Experimental Studies

As in human subjects, pathological changes in experimental animals were also found primarily in the cerebellum, in the calcarine cortex, and in the dorsal root ganglia. Pathological changes in the cerebellum and in the dorsal root ganglia were most extensively studied.

Brown and Yoshida[123] investigated cerebellar changes in young chicks following organomercurial exposure and described alterations in the cytoplasmic reticulum in both the granule cells and Purkinje neurons. They postulated an interference with protein synthesis in the nerve cells as a result of methylmercury; this hypothesis was subsequently supported by other investigators.[124-131]

Investigators with cats and rats exposed to MeHg both demonstrated prominent degeneration of the cerebellar granule cells.[107,132,133] It was found that the disappearance of granule cells first occurred beneath the Purkinje cell layer and was most prominently found at the depth of the sulci (Figures 12 and 13). Ultrastructural studies revealed accumulation of lysosomes, membraneous

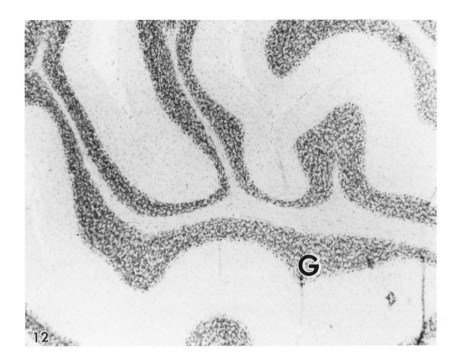

FIGURE 12. Cerebellum, hamster. The cerebellar cortex of normal animal showed prominent and consistent neuronal populations throughout the granule cell layer (G). H & E, magnification × 250.

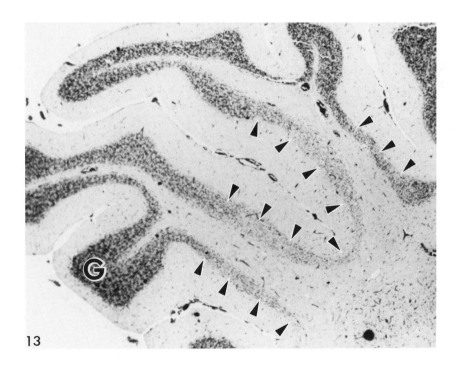

FIGURE 13. Cerebellum, hamster. Methylmercury, daily injection 1.0 mg/kg, 2 weeks. Extensive cell loss in the granule cell layer (G), particularly at the depth of the sulci (arrowheads), was evident (compare with Figure 12). H & E, magnification × 250.

degeneration, degranulation and destruction of rough endoplasmic reticulum, and cytoplasmic coagulation.[107] By means of an electron microscopic histochemical technique, Chang and co-workers[129] first demonstrated visually the intracellular location of mercury. These investigators demonstrated beyond doubt that methylmercury has a high affinity to cytoplasmic membrane-ous components, including the endoplasmic reticulum, nuclear envelope, Golgi complex, and mitochondria.

Chang and co-workers[107,108] also demonstrated that the dorsal root ganglia, at least in rats, were extremely sensitive to the toxicity of methylmercury (Figures 14 and 15) and suggested that these sensory ganglia could be used as a model for the study of methylmercury. These observations were later confirmed by other investigators.[134]

Disintegration of the endoplasmic reticulum and degranulation of ribosomes in neurons after MeHg poisoning (Figure 16) strongly suggest alterations in both RNA and protein metabolism in the nerve cells. These observations are consistent with those findings showing a reduction in neuronal RNA[130,131] and a decrease in protein synthesis by the cells after mercury intoxication.[125,128]

Induction of axonal degeneration and disintegration of myelin in peripheral nerves were also observed.[108,134-136] Dorsal root fibers were found to be particularly sensitive to the toxicity of methylmercury.[108]

Methylmercury was found to induce gross abnormalities of the developing brain including anencephaly, encephalocele, and hydrocephalus in animals exposed *in utero*.[137,139,141,142] Impacts of methylmercury on the fetus were found to be influenced not only by the dosage of exposure, but also by the species/strain of animals employed,[137,138] duration and periods of exposure during pregnancy,[139] and by the routes of exposure.

In animals exposed to methylmercury *in utero*, loss of nerve cells together with cytoarchitectural abnormalities are induced. Small hemorrhages and cytic changes in both the cortex and white matter are also observed.[143,144] Careful examination revealed neuronal destruction, disruption in neuronal migrations, and alterations in cerebral and cerebellar cytoarchitecture. Hypoplastic and ectopic neurons, delayed migration of cerebellar external granule layer, and incomplete cerebellar granule cell layer formation have been described both in cats and rats.[120,145-149]

By means of electron microscopy, Chang and co-workers[147-151] further elucidated the pathological changes in the nerve cells of mice, rats, and hamsters exposed to methylmercury *in utero*. Besides the relatively non-specific cytological changes such as lysosomal accumulation, disintegration of endoplasmic reticulum, and cytoplasmic degeneration, these investigators also reported aggregation of tubular saccules in the neuronal process, segmental incomplete myelination (hypomyelination) of axons, abnormal formation of myelin sheaths (dysmyelination), and abnormal dendritic process and synaptic developments. Some of these changes were found to persist well into adult lives of these animals.

D. ALUMINUM

The neurological entities associated with aluminum may be divided into three major categories: (1) experimental aluminum intoxication, (2) Alzheimer's disease, and (3) dialysis dementia. Each of these entities presents its own unique neurological phenomena.

1. Experimental Aluminum Intoxication

Ample evidence indicates that aluminum readily crosses the blood brain barrier.[152,153] Upon exposure, aluminum is found to accumulate in various brain areas, most noticeably in the cerebral cortex, midbrain, medulla, and cerebellum,[153,154] with much higher concentration in the gray than in the white matter.[155-159] Intracellularly, aluminum is found to a large extent localized within the nuclei of cells.[152,160-162] Neurofibrillary tangles have also been demonstrated to have a high concentration of aluminum.[153,163]

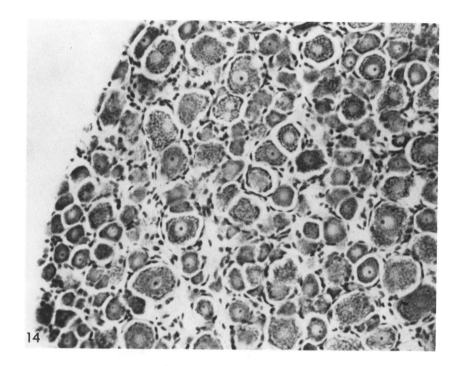

FIGURE 14. Dorsal root ganglion, rat. The dorsal root ganglion consisted of a dense population of neurons. These neurons were rich in Nissl substance and had prominent, centrally located nuclei. Gallocyanine stain, magnification × 350.

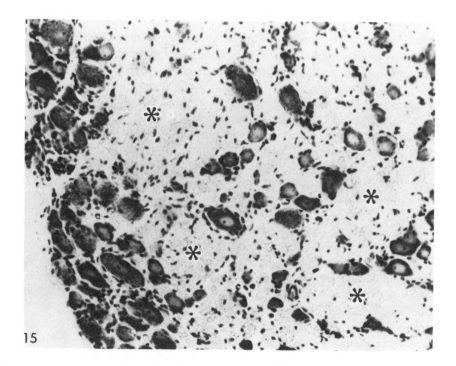

FIGURE 15. Dorsal root ganglion, rat. Methylmercury, daily injection, 1.0 mg/kg, 2 weeks. Large areas of the ganglion (*) showed a loss of nerve cells. The larger neurons appeared to be more affected (compare with Figure 14). Gallocyanine stain, magnification × 350.

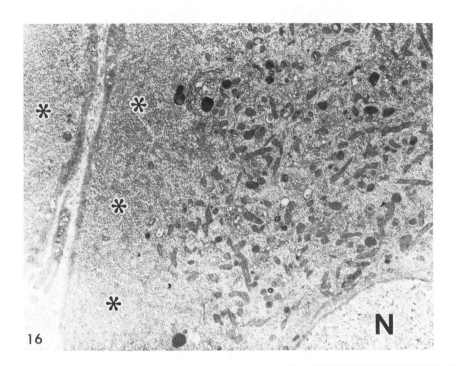

FIGURE 16. Dorsal root ganglion neuron, rat. Methylmercury, daily injection, 1.0 mg/kg, 2 weeks. Large areas (*) of the neuronal cytoplasm were void of rough endoplasmic reticulum; this could be correlated with chromatolysis and reduced protein synthesis. N, nucleus. Magnification × 8500.

Increased brain aluminum accumulation can induce proliferation of neurofibrillary elements in the nervous system of some species, such as rabbits and cats. Brain aluminum concentrations when exceeding 6 μg/g dry weight induce in rabbits development of such fibrillary changes.[152,164] Rats, however, are extremely insensitive to aluminum toxicity. Concentrations of up to 40 μg/g in rat brain still do not induce any observeable neuropathological effects.

Neurofibrillary tangles can be demonstrated with silver stains such as Bielschowsky's and Bodian stains. Depending on the species studied, neurofibrillary tangles develop rapidly in neurons of the spinal cord, brain stem, neocortex, Purkinje layer of cerebellum, and hippocampal Ammon's horn.[165-168] In general, nerve cells with prominent dendritic structures, large to moderate cell bodies, and large proximal dendrites are more prone to tangle formation.[162,169]

Electron microscopic examination reveals areas of cytoplasmic clearing in affected neurons; these are filled with bundles of random accumulations of 10 nm neurofilaments.[163,168,170,171] The tangles are usually sharply demarcated from the rest of the cytoplasmic matrix and organelles. An associated reduction in normal neurotubules was also reported in these affected neurons.[168]

Neurofibrillary tangles are generally found in the cytoplasm adjacent to the nucleus.[171] Dendritic involvements are also noted in some neurons.[170,173] Aluminum also induces progressive reduction of dendritic branches, spines, and beadings in the nerve cells.[170,174] These changes are most prominent at the distal portion of the dendritic tree,[169,170] suggesting that a reduced neuroplasmic transport in the dendrites may be responsible for this "distal dendritic atrophy". A reduction in neurotubules with a concomitant neurofibrillary accumulation is believed to create a disturbance in the normal neuroplasmic flow and transport in the dendrites. By means of a radioactive tracer technique, Liwnicz et al.[175] demonstrated a disruption of retrograde axonal transport in aluminum intoxicated rabbits. Since axons were not known to accumulate neurofibrillary tangles, no antrograde transport alteration was observed in the axons. The conditions of neuroplasmic transport in dendrites (dendritic flow) under the influence of aluminum are yet to be tested.

2. Alzheimer's Disease

Alzheimer's disease is a neurological problem found in human populations and is characterized by progressive loss of higher mental function.[176] Because of the similarities of Alzheimer's disease and experimental aluminum intoxication (such as accumulation of neurofibrillary materials in certain neurons, an elevation of aluminum concentration in the brain, and certain behavioral and neurological changes), the possibility of aluminum's association with Alzheimer's disease has attracted much attention.

It must be emphasized, however, despite the accumulation of neurofibrillary materials in both aluminum intoxicated animals and in Alzheimer's disease, that these neurofibrillary materials are not identical. While the neurofibrillary tangles in experimental aluminum intoxication are composed of bundles of single units of neurofilaments of approximately 10 nm in thickness,[163,168,170,172] the neurofibrillary tangle materials found in Alzheimer's disease are helical filaments composed of paired 10 nm neurofilaments inter-twisted in the form of double helix displaying a periodicity of approximately 80 nm.[172,176]

An elevation of brain aluminum has been reported in Alzheimer's patients by various investigators.[154,162,177] Other studies, however, revealed that elevations of brain aluminum were also found in age-matched but apparently normal individuals.[178,179] Some degree of accumulation of neurofibrillary tangle has also been demonstrated in elderly but asymptomatic individuals.[176] Crapper et al.[162,180] reported that Alzheimer's patients have high brain aluminum concentrations and most of the aluminum was found in the nucleus and heterochromatin fractions. High concentrations of aluminum, however, are demonstrable within the nucleus and in the nuclear region of neurofibril-containing neurons of both Alzheimer's patients and elderly asymptomatic individuals.[181,182] Thus, accumulation of aluminum in neurons may simply be an aging phenomenon and not necessarily related to Alzheimer's disease.

In vitro study by DeBoni and co-workers[152] demonstrated that when aluminum was introduced into human dorsal root ganglion and neocortical cell cultures, neurofibrillary tangles were induced only in the cortical neurons. Furthermore, the neurofibrillary materials induced were similar to those obtained in animal models (experimental aluminum intoxication) and not to those in human Alzheimer's disease. In a separate study, these investigators successfully induced formations of the double helical filaments, like those seen in Alzheimer's disease, in human neurons *in vitro* by means of Alzheimer brain extracts without aluminum.[183] A similar study with Alzheimer brain extracts and animal neuronal cultures, however, has not yet been performed.

Although there are certain similarities between Alzheimer's disease and experimental aluminum intoxication, there are also many unique pathological changes, such as accumulation of "brain amyloids", formation of senile plaques, as well as granulovacuolar degeneration of certain neurons, which are not found in aluminum intoxications. In view of the information available at the present time, aluminum accumulation in neurons of Alzheimer's patients should be considerd only as a secondary or associated event, rather than an underlying factor, for the pathogenesis of the disease.

3. Dialysis Dementia

Dialysis dementia is a neurological syndrome developed in some patients undergoing renal dialysis. It is characterized by speech disturbance, confusion, paranoia, personality changes, loss of muscular and motor coordination, as well as seizure and convulsion.[155,184,185] An elevation of aluminum was detected in the cerebral gray, muscle, and bone of these patients.[155] The elevation of tissue aluminum was believed to be associated with high aluminum content in the water of the dialyzate.[185,186]

Unlike in experimental aluminum intoxication and in Alzheimer's disease, where the neuronal aluminum is mainly nuclear,[180] the neuronal aluminum in dialysis encephalopathy is mainly cytoplasmic.[156,187] Furthermore, even with aluminum concentrations several-fold higher than those in Alzheimer's disease, none of the neurological signs or pathology are found in patients with dialysis encephalopathy.[155] These findings further affirm that, while aluminum maybe neurotoxic and induce certain behavioral, neurological, and pathological changes in animal models and in human subjects, Alzheimer's disease, which may be associated with an elevated aluminum content in the brain, probably has no direct etiological relationship with aluminum exposure.

E. CADMIUM

Data on cadmium effects on the human nervous system are very limited. In a study of 160 workers exposed to an unspecified amount of cadmium, Vorobjeva[188] reported headache, vertigo, sleep disturbance, tremor, sweating, and sensory disturbance. Postmortem study of a cadmium-exposed worker showed regressive changes and destruction of ganglion cells in the gut, stomach, esophagus, and bronchial and tracheal walls.[189]

In 1966, Gabbiani[190] first described hemorrhagic lesions in the dorsal root and Gasserian ganglia of rats with neuronal necrosis in some of the ganglia. Similar pathological lesions were seen in Guinea pigs, hamsters, and mice.[191-193] Vascular lesions involving arterioles, capillaries, and venules were observed.[194-195] Cadmium does not cross the blood brain barrier significantly but tends to accumulate in peripheral ganglia.[196] The high concentration of cadmium in ganglia, related perhaps to the highly fenestrated and permeable ganglionic vessels, may contribute to the hemorrhagic and neuronal changes in peripheral ganglia.

Many of the neuronal changes in the ganglia may be induced as secondary changes following hemorrhagic and vascular lesions[193-195] and show dilatation of endoplasmic reticulum, mitochondrial condensation, and other non-specific pathological chnages. Such changes are also demonstrated in the cerebellum during early developmental life of the animals exposed to cadmium.[197,198] While granule cells of the cerebellum appear to be most vulnerable, Purkinje cell damage was also evident. Such neuronal damage is not seen in adult CNS where the blood brain barrier is well developed.

Direct toxicity of cadmium to nerve cells has also been demonstrated *in vitro* without the complication of vascular effects.[199] Accumulation of glycogen, lipid droplets, and neurofilaments have been reported in neurons exposed *in vitro* to cadmium.[199] Nerve fibers are found to be even more sensitive to cadmium, at least *in vitro,* than neuronal bodies.[200] In animals, myelin degeneration and axoplasmic degeneration with accumulation of glycogen particles in large glycogenosomes have also been observed.[201]

Although cadmium does not effectively cross the blood brain barrier, certain regions of the central nervous system have weak or no blood brain barrier system; continuous exposure to cadmium may constitute a neurological risk to these regions. Furthermore, the reason for differential damage to vasculatures in different regions of the brain is still obscure. Future investigations on endothelial cell responses to cadmium are needed to elucidate these issues.

III. NUTRITIONAL INFLUENCES ON METAL NEUROTOXICITY

Present evidence indicates that among some 90 naturally occurring elements, 26 appear to be essential to life. These consist of 11 major elements (C, H, N, O, S, Na, K, Ca, Mg, Cl, and P) and 15 trace elements (As, Cr, Co, Cu, F, I, Fe, Mn, Mo, Ni, Se, Si, Sn, V, and Zn). Nonessential and perhaps "toxic" elements, such as mercury, lead, and cadmium, may interact either directly or indirectly with the essential mineral elements, creating an alteration in the well-balanced microenvironment in the biological system. Such disturbance in the balance and normal metabolism of essential elements may become the basis of some of the "toxic" phenomena observed in animals or man exposed to heavy metals.

Because of heavy industrial use, cadmium, lead, and mercury have attracted the most attention. Experimental evidence for metal-metal antagonism suggests that dietary presence or supplementation of certain mineral elements may contribute to the reduction of adverse risks from heavy metal exposures. On the other hand, deficient intake of certain mineral elements may potentiate the toxic effects of these heavy metals. Therefore the understanding of mineral element/metal interaction becomes important not only to nutritionists but also to toxicologists. The clarification of such interactions may also help to elucidate the toxic effects and mechanisms of some heavy metals. We will shortly review instances where metal-metal interaction influences the neurotoxicology of heavy metals.

A. CADMIUM
1. Interaction with Calcium

The interrelationship between cadmium and calcium was first noticed in Itai-itai disease (cadmium poisoning) in Japan. Patients of Itai-itai disease developed severe osteomalacia, with severely reduced skeletal calcification and pathological fractures.[202] Experimental studies showed that cadmium and calcium play an antagonistic role in the gastrointestinal tract. Animals fed a calcium-deficient diet showed an increase in cadmium absorption.[206]

An increase of cadmium absorption, under the condition of a calcium-deficient diet, would certainly enhance the overall toxicity of cadmium, including those to the nervous system.[205] The suppression of calcium absorption by cadmium[207,208] may also promote secondary cellular and neuronal functional changes. Thus, the awareness of interaction between cadmium and calcium becomes of prime importance in the study of cadmium toxicity.

2. Interaction with Copper

Cadmium also has an antagonistic effect on copper metabolism. Inhibition of copper absorption by cadmium has been demonstrated.[209,210] Placental transport of copper is also significantly reduced by cadmium,[211] inducing "copper-deficiency" in the fetal animals. Indeed, even in adults, when the dietary copper intakes are marginal, a disturbance of copper metabolism by cadmium would lead to the onset of copper deficiency syndromes which represent one of the earliest manifestations of cadmium toxicosis.[210,211]

Since copper is an essential mineral element and plays an important role in neuronal functions, disturbance of copper metabolism by cadmium would most likely have an adverse effect on the nervous system. Indeed, sheep fed with cadmium-contaminated grass showed an unusually low tissue copper level and their offspring developed a demyelinating disease.[212]

3. Interaction with Zinc and Selenium

Cadmium and zinc share many similarities in both physical and chemical properties and most likely may compete for various biological ligands. Another element known to interact with Cd is Se. No role of Cd-Zn or Cd-Se interaction in the neurotoxicity of Cd has been reported.

B. LEAD

The toxicity of lead can be modified by mineral elements such as calcium, zinc, copper, selenium, iron, and chromium. In general, excess of these mineral elements appears to have a protective effect against lead toxicity, and a dietary deficiency of one or more of these elements increases lead absorption and toxicity.

In the central nervous system, copper and zinc alter the deposition of lead in various regions of the brain. With rabbits, Rehman and Chandra[213] reported a reduction of lead deposit in most regions of the brain (except in brain stem) by copper and an elevation of lead deposit in CNS (except in hypothalamus) by zinc. Lead also disturbs the brain copper and zinc concentrations. With lead administration, there was an increase in copper and a decrease in zinc in hippocampus, cerebral cortex, and cerebellum; however, a decrease in copper and an elevation of zinc were observed in the amygdala and hypothalamus.[213] When lead was administered together with copper, there was a greater copper increase in the hippocampus, cerebellum, and brain stem than if copper was administered alone. In the event that zinc was given together with lead, hypothalamus and amygdala showed a higher zinc concentration while cerebellum and blood displayed a lower level of zinc than in animals receiving zinc injections alone. It appears, therefore, that the interactions between lead and zinc or lead and copper are extremely complex and probably vary with the tissues and cells involved. The basic function of the cells and their primary need for the mineral elements involved would probably also influence the outcome of the interaction.

C. MERCURY

Certain "antioxidation" agents, most noticeably selenium, vitamin E, and vitamin C (ascorbic acid), have displayed antagonistic effects on mercury toxicity. The element selenium has received the most attention.

1. Interaction with Selenium

The counteraction of mercury by selenium has been demonstrated in various animal species for both organic and inorganic mercurial compounds.[214-217] Among different selenium compounds, selenite appears to be most effective in this interaction.

Selenium was found to reduce the toxic effects of mercury, particularly in the nervous system.[217,218] As seen with cadmium, the protective effect of selenium was exerted by a diversion of mercury binding to a high molecular weight, selenium-containing protein.[217] A reduction of mercury toxicity can thus be achieved without actual reduction of mercury concentration in tissues.[217] On a molar basis, selenium and mercury accumulate in an approximately 1:1 ratio in tissues.[219] Diversion of mercury to a selenoprotein complex as a protective mechanism has been proposed.[218,220]

Vitamin E, an antioxidant, also has similar protective effects against methylmercury in the nervous system.[221-223] The precise mechanism of the effects of selenium and vitamin E on mercury toxicity is still obscure. Further study is required to elucidate this phenomenon.

D. CONCLUDING REMARKS

Environmental contamination and occupational exposures to heavy metals are an unavoidable situation that man must contend with. Besides avoiding "over-exposure" to the metals, and seeking to understand their toxic effects, it is also important to acquire the knowledge of factors which may substantially modify toxicity. It must be emphasized that the ultimate effects of these toxic metals are not only their direct "toxic impacts" on cell components alone, but also on important nutrients, such as the essential mineral elements needed for the functioning of these cells. Essential dietary elements and metals play an extremely important role in the reduction or enhancement of injuries induced by the toxic metals.

REFERENCES

1. **Singhal, R. L. and Thomas, T. A., Eds.,** *Lead Toxicity,* Urban & Schwarzenberg, Baltimore-Munich, 1980.
2. **McConnell, P.,** Neurotoxic effects of lead, in *Neurobiology of the Trace Elements,* Vol. 2, Dreosti, I. E. and Smith, R. M., Eds., Humana Press, Clifton, NJ, 1983, chap. 5.

3. **Chang, L. W.,** Pathological effects of mercury poisoning, in *Biogeochemistry of Mercury,* Nriagu, J. O., Ed., Elsevier, New York, 1979, 519.

4. **Chang, L. W.,** Neurotoxic effects of mercury, in *Experimental and Clinical Neurotoxicology,* Spencer, P. S. and Schaumberg, H. H., Eds., William & Wilkins, 1980, 508.

5. **Chang, L. W., Reuhl, K. R., and Wade, P. R.,** Pathological effects of cadmium, in *Biogeochemistry of Cadmium. II. Health Effects,* Nriagu, J. O., Ed., Elsevier, New York, 1981, 783.

6. **Petit, T. L.,** Aluminum neurobehavioral toxicology, in *Neurobiology of the Trace Elements,* Vol. 2, Dreosti, I. E. and Smith, R. M., Eds., Humana Press, Clifton, NJ, 1983, chap. 8.

7. **World Health Organization,** Environmental Health Criteria, Vol. 15, Tin and Organotin Compounds, A Preliminary Review, WHO, Geneva, 1980.

8. **Chang, L. W., Tiemeyer, T. M., Wenger, G. R., McMillan, D. E., and Reuhl, K. R.,** Neuropathology of trimethyltin intoxication. I. Light microscopy study, *Environ. Res.,* 29, 435, 1982.

9. **Chang, L. W., Tiemeyer, T. M., Wenger, G. R., McMillian, D. E., and Reuhl, K. R.,** Neuropathology of trimethyltin intoxication. II. Electron microscopy study of the hippocampus, *Environ. Res.,* 29, 445, 1982.

10. **Chang, L. W. and Dyer, R. S.,** A time-course study of trimethyltin induced neuropathology in rats, *Neurobehav. Toxicol. Teratol.,* 5, 443, 1983.

11. **Chang, L. W. and Dyer, R. S.,** Trimethyltin induced pathology in sensory neurons, in disease and chemically induced neurological dysfunction, *Neurobehav. Toxicol. Teratol.,* 5, 673, 1983.

12. **Chang, L. W., Tiemeyer, T. M., Wenger, G. R., and McMillan, D. E.,** Neuropathology of trimethyltin intoxication. III. Changes in the brain stem neurons, *Environ. Res.,* 30, 399, 1983.

13. **Chang, L. W., Wenger, G. R., McMillan, D. E., and Dyer, R. S.,** Species and strain comparison of acute neurotoxic effects of trimethyltin in mice and rats, *Neurobehav. Toxicol. Teratol.,* 5, 337, 1983.

14. **Chang, L. W.,** Trimethyltin induced hippocampal lesions at various neonatal ages, *Bull. Environ. Contam. Toxicol. Teratol.,* 3, 295, 1984.

15. **Chang, L. W. and Dyer, R. S.,** Septotemporal gradients of trimethyltin-induced hippocampal lesions, *Neurobehav. Toxicol.,* 7, 43, 1985.

16. **Chang, L. W. and Dyer, R. S.,** Early effects of trimethyltin on the dentate gyrus basket cells: a morphological study, *J. Toxicol. Environ. Health,* 16, 641, 1985.

17. **Chang, L. W.,** Neuropathology of trimethyltin: a proposed pathogenetic mechanism, *Fund. Appl. Toxicol.,* 6, 217, 1986.

18. **McMillan, D. E., Chang, L. W., Ideumdia, S. O., and Wenger, G. R.,** Effects of trimethyltin and triethyltin on lever pressing, water drinking and running in an activity wheel: associated neuropathology, *Neurobehav. Toxicol. Teratol.,* 8, 499, 1986.

19. **Wenger, G. E., McMillan, D. E., and Chang, L. W.,** Effects of triethyltin on responding of mice under a multiple schedule of food presentation, *Toxicol. Appl. Pharmacol.,* 8, 659, 1986.

20. **Chisolm, J. J. and Barltrop, D.,** Recognition and management of children with increased lead absorption, *Arch. Dis. Child.,* 54, 249, 1979.

21. **Grandjean, P.,** Widening perspectives of lead toxicity: a review of health effects of lead exposure in adults, *Environ. Res.,* 17, 303, 1978.

22. **Morris, C. E., Heyman, A., and Pozefsky, T.,** Lead encephalopathy caused by ingestion of illicitly distilled whiskey, *Neurology,* 14, 493, 1964.

23. **National Bureau of Standards,** Survery manual for estimating the incidence of lead paint in housing, NBS Technical Note No. 921. Washington, D.C., 1976.

24. **Posner, H. S., Damstra, T., and Nriagu, J. O.,** Human health effects of lead, in the *Biogeochemistry of Lead in the Environment,* Nriagu, J. O., Ed., Elsevier/North Holland, Amsterdam, 1978, 173.

25. **Alfano, D. P. and Petit, T. L.,** Neonatal lead exposure alters the dendritic development of hippocampal dentate granule cells, *Exp. Neurol.,* 75, 275, 1982.

26. **Krigman, M. R., Druse, M. J., Traylor, T. D., Wilson, M. H., Newell, L. R., and Hogan, E. L.,** Lead encephalopathy in the developing rat: effect on myelination, *J. Neuropath. Exp. Neurol.,* 33, 58, 1974.

27. **Krigman, M. R., Druse, M. J., Traylor, T. D., Wilson, M. H., Newell, L. R., and Hogan, E. L.,** Lead encephalopathy in the developing rat: effect on cortical ontogenesis, *J. Neuropath. Exp. Neurol.,* 33, 671, 1974.

28. **Krigman, M. R. and Hogan, E. L.,** Effect of lead intoxication on the postnatal growth of the rat nervous system, *Environ. Health Perspect.,* 7, 187, 1974.

29. **Petit, T. L. and Alfano, D. P.,** Differential experience following developmental lead exposure: effects on brain and behavior, *Pharmacol. Biochem. Behav.,* 11, 156, 1979.

30. **Petit, T. L. and LeBoutillier, J. C.,** The effects of lead exposure during development on neocortical dendritic and synaptic structure, *Exp. Neurol.,* 64, 487, 1979.

31. **Clasen, R. A., Hartman, J. F., Coogan, P. S., Pandolfi, S., Laing, I., and Becker, R. A.,** Experimental acute lead encephalopathy in the juvenile rhesus monkey, *Environ. Health Perspect.,* 7, 175, 1974.

32. **Holtzman, D., Herman, M. M., Shen-Hsu, J., and Mortell, P.,** The pathogenesis of lead encephalopathy, *Virchows Arch. A,* 387, 147, 1980.

33. **Michaelson, I. A. and Sauerhoff, M. W.,** Animal models of human disease: severe and mild encephalopathy in the neonatal rat, *Environ. Health Perspect.,* 7, 201, 1974.
34. **Niklowitz, W. J.,** Ultrastructural effects of acute tetraethyllead poisoning on nerve cells of the rabbit brain, *Environ. Res.,* 8, 17, 1974.
35. **Pentschew, A. and Garrow, F.,** Lead encephalo-myelopathy of the suckling rat and its implications on the porphyrinopathic nervous diseases, with special reference to the permeability disorders of the nervous system capillaries, *Acta Neuropathol.,* 6, 266, 1966.
36. **Press, M.,** Lead encephalopathy in neonatal Long Evans rat Morphologic studies, *J. Neuropath. Exp. Neurol.,* 34, 169, 1977.
37. **Louis-Ferdinand, R. T., Brown, D. R., Fiddler, S. F., Daughtrey, W. C., and Klein, A. W.,** Morphometric and enzymatic effects of neonatal lead exposure in the rat brain, *Toxicol. Appl. Pharmacol.,* 43, 531, 1978.
38. **Thomas, J. A., Dallenback, F. D., and Thomas, M.,** The distribution of radioactive lead ^{210}(Pb) in the cerebellum of developing rats, *J. Pathol.,* 109, 49, 1973.
39. **Toews, A. D., Kolber, A., Haywood, J., Krigman, M. R., and Morell, P.,** Experimental lead encephalopathy in the suckling rat: concentration of lead in the cellular fractions enriched in brain capillaries, *Brain Res.,* 147, 131, 1978.
40. **Windebank, A. J. and Dyck, P. J.,** Kinetics of ^{210}Pb entry into the endoneurium, *Brain Res.,* 225, 67, 1981.
41. **Vistica, D. T. and Ahrens, F. A.,** Microvascular effects of lead in the neonatal rat. II. An ultrastructural study, *Exp. Mol. Pathol.,* 26, 139, 1977.
42. **Thomas, J. A., Dallenbach, F. D., and Thomas, M.,** Consideration of the development of experimental lead encephalopathy, *Virchows Arch. A,* 352, 61, 1971.
43. **Popoff, N., Weinberg, S., and Feigin, I.,** Pathologic observation in lead encephalopathy, *Neurology,* 13, 101, 1963.
44. **Tennekoon, G., Aitchison, C. S., Frangia, J., Price, D. L., and Goldberg, A. M.,** Chronic lead intoxication: effect on developing optic nerve, *Ann. Neurol.,* 5, 558, 1979.
45. **Brashear, C. W., Koop, V. J., and Krigman, M. R.,** Effect of lead on the developing peripheral nervous system, *J. Neuropathol. Exp. Neurol.,* 38, 414, 1978.
46. **Fullerton, P. M.,** Chronic peripheral neuropathy produced by lead poisoning in guinea pigs, *J. Neuropathol. Exp. Neurol.,* 25, 214, 1966.
47. **Lampert, P. W. and Schochet, S. S.,** Demyelination and remyelination in lead neuropathy, *J. Neuropathol. Exp. Neurol.,* 27, 527, 1966.
48. **Sauer, R. M., Zook, B. C., and Garner, F. M.,** Demyelinating encephalopathy associated with lead poisoning in nonhuman primates, *Sciences,* 169, 1091, 1970.
49. **Windebank, A. J., McCall, J. T., Hunder, H. G., and Dyck, P. J.,** The endoneurial content of lead related to the onset and severity of segmental demyelination, *J. Neuropathol. Exp. Neurol.,* 39, 692, 1980.
50. **Dyck, P. J., O'Brien, P. C., and Ohnishi, A.,** Lead neuropathy: 2. Random distribution of segmental demyelination among old internodes of myelinated fibres, *J. Neuropathol. Exp. Neurol.,* 36, 570, 1977.
51. **Cavanagh, J. B.,** Metallic toxicity and the nervous system, in *Recent Advances in Neuropathology,* Smith, W. Thomas and Cavanagh, J. B., Eds., Churchill Livingstone, Edinburgh, 1977, 247.
52. **Myers, R. R., Powell, H. C., Shapiro, H. M., Costello, M. L., and Lampert, P. W.,** Changes in endoneurial fluid pressure, permeability and peripheral nerve ultrastructure in experimental lead neuropathy, *Ann. Neurol.,* 8, 392, 1980.
53. **Lampert, P., Garro, F., and Pentschew, A.,** Lead encephalopathy in sucking rats - an electron microscopic study, in *Brain Edema,* Klatzo, I. and Seitelberg, F., Eds., Springer-Verlag, New York, 1967, 207.
54. **McConnell, P. and Berry, M.,** The effects of postnatal lead exposure on Purkinje cell dendritic development in the rat, *Neuropathol. Appl. Neurobiol.,* 5, 115, 1979.
55. **Petit, T. L. and LeBoutillier, J. C.,** Effects of lead exposure during development on neocortical dendritic and synaptic structure, *Exp. Neurol.,* 64, 482, 1979.
56. **Press, M. F.,** Neuronal development in the cerebellum of lead poisoned neonatal rats, *Acta Neuropathol.,* 40, 259, 1977.
57. **Takeichi, M. and Noda, Y.,** Electron microscopy of experimental lead encephalopathy-consideration on the development mechanism of brain lesions, *Folia Psychiatr. Neurol. Jpn.,* 28, 217, 1974.
58. **Thomas, J. A. and Thomas, M.,** The pathogenesis of lead encephalopathy, *Ind. J. Med. Res.,* 62, 36, 1974.
59. **Zook, B. C., London, W. T., Wilpizeski, C. R., and Sever, J. L.,** Experimental lead paint poisoning in non-human primates. III. Pathologic findings, *J. Med. Primatol.,* 9, 343, 1980.
60. **Michaelson, I. A.,** Effects of inorganic lead on RNA, DNA and protein content of the developing neonatal rat brain, *Toxicol. Appl. Pharmacol.,* 26, 539, 1973.
61. **Alfano, D. P., Le Boutillier, J. C., and Petit, T. L.,** Hippocampal mossy fiber pathway development in normal and postnatally lead exposed rats, *Exp. Neurol.,* 75, 308, 1982.
62. **Averill, D. R. and Needleman, H. L.,** Neonatal lead exposure retards cortical synapatogenesis in the rat, in *Low Level Lead Exposure: The Clinical Implications of Current Research,* Needleman, H. L., Ed., Raven Press, New York, 201, 1980.

63. **McCauley, P. T., Bull, R. J., and Lutkenhoff, S. D.,** Association of alterations in energy metabolism with lead-induced delays in rat cerebral cortical development, *Neuropharmacology,* 18, 1979.

64. **Davis, R. K., Horton, A. W., Larson, E. E., and Stemmer, K. L.,** The acute effects of lead alkyls, *Arch. Environ. Health,* 6, 467, 1963.

65. **Schepers, G. W. H.,** Tetraethyllead and tetramethyllead. Comparative experimental pathology. I. Lead absorption and pathology, *Arch. Environ. Health,* 8, 277, 1964.

66. **Niklowitz, W. J.,** Ultrastructural effects of acute tetraethyllead poisoning on nerve cells in the rabbit brain, *Environ. Res.,* 8, 17, 1974.

67. **Cremer, J. E.,** Biochemical studies on the toxicity of tetraethyllead and other organolead compounds, *Br. J. Ind. Med.,* 16, 191, 1959.

68. **Stoner, H. B., Barnes, J. M., and Duff, J. I.,** Studies on the toxicity of alkyltin compounds, *Br. J. Pharmacol.,* 10, 16, 1955.

69. **Casida, J. E., Kimmel, E. C., Holm, B., and Widmark, G.,** Oxidative dealkylation of tetra-, tri- and dialkyltins and tetra- and trialkylleads by liver microscomes., *Acta Chem. Scand.,* 25, 1494, 1971.

70. **Seawright, A. A., Brown, A. W., Ng, T. C., and Hrdlicka, T.,** Experimental pathology of short-chain alkyllead compounds, in *Biological Effects of Organolead Compounds,* Grandjean, P., Ed., CRC Press, Boca Raton, FL, 1984, chap. 14.

71. **Walsh, T. J., McLamb, R. L., Bondy, S. C., Tilson, H. A., and Chang, L. W.,** Triethyl and trimethyl lead: effects on behavior, central nervous system morphology and concentrations of lead in blood and brain of rat, *Neurotoxicology,* 7, 21, 1986.

72. **Chang, L. W.,** Central nervous system changes: selective and non-selective effects, in *Structural and Functional Effects of Neurotoxicants: Organometals,* Tilson, T. A., and Sparber, S. B., Eds., John Wiley & Sons, New York, 1987, 82.

73. **Watanabe, I.,** Effect of triethyltin on the developing brain of the mouse, in *Neurotoxicology,* Roisin, L., Shiraki, H., and Grcevic, N., Eds., Raven Press, New York, 1977, 317.

74. **Foncin, J. R. and Gruner, J. E.,** Tin neurotoxicity, in *Handbook of Clinical Neurology, Intoxications of the Nervous System,* Part I, Vinken, P. J. and Bruyn, G. W., Eds., North-Holland, New York, 1979, 279.

75. **Gruner, J. E.,** Lesions du nevraxe secondaires a l'ingestion d'ethyltain (stalinon), *Rev. Neurol.,* 98, 109, 1958.

76. **Magee, P. N., Stoner, H. B., and Barnes, J. M.,** The experimental production of edema in the central nervous system of the rat by triethyltin compounds, *J. Pathol. Bacteriol.,* 73, 102, 1957.

77. **Torak, R. M., Gordon, J., and Prokop, J.,** Pathobiology of acute triethyl tin intoxication, *Int. Rev. Neurobiol.,* 12, 45, 1970.

78. **Torak, R. M., Terry, R. D., and Zimmerman, H. M.,** The fine structure of cerebral fluid accumulation. II. Swelling produced by triethyltin poisoning and its comparison with that in human brain, *Am. J. Pathol.,* 36, 273, 1960.

79. **Aleu, F. P., Lkatzman, R., and Terry, R. D.,** Fine structure and electrolyte analyses of cerebral edema induced by alkyltin intoxication, *J. Neuropathol. Exp. Neurol.,* 22, 403, 1963.

80. **Hirano, A., Zimmerman, H. M., and Levine, S.,** Intramyelinic and extracellular space in triethyltin intoxication, *J. Neuropathol. Exp. Neurol.,* 27, 571, 1968.

81. **Graham, D. I. and Gonatas, N. K.,** Triethyltin sulfate-induced splitting of peripheral myelin in rats, *Lab. Invest.,* 29, 628, 1973.

82. **Jacobs, J. M., Cremer, J. E., and Cavanagh, J. B.,** Acute effects of triethyltin on the rat myelin sheath, *Neuropathol. Appl. Neurobiol.,* 3, 169, 1977.

83. **Alea, F. P., Katzman, R., and Terry, R. D.,** Fine structure and electrolyte analyses of cerebral oedema induced by alkyltin intoxication, *J. Neuropathol. Exp. Neurol.,* 22, 403, 1963.

84. **Jacobs, J. M., Cremer, J. E., and Cavanaugh, J. B.,** Acute effects of triethyltin on the rat myelin sheath, *Neuropathol. Appl. Neurobiol.,* 3, 169, 1977.

85. **Magee, P. N., Stoner, H. B., and Barnes, J. M.,** The experimental production of oedema in the central nervous system of the rat by triethyltin compounds, *J. Pathol. Bacteriol.,* 73, 107, 1957.

86. **Suzuki, K.,** Some new observations in triethyltin intoxication of rats, *Exp. Neurol.,* 31, 207, 1971.

87. **Watanabe, I.,** Organotins, in *Experimental and Clinical Neurotoxicology,* Spencer, P. S. and Schaumburg, H. H., Eds., Williams & Wilkins, Baltimore, 1980, 545.

88. **Padilla, S. and Veronesi, B.,** Triethyltin induced encephalopathy in perinatally exposed rats: effects on CNS myelin development, *Neurotoxicology,* 3, 131, 1982.

89. **Veronesi, B., Brady, A., and Reiter, L. W.,** Triethyltin induced encephalopathy in perinatally exposed rodents: effects on cortical neurons, *Neurotoxicology,* 3, 136, 1982.

90. **Veronesi, B. and Chang, L. W.,** A comparative study of the pathological effects of TMT and TET on the developing nervous system, *Int. Neurotoxicol. Symp.,* Little Rock, AR, 1985.

91. **Wenger, G. R., McMillan, D. E., and Chang, L. W.,** Behavioral toxicology of acute trimethyltin exposure in the mouse, *Neurobehav. Toxicol. Teratol.,* 4, 157, 1982.

92. **Wenger, G. R., McMillan, D. E., and Chang, L. W.,** Behavioral effects of TMT in two strains of mice. I. Spontaneous motor activity, *Toxicol. Appl. Pharmacol.,* 73, 78, 1984.

93. **Wenger, G. R., McMillan, G. R., and Chang, L. W.,** Behavioral effects of TMT in two strains of mice. II. Multiple fixed-ratio, fixed-interval, *Toxicol. Appl. Pharmacol.,* 73, 89, 1984.

94. **Chang, L. W., Wenger, G. R., and McMillan, D. E.,** Neuropathology of trimethyltin intoxication. IV. Changes in the spinal cord, *Environ. Res.,* 34, 123, 1984.

95. **Brown, A. W., Aldridge, W. N., Street, B. W., and Verschoyle, R. D.,** The behavioral and neuropathologic sequelae of intoxication by trimethyltin compounds in the rat, *Am. J. Pathol.,* 97, 59, 1979.

96. **Bouldin, T. W., Goines, N. D., Bagnell, C. R., and Krigman, M. R.,** Pathogenesis of trimethyltin neuronal toxicity, *Am. J. Pathol.,* 104, 237, 1981.

97. **Valdes, J. J., Mactutus, C. F., Santos-Anderson, R. M., Dawson, R., Jr., and Annau, Z.,** Selective neurochemical and histological lesions in rat hippocampus following chronic trimethyltin exposure, *Neurobehav. Toxicol. Teratol.,* 5, 357, 1983.

98. **Dyer, R. S., Wonderlin, W. F., and Deshields, T. L.,** Trimethyltin-induced changes in gross morphology of the hippocampus, *Neurobehav. Toxicol. Teratol.,* 4, 141, 1982.

99. **Chang, L. W., Tiemeyer, T. M., Wenger, G. R., and McMillan, D. E.,** Neuropathology of mouse hippocampus in acute trimethyltin intoxication, *Neurobehav. Toxicol. Teratol.,* 4, 149, 1982.

100. **Chang, L. W.,** Hippocampal lesions induced by TMT in the neonatal rat brain, *Neurotoxicology,* 5, 205, 1984.

101. **Chang, L. W.,** A proposed pathogenic mechanism on trimethyltin induced lesions in the hippocampus of adult and neonatal rats, in *Biological Trace Element Research,* Vol. 13, Schrauzer, Ed., Humana Press, New York, 1987, 77.

102. **Chang, L. W. and Dyer, R. S.,** Trimethyltin induced zinc depletion in rat hippocampus, in *Neurobiology of Zinc,* Vol. 2, Frederickson, C. and Howell, G., Eds., Alan R. Liss, New York, 1984, 275.

103. **Dyer, R. S., Walsh, T. J., Wonderlin, W. F., and Bercegeay, M.,** The trimethyltin syndrome in rats, *Neurobehav. Toxicol. Teratol.,* 4, 127, 1982.

104. **Doctor, S. V., Costa, L. G., Kendall, D. A., and Murphy, S. D.,** Trimethyltin inhibits uptake of neurotransmitters into mouse forebrain synaptosomes, *Toxicology,* 25, 213, 1982.

105. **Cremer, T. E.,** Possible mechanisms for the selective neurotoxicity, in *Biological Effects of Organolead Compounds,* Grandjean, P., Ed., CRC Press, Boca Raton, FL, 1984, chap. 15.

106. **Chang, L. W. and Hartmann, H. A.,** Blood-brain barrier dysfunction in experimental mercury intoxication, *Acta Neuropathol.,* 21, 179, 1972.

107. **Chang, L. W. and Hartmann, H. A.,** Ultrastructural studies of the nervous system after mercury intoxication. I. Pathological changes in the nerve cell bodies, *Acta Neuropathol.,* 20, 122, 1972.

108. **Chang, L. W. and Hartmann, H. A.,** Ultrastructural studies of the nervous system after mercury intoxication. II. Pathological changes in the nerve fibers, *Acta Neuropathol.,* 20, 316, 1972.

109. **Chang, L. W. and Annu, Z.,** Developmental neuropathology and behavioral teratology of methylmercury, in *Neurobehavioral Teratology,* Yanai, J., Ed., Elsevier, Amsterdam, 1984, chap. 18.

110. **Chang, L. W.,** Developmental toxicology of methylmercury, in *Toxicology and the Newborn,* Kacew, S. and Reasor, M. H., Eds., Elsevier, Amsterdam, 1984, 175.

111. **Takeuchi, T.,** Pathology of Minamata disease, in *Study Group of Minamata Disease,* Kutsuma, M., Ed., Kumamoto University, Japan, 1968, 141.

112. **Takeuchi, T., Kambara, T., Morikawa, N., Matsumoto, H., Shiraishi, Y., and Ito, H.,** Pathologic observations of the Minamata disease, *Acta Pathol. Jpn.,* 9, 768, 1959.

113. **Takeuchi, T., Matsumoto, H., Sasaki, M., Kambara, T., Shiraishi, Y., Hirata, Y., Nobuhiro, M., and Ito, H.,** Pathology of Minamata disease, *Kumamoto Med. J.,* 34, 521, 1968.

114. **Takeuchi, T., Morikawa, N., Matsumoto, H., and Shiraishi, Y.,** A pathological study of Minamata disease in Japan, *Acta Neuropathol.,* 2, 40, 1962.

115. **Amin-Zaki, L., Elhassani, S., Majeed, M. A., Clarkson, T. W., Doherty, R. A., and Greenwood, M. R.,** Intra-uterine methylmercury poisoning in Iraq, *Pediatrics,* 54, 587, 1974.

116. **Amin-Zaki, L., Elhassani, S., Majeed, M. A., Clarkson, T. W., Doherty, R. A., and Greenwood, M. R.,** Studies of infants postnatally exposed to methylmercury, *J. Pediatrics,* 85, 81, 1974.

117. **Amin-Zaki, L., Elhassani, S., Majeed, M. A., Clarkson, T. W., Doherty, R. A., Greenwood, M. R., and Giovanoli-Jakubczak, T.,** Perinatal methylmercury poisoning in Iraq, *Am. J. Dis. Child.,* 130, 1070, 1976.

118. **Harada, Y.,** Clinical investigations of Minamata disease. Congenital (or fetal) Minamata disease, in *Minamata Disease,* Kutsuna, M., Ed., Study group of Minamata disease, Kumamoto University, Japan, 1968, 70.

119. **Harada, Y.,** Intrauterine poisoning, Clinical and epidemiological studies and significance of the problem, Bulletin of the Institute of Constitutional Medicine, *Kumamoto Univ. Suppl.,* 25, 1, 1976.

120. **Harada, Y.,** Congenital Minamata disease, in *Minamata Disease: Methylmercury Poisoning in Minamata and Niigata, Japan,* Tsubaki, T. and Irukayama, K., Eds., Elsevier, New York, 1977, 86.

121. **Takeuchi, T.,** Pathology of fetal Minamata disease, *Pediatrician,* 6, 69, 1977.

122. **Choi, B. H., Lapham, L. W., Amin-Zaki, L., and Saleem, T.,** Abnormal neuronal migration, deranged cerebral cortical organization, and diffuse white matter astrocytosis of human fetal brain: a major effect of methylmercury poisoning *in utero, J. Neuropathol. Exp. Neurol.,* 37, 719, 1978.

123. **Brown, W. J. and Yoshida, N.,** Organic mercurial encephalopathy: an experimental electron microscopy study, *Adv. Neurol. Sci. (Tokyo)*, 9, 34, 1965.

124. **Yoshino, Y., Mozai, T., and Nakao, K.,** Distribution of mercury in the brain and its subcellular units in experimental organic mercury poisoning, *J. Neurochem.*, 13, 397, 1966.

125. **Yoshino, Y., Mozai, T., and Nakao, K.,** Biochemical changes in the brain of rats poisoned with an alkyl mercuric compound, with special reference to the inhibition of protein synthesis in brain cortex slice, *J. Neurochem.*, 13, 1223, 1966.

126. **Steinwall, O.,** Brain uptake of Se[75]. Selenomethionine after damage to blood-brain barrier bay mercuric ions, *Acta Neurol. Scand.*, 45, 362, 1969.

127. **Steinwall, O. and Snyder, H.,** Brain uptake of C[14]-cycloleucine after damage to blood-brain barrier by mercuric ions, *Acta Neurol. Scand.*, 45, 369, 1969.

128. **Cavanagh, J. B. and Chen, F. C. K.,** Amino acid incorporation in protein during the "silent phase" before organomercury and p-bromophenylacetylurea neuropathy in the rat, *Acta Neuropathol. (Berlin)*, 19, 216, 1969.

129. **Chang, L. W. and Hartmann, H. A.,** Electron microscopic histochemical study on the localization and distribution of mercury in the nervous system after mercury intoxication, *Exp. Neurol.*, 35, 122, 1972.

130. **Chang, L. W., Desnoyers, P. A., and Hartmann, H. A.,** Quantitative cytochemical studies of RNA in experimental mercury poisoning. I. Changes in RNA content, *J. Neuropathol. Exp. Neurol.*, 32, 489, 1972.

131. **Chang, L. W., Martin, A. H., and Hartmann, H. A.,** Quantitative autoradiographic study of the RNA synthesis in the neurons after mercury intoxication, *Exp. Neurol.*, 37, 62, 1972.

132. **Morikawa, N.,** Pathological studies on organic mercury poisoning, *Kumamoto Med. J.*, 14, 71, 1961.

133. **Miyakawa, T. and Deshimaru, M.,** Electron microscopic study of experimentally induced poisoning due to organic mercury compund, *Acta Neuropathol. (Berlin)*, 14, 126, 1969.

134. **Herman, S. P., Klein, R., Talley, F. A., and Krigman, M. R.,** An ultrastructural study of methylmercury-induced primary sensory neuropathy in rats, *Lab. Invest.*, 28, 104, 1973.

135. **Miyakawa, T., Deshimaru, M., Sumiyoshi, S., Tersoka, A., Udo, N., Hattori, E., and Tatetsu, L.,** Experimental organic mercury poisoning: pathological changes in peripheral nerves, *Acta Neuropathol. (Berlin)*, 15, 45, 1970.

136. **Cavanagh, J. B. and Chen, F. C. K.,** The effects of methyl-mercury-dicyanidamide on the peripheral nerves and spinal cord of rats, *Acta Neuropathol (Berlin)*, 19, 208,.1971.

137. **Spyker, J. M. and Smithberg, M.,** Effects of methylmercury on prenatal development in mice, *Teratology*, 5, 181, 1972.

138. **Su, M. Q. and Okita, G. T.,** Embryocidal and teratogenic effects of methylmercury in mice, *Toxicol. Appl. Pharmacol.*, 38, 207, 1976.

139. **Harris, S. B., Wilson, J. G., and Printz, R. H.,** Embryotoxicity of methylmercuric chloride in golden hamster, *Teratology*, 6, 139, 1972.

140. **Lown, B. A., Morganti, J. B., Stineman, C. H., and Massaro, E. J.,** Differential effects of acute methylmercury poisoning in the mouse arising from route of administration: LD$_{50}$, tissue distribution and behavior, *Gen. Pharm.*, 8, 97, 1977.

141. **Murakami, U.,** The effect of organic mercury on intrauterine life, *Adv. Exp. Biol.*, 27, 301, 1972.

142. **Gilani, S. H.,** Congenital abnormalities in methylmercury poisoning, *Environ. Res.*, 9, 128, 1975.

143. **Tatetsu, S., Takagi, M., and Miyakawa, T.,** *Psychiat. Neurol. Jpn.*, 70, 162, 1968.

144. **Fuyuta, M., Fujimoto, T., and Hirata, S.,** Embryotoxic effects of methylmercuric chloride administered to mice and rats during organogenesis, *Teratology*, 18, 353, 1978.

145. **Khera, K. S.,** Teratogenic effects of methylmercury in the cat: note on the use of this species as a model for teratogenicity studies, *Teratology*, 8, 293, 1978.

146. **Khera, K. S. and Tobacova, S. A.,** Effects of methylmercuric chloride on the progency of mice and rats treated before or during gestation, *Fd. Cosmet. Toxicol.*, 11, 245, 1973.

147. **Reuhl, K. R. and Chang, L. W.,** Effects of methylmercury on the development of the nervous system: a review, *Neurotoxicology*, 1, ?1, 1979.

148. **Reuhl, K. R., Chang, L. W., and Townsend, J. W.,** Pathological effects of *in utero* methylmercury exposure on the cerebellum of the golden hamster. I. Early effects upon the neonatal cerebellar cortex, *Environ. Res.*, 26, 281, 1981.

149. **Reuhl, K. R., Chang, L. W., and Townsend, J. W.,** Pathological effects of *in utero* methylmercury exposure on the cerebellum of the golden hamster. II. Residual effects on the adult cerebellum, *Environ. Res.*, 26, 307, 1981.

150. **Chang, L. W., Reuhl, K. R., and Spyker, J. M.,** Ultrastructural study of the long-term effects of methylmercury on the nervous system after prenatal exposure, *Environ. Res.*, 13, 171, 1977.

151. **Chang, L. W., Reuhl, K. R., and Lee, G. W.,** Electron microscopic evidence of degenerative changes in the developing nervous system as a result of in utero exposure to methylmercury, *Environ. Res.*, 14, 414, 1977.

152. **DeBoni, U., Otvos, A., Scott, J. W., and Crapper, D. R.,** Neurofibrillary degeneration induced by systemic aluminum, *Acta Neuropathol.*, 35, 285, 1976.

153. **Klatzo, I., Wisniewski, H., and Streicher, E.,** Experimental production of neurofibrillary degeneration. I. Light microscopic observations, *J. Neuropathol. Exp. Neurol.,* 24, 187, 1965.

154. **Crapper, D. R., Krishnan, S. S., and Dalton, A. J.,** Brain aluminum distribution in Alzheimer's disease and experimental neurofibrillary degeneration, *Science,* 180, 511, 1973.

155. **Alfrey, A. C., LeGendre, G. R., and Kaehny, W. D.,** The dialysis encephalopathy syndrome. Possible aluminum intoxication, *N. Eng. J. Med.,* 294, 184, 1976.

156. **Arieff, A. I., Cooper, J. D., Armstrong, D., and Lazarowitz, V. C.,** Dementia renal failure, and brain aluminum, *Ann. Intern. Med.,* 90, 741, 1979.

157. **Ball, J. H., Butkus, D. E., and Madison, D. S.,** Effect of subtotal parathyroidectomy on dialysis dementia, *Nephron,* 18, 151, 1972.

158. **Berlyne, G. M.,** Plasmaphoresis, aluminum, and dialysis dementia, *Lancet,* 1, 1155, 1978.

159. **Berlyne, G. M., Ben-Ari, J., Knopf, E., Yagil, R., Weinberger, G., and Danovitch, G. M.,** Aluminum toxicity in rats, *Lancet,* 1, 564, 1972.

160. **DeBoni, U., Scott, J. W., and Crapper, D. R.,** Intracellular aluminum binding: a histochemical study, *Histochemistry,* 40, 31, 1974.

161. **Crapper, D. R. and DeBoni, U.,** Aluminum and the genetic apparatus in Alzheimer disease, in *The Aging Brain and Senile Dementia,* Nandy, K. and Sherwin, I., Eds., Plenum Press, New York, 1977, 229.

162. **Crapper-McLachlan, D. R. and DeBoni, U.,** Aluminum in human brain disease - an overview, *Neurotoxicology,* 1, 3, 1980.

163. **Terry, R. D. and Pena, C.,** Experimental production of neurofibrillary degeneration. Electron microscopy phosphatase histochemistry and electron probe analysis, *J. Neuropathol. Exp. Neurol.,* 24, 200, 1965.

164. **Crapper, D. R. and Tomko, G. J.,** Neuronal correlates of an encephalopathy associated with aluminum neurofibrillary degeneration, *Brain Res.,* 97, 253, 1975.

165. **Wisniewski, H. M., Iqbal, K., and McDermott, J. R.,** Aluminum-induced neurofibrillary changes: its relationship to senile dementia of the Alzheimer's type, *Neurotoxicology,* 1, 121, 1980.

166. **Crapper, D. R.,** Experimental neurofibrillary degeneration and altered electrical activity, *Electroencephal. Clin. Neurophysiol.,* 35, 575, 1973.

167. **Crapper, D. R.,** Functional consequences of neurofibrillary degeneration, in *Neurobiology of Aging,* Terry, R. D. and Gershon, S., Eds., Raven Press, New York, 1976, 405.

168. **Crapper, D. R. and Dalton, A. J.,** Aluminum induced neurofibrillary degeneration, brain electrical activity, and alterations in acquisition and retention, *Physiol. Behav.,* 10, 935, 1973.

169. **Wisniewski, H., Narkiewicz, O., and Wisniewska, K.,** Topography and dynamics of neurofibrillar degeneration in aluminum encephalopathy, *Acta Neuropathol.,* 9, 127, 1967.

170. **Petit, T. L., Biederman, G. B., and McMullen, P. A.,** Neurofibrillary degeneration, dendritic dying back, and learning-memory deficits after aluminum administration: implications for brain aging, *Exp. Neurol.,* 67, 152, 1980.

171. **Selkoe, D. J., Liem, R. K. H., Yen, S. H., and Shelanski, M. I.,** Biochemical and immunological characterization of neurofilaments in experimental neurofibrillary degeneration induced by aluminum, *Brain Res.,* 163, 235, 1979.

172. **Wisniewski, H., Terry, D., and Hirano, A.,** Neurofibrillary pathology, *J. Neuropathol. Exp. Neurol.,* 29, 163, 1970.

173. **Crapper, D. R. and Dalton, A. J.,** Alterations in short-term retention conditioned avoidance response acquisition and motivation following aluminum induced neurofibrillary degeneration, *Physiol. Behav.,* 10, 925, 1973.

174. **Westrum, L. E., White, L., and Ward, A.,** Morphology of the experimental epileptic focus, *J. Neurosurg.,* 21, 1033, 1964.

175. **Liwnicz, B. H., Kristensson, K., Wisniewski, H. M., Shelanski, M. L., and Terry, R. D.,** Observations on axoplasmic transport in rabbits with aluminum induced neurofibrillary tangles, *Brain Res.,* 80, 413, 1974.

176. **Petit, T. L.,** Neuroanatomical and clinical neuropsychological changes in aging and senile dementia, in *Aging and Cognitive Processes,* Craik, F. I. M. and Trehub, S., Eds., Plenum Press, New York, 1, 1982.

177. **Crapper, D. R., Krishnan, S. S., and Quittkat, S.,** Aluminum neurofibrillary degeneration and Alzheimer's disease, *Brain,* 99, 67, 1976.

178. **McDermott, J. R., Smith, A. I., Iqbal, K., and Wisniewski, H. M.,** Aluminum and Alzheimer's disease, *Lancet,* 2, 710, 1977.

179. **McDermott, J. R., Smith, A. I., Iqbal, K., and Wisniewski, H. M.,** Brain aluminum in aging and Alzheimer disease, *Neurology,* 29, 809, 1979.

180. **Crapper, D. R. and DeBoni, U.,** Brain aging and Alzheimer's disease, *Can. Psychiatr. Assoc. J.,* 23, 229, 1978.

181. **Perl, D. P. and Brody, A. R.,** Alzheimer's disease: x-ray spectrometric evidence of aluminum accumulation in neurofibrillary tangle-bearing neurons, *Science,* 208, 297, 1980.

182. **Perl, D. P. and Brody, A. R.,** Detection of aluminum by SEM x-ray spectrometry within neurofibrillary tangle-bearing neurons of Alzheimer's disease, *Neurotoxicology,* 1, 133, 1980.

183. **DeBoni, U. and Crapper, D. R.,** Paired helical filaments of the Alzheimer type in cultured neurons, *Nature,* 271, 566, 1978.
184. **Chokroverty, S., Bruetman, M. E., Berger, V., and Reyes, M. D.,** Progressive dialytic encephalopathy, *J. Neurol. Neurosurg. Psychiat.,* 39, 411, 1976.
185. **Rozas, V. V., Port, F. K., and Rutt, W. M.,** Progressive dialysis encephalopathy from dialysis aluminum, *Arch. Int. Med.,* 138, 1375, 1978.
186. **Dunea, G., Mahurkar, S. D., Mamdani, B., and Smith, E. C.,** Role of Aluminum in dialysis dementia, *Ann. Int. Med.,* 88, 502, 1978.
187. **McDermott, J. R., Smith, A. I., Ward, M. K., Parkinson, I. S., and Kerr, D. N. S.,** Brain aluminum concentration in dialysis encephalopathy, *Lancet,* 1, 901, 1978.
188. **Vorobjeva, R. S.,** Investigations of the nervous system function in workers exposed to cadmium oxide, *In. Neuropathol. Psikhiat,* 57, 385, 1975.
189. **Baader, E. W.,** Chronic cadmium poisoning, *Ind. Med. Surg.,* 21, 427, 1952.
190. **Gabbiani, G.,** Action of cadmium chloride on sensory ganglia, *Experientia,* 22, 261, 1966.
191. **Arvidson, B.,** Regional differences in severity of cadmium induced lesions in the peripheral nervous system in mice, *Acta Neuropathol.,* 49, 213, 1980.
192. **Gabbiani, G., Baic, D., and Deziel, C.,** Studies on tolerance and ionic antagonism for cadmium and mercury, *Can. J. Physiol. Pharmacol.,* 45, 443, 1967.
193. **Gabbiani, G., Gregory, A., and Baic, D.,** Cadmium induced selective lesions of sensory ganglia, *J. Neuropathol. Exp. Neurol.,* 26, 498, 1967.
194. **Gabbiani, G., Badonnel, M. C., Mathewson, S. M., and Ryan, G. B.,** Acute cadmium intoxication: early selective lesions of endothelial clefts, *Lab. Invest.,* 30, 686, 1974.
195. **Schlaepfer, W. W.,** Sequential study of endothelial changes in acute cadmium intoxication, *Lab Invest.,* 25, 556, 1971.
196. **Arvidson, B. and Tjalve, H.,** Distribution of ^{109}Cd in the nervous system of rats after intravenous injection, *Neurotoxicol. Neurobiol.,* 20, 110, 1983.
197. **Gabbiani, G., Baic, D., and Deziel, C.,** Toxicity of cadmium for the central nervous system, *Exp. Neurol.,* 18, 154, 1967.
198. **Rohrer, S. R., Shaw, S. M., and Lamar, C. H.,** Cadmium induced endothelial cell alteration in the fetal brain from prenatal exposure, *Acta Neuropathol.,* 44, 147, 1978.
199. **Tischner, K. H. and Schroder, J. M.,** The effects of cadmium chloride on organotypic cultures of rat sensory ganglia, *J. Neurol. Sci.,* 16, 383, 1972.
200. **Sharma, R. P. and Obersteiner, E. J.,** Metals and neurotoxic effects: cytotoxicity of selected metallic compounds on chick ganglia cultures, *J. Comp. Pathol.,* 91, 235, 1981.
201. **Sato, K., Iwamasa, T., Tsuru, T., and Takeuchi, T.,** An ultrastructural study of chronic cadmium-induced neuropathy, *Acta Neuropathol.,* 41,185, 1978.
202. **Hagino, N. J.,** About investigations on Itai-itai disease, *J. Toyama Med. Assoc.,* Suppl. 21, 60, 1957.
203. **Nath, R., Prasad, R., Palinal, V. K., and Chopra, R. K.,** Molecular basis of cadmium toxicity, *Prog. Food Nutr. Sci.,* 8, 109, 1984.
204. **Larsson, S. E. and Piscator, M.,** Effect of cadmium on skeletal tissue in normal and calcium-deficient rats, *Isr. J. Med. Sci.,* 7, 495, 1971.
205. **Washko, P. W. and Cousins, R. J.,** Effect of low dietary calcium on chronic cadmium toxicity in rats, *Nutr. Rep. Int.,* 11, 113, 1975.
206. **The Task Group on Metals Interaction,** Factors influencing metabolism and toxicity of metals: a concensus report, *Environ. Health Perspect.,* 25, 3, 1978.
207. **Yuhas, E. M., Miya, T. S., and Schnell, R. C.,** Influence of cadmium on calcium absorption from the rat intestine, *Toxicol. Appl. Pharmacol.,* 43, 23, 1978.
208. **Chertok, R. J., Sasser, L. B., Callahan, M. F., and Jarboe, G. E.,** Influence of cadmium on the intestinal uptake and absorption of calcium in the rat, *J. Nutr.,* 111, 631, 1981.
209. **Van Campen, D. R.,** Effects of zinc, cadmium, silver and mercury on the absorption and distribution of copper-64 in rats, *J. Nutr.,* 88, 125, 1966.
210. **Davies, N. T. and Campbell, J. K.,** The effect of cadmium on intestinal copper absorption and binding in the rat, *Life Sci.,* 20, 955, 1977.
211. **Bremner, I. and Campbell, J. K.,** The influence of dietary copper intake on the toxicity of cadmium, *Ann. N.Y. Acad. Sci.,* 355, 319, 1980.
212. **Grun, M., Anke, M., and Partschefeld, M.,** Cadmium toxicity, in *Kadmium-Symposium,* Friedrich-Schiller-Universitat, Jena, German Democratic Republic, 1977, 253.
213. **Rehman, S. and Chandra, O.,** Regional interrelationships of zinc, copper, and lead in the brain following lead intoxication, *Bull. Environ. Contam. Toxicol.,* 32, 157, 1984.
214. **Johnson, S. L. and Pond, W. G.,** Inorganic vs. organic Hg toxicity in growing rats: protection by dietary Se but not Zn, *Nutr. Rep. Int.,* 9, 135, 1974.

215. **Potter, S. and Matrone, G.,** Effect of selenite on the toxicity of dietary methylmercury and mercuric chloride in the rat, *J. Nutr.,* 104, 638, 1974.

216. **Skerfving, S.,** Interaction between selenium and methylmercury, *Environ. Health Perspect.,* 25, 57, 1978.

217. **Chang, L. W.,** Protective effects of selenium against methylmercury neurotoxicity: a morphological and biochemical study, *Exp. Pathol.,* 23, 143, 1983.

218. **Chang, L. W., Dudley, A. W., Jr., Dudley, M. A., Ganther, H. E., and Sunde, M. L.,** Modification of the neurotoxic effects of methylmercury by selenium, in *Neurotoxicology,* Roizin, L., Shiraki, H., and Grcevic, N., Eds., Raven Press, New York, 1977, 275.

219. **Burk, R. F., Foster, K. A., Greenfield, P. M., and Kiker, K. W.,** Binding of simultaneously administered inorganic selenium and mercury to a rat plasma protein, *Proc. Soc. Exp. Biol. Med.,* 145, 782, 1974.

220. **Chang, L. W. and Suber, R.,** Protective effect of selenium on methylmercury toxicity: a possible mechanism, *Bull. Environ. Contam. Toxicol.,* 29, 285, 1982.

221. **Kasuya, M.,** The effect of vitamin E on the toxicity of alkyl mercurials on nervous tissue in culture, *Toxicol. Appl. Pharmacol.,* 32, 347, 1975.

222. **Chang, L. W., Gilbert, M. M., and Sprecher, J. A.,** Modification of the neurotoxic effects of methylmercury by vitamin E, *Environ. Res.,* 17, 356, 1978.

223. **Yip, R. K. and Chang, L. W.,** Protective effects of vitamin E on methylmercury toxicity in the dorsal root ganglia, *Environ. Res.,* 28, 84, 1982.

Chapter 5

NEUROTOXICOLOGY OF ORGANOTIN COMPOUNDS

George G. Bierkamper and Iain L. O. Buxton

TABLE OF CONTENTS

I. INTRODUCTION

Tin is an abundant metal which was known to the early Egyptians, Romans and Phoenicians. Its symbol, Sn, is derived from the Latin name *stannum*. The biological function of tin, if any, has yet to be established.[6] Inorganic tin is relatively nontoxic. The poor oral absorption of inorganic tin appears to limit systemic toxicity. In the 19th century the increasing practice of storing food in tinned cans raised concern over the possible toxicology of tin.[380,407] Tin contamination in food from the soil or tinned cans does not, however, constitute a serious toxicological threat. Oral ingestion of significant amounts of inorganic tin salts may produce gastrointestinal disturbances due to local corrosive damage.[38] Chronic inhalation of dust, particles or fumes of tin or tin oxides may produce stannosis, a benign pneumoconiosis.

When tin has at least one bond to a carbon atom it becomes an organotin compound. These compounds represent a versatile family possessing many useful commercial, agricultural, and industrial applications. Despite their uses, organotins, particularly the trialkyltins, are highly neurotoxic. This review will, therefore, focus on the toxicity of trialkyltins with emphasis on their neurotoxicology. The objective is to review, in a comprehensive manner, the substantive literature which will suggest the underlying mechanism(s) of neurotoxicity as well as general systemic toxicity of the organotin compounds.

A. HISTORY

Organotin synthesis was first reported by Löwig in 1852.[227] In 1858 Buckton[58] described the preparation of triethyltin (TET) chloride and reported its ability to produce headache and cause nasal irritation. The toxicity of organotin compounds in dogs and frogs was reported in 1869.[193] TET induced a progressive muscular weakness; diethyltin caused G.I. irritation and disturbance.[193] In 1881, White[407] reported an incident of human neurotoxicity by organotin compounds involving a Strasbourg chemist named Harnack. Prostration and encephalopathy were reported in dogs and cats intoxicated with TET.[34] Seifter[327] studied the mammalian toxicity of tetramethyltin in 1939.

The so-called "Stalinon Affair" of 1953 to 1954 in France introduced the clinical community to the devastatingly toxic effects of alkyltins. A proprietary preparation used to treat skin infections known as Stalinon, was inadvertently compounded with approximately 10% TET. The capsules intentionally contained the less toxic diethyltin diiodide as the antibacterial agent against boils and other cutaneous staphylococcal infections. More than 100 people died as a result of ingestion of the adulterated medication.[3,35] The primary neurological symptoms were persistent headache, vertigo, visual disturbances, abdominal pain and cramping, urinary retention, psychic disturbances, muscle weakness, with signs that included electroencephalographic (EEG) changes, increased cerebral spinal fluid (CSF) pressure, and convulsions.[3] The full clinical picture took days to develop and in some cases progressed to a flaccid type paraplegia, sensory loss, absence of reflexes, severe psychiatric disturbances, convulsions, coma and death. EEG recordings of the Stalinon victims revealed transient and unfocused slow activities.[93] Autopsies revealed an unusual edema of the white matter of the brain. Neurological recovery was not always complete in those who survived the intoxication. Memory dysfunction, flaccid paraplegia, asthenia, depression, and visual disturbances persisted in many patients, emphasizing the irreversible nature of the neurological damage.

An insightful report by Stoner et al.[358] bridges the historical organotin experimentation with modern times. It's timely submission preceded the Stalinon tragedy in France by several months. In addition, Barnes and Stoner[35] presented a good review of general mammalian toxicology of many different tin species. A number of other authors have also provided interesting historical reviews or introductions regarding organotin compounds.[35,149,358,395]

Scattered incidences of human organotin poisoning have occurred since the Stalinon affair. Aldridge and his collaborators published a warning in Lancet[7] concerning the irreversible brain damage possible from exposure to trimethyltin (TMT) which is produced during the manufacture of dimethyltin dichloride. Fortemps et al.[150] and Ross et al.[316] have described several incidents of human exposure to TMT. Two chemists exposed chronically to dimethyltin and TMT suffered mental confusion, generalized seizures, headaches, and numerous psychic disturbances.[150] The effects were apparently reversible upon removal from exposure. In 1981, six German industrial workers suffered serious intoxication with TMT resulting in one fatality.[299] Moreover, 22 workers were examined for the toxic effects of exposure to TMT in an industrial setting in 1978.[316] A wide range of psychomotor symptoms have been reported for trialkyltin intoxication including irritability, personality changes, depression, memory defects, insomnia, aggressiveness, headaches, tremors, convulsions, and changes in libido. Most of the symptoms appear to be reversible although follow-up studies have not been thorough. A paucity of histopathological data is available on humans since there have been few organotin-related deaths in recent years.

B. CHEMISTRY

The reader is directed to an exhaustive review of the chemistry of organotin compounds by

Ingham et al.[181] The alkyltin derivatives form four main groups according to structure: R-Sn-X_3, R_2-Sn-X_2, R_3-Sn-X, and R_4-Sn where R is the alkyl group and X is the anion. As a general rule, increasing the carbon chain length increases the lipid solubility of the compound and also lowers the human toxicity. Triorganotin compounds break down at various rates in the environment to simpler non-toxic forms of tin including elemental tin. For example, triphenyltin compounds break down in the aqueous environment by hydrolysis and photolysis, but without volatilization.[351] Tricyclohexyltin hydroxide, the active ingredient of the miticide Plictran appears to break down in the environment (e.g., soil) through a metabolic pathway which removes each cyclohexyl group in a stepwise manner from the tin.[44] The remaining elemental Sn^{4+} is a natural component of soils. Thus, environmental distribution of organotin compounds may not pose a long-term threat under certain conditions (see below).

A wide variety of methods are now available for extracting and measuring organotin compounds in many media. Mushak[254] reviewed the available methods in 1984. In general, these procedures require relatively expensive instrumentation, the cost of which varies depending on the limit of detection and the capability to speciate the tin compounds. In earlier toxicological studies, the limited availability of methods and the expense of the instrumentation appears to have greatly curtailed the reporting of tissue levels of tin coincident with *in vivo* exposure to organotin compounds. It is imperative that all future studies include the level and, if possible, the speciation of the organotins in tissue at the time of testing. This practice would greatly simplify the cross-comparison of studies involving the neurotoxicity of organotin compounds.

Sherman and Carlson[332] introduced a modified phenylfluorone method for determining organotin compounds in environmental samples (e.g., fish, water, soil, etc.) with a limit of sensitivity of 0.1 µg tin (sub-ppb range) per sample. While reliable, this spectrophotometric method has an inadequate resolution for most neurotoxicity studies. Aldridge and Street[14] provided a more sensitive method in 1981 for the spectrophotometric (absorbance and fluorescence) determination of organotin compounds. The maximum sensitivities for TMT and triphenyltin (TPT) were 0.5 and 0.13 nmol, respectively. Differentiation by spectral separation appears to give reasonable separation.

A biochromatographic method for the quantitative estimation of TPT fungicides in the ng range has been described.[1] In addition, a method to determine the *in vitro* cytotoxicity of organotin compounds has been reported utilizing neuroblastoma and fibroblast cell lines.[48]

Nanogram concentrations of mono-, di-, and triphenyltin and tributyltin may be detected in biological and sediment samples by electron-capture gas chromatography (GC).[379] GC speciation is a common method with limited resolution unless coupled to another instrument for quantification.[185,322,361] Means and Hulebak[244] published a GC-mass spectrometric (MS) method for speciation of methyltins in biological tissues. The sensitivity range was 1ng to 30 µg in brain homogenates with better than 90% recovery. Burns et al.[61] reported a comparative study on organotin detection instrumentation using flame atomization (µg range), direct pyrolysis (1 µg) and hydride reduction (2 to 20 pg) with gas-liquid and HPLC separation prior to atomic absorption (AA) spectrometry.[253]

Ion-exchange HPLC coupled to an element-specific graphite furnace AA spectrometer provides a detection limit of 5 ng in field samples.[189] Orren et al.[273] have reported a moderately sensitive method for measuring ethyltin compounds in mammalian tissue using HPLC-flameless AA spectrometry. Pinel and Madiec[284] published a procedure for assessing "heavy" organotin pollution of water and shellfish by modified hydride AA spectrometry. Han and Weber[165] have also reported the use of AA spectrometry with hydride generation for speciated methyl and butyltin compounds at detection levels of 1 ng/100 mg (wet weight) of sample.

Brinckman and colleagues have published a number of methods for measuring organotin compounds in various media, at various levels of resolution and on different instrumentation[45,54] A clever method was recently published by this group describing the use of microbore and capillary HPLC techniques coupled to a UV epifluorescence microscope for the relatively high resolution (pg) characterization of organotin species.[45]

C. USES

Ross[315] and Piver[285] and more recently, Wilkinson[410] have published interesting reviews of the versatile industrial and commercial applications of organotin and tin compounds. Additionally, Wilkinson[410] recently reviewed the major industrial and commercial uses of organotin and tin compounds. In the late 1920s, organotin compounds were used as moth proofing agents and for protection of woolen stockpiles at mills.[181] Organotin compounds were introduced in the 1930s as stabilizers in transformer oils (e.g., tetraalkyltin compounds).[285,315] Later their usefulness as fungicides,[383] algicides,[325] and antifouling paint[181] was discovered. Antifouling paint issues were addressed in a fine introduction by Walsh et al.[388] in a study of alkyltin polymers prepared for the U.S. Navy. The action of organotin antifouling agents on barnacles and marine life has been described elsewhere.[50] Prior to and during World War II research was conducted on organotin compounds as potential chemical warfare agents; however, they neither acted rapidly enough nor were sufficiently potent for further consideration.[181] Part of the war-induced research resulted in numerous publications on the biological effects of organotin compounds.[327] Many uses have been discovered for the organotin family. Dibutyltin (DBT) is used as an anti-wear additive to high pressure lubricating oils.[27] The widespread use of polyvinylchloride (PVC) materials and silicon products has greatly increased the use of organotin as stabilizers and cold vulcanizers over the last 3 decades.[285] Disinfectants containing organotin compounds have been tried in hospitals in an effort to fight resistant strains of microorganisms.[179,220] Organotin oxides have utility as anthelmintics.[59] Dibutyltin compounds are very effective in the treatment of cestode-infected hens, but have proven to be too toxic to treat cestode infected rats, illustrating species differences in the toxicity of the organotins.[198] Organotin compounds have become more popular in recent years as pesticides, insecticides, and acaricides. Since organotin compounds eventually degrade to non-toxic elemental tin, the advantages over organomercury compounds which break down to toxic inorganic mercury are obvious. Organotins such as cyhexatin (Plictran) and fenbutatin oxide (Vendex) have been used against spider mites,[178] cotton leafworms,[196] and other parasitic arthropods[182] in, for example, California citrus and almond orchards.[283]

Organotin compounds have provided intriguing tools for biomedical researchers and biochemists. The Stalinon incident raised much interest in TET intoxication as a model of CNS edema. In the 1970s, renewed interest in trialkyltin compounds emerged as the field of neurotoxicology began to develop. Behaviorists, in particular, were interested in the selective hippocampal damage induced by TMT. For biochemists, the trialkyltin compounds have been widely used in studies on energy metabolism as uncouplers of mitochondrial oxidative phosphorylation. Triorganotin compounds are currently being used a tools to study immunotoxicological mechanisms.[349] Acetoxytriphenylstanane has been used as an experimental carcinogen.[325] Paradoxically, current experimentation is examining organotin compounds as possible anticancer agents.[33,71,188] Dibutyltin dichloride shows promise against pancreatic adenocarcinoma in experimental animals.[188] These recent reports supersede a 1929 report cited by Barnes and Stoner, in which several organotin compounds were found to be ineffective against tumors in mice.[35]

The major current uses of the organotins include heat stabilizers for PVC products (e.g., pipes, tubing, siding, window casings, etc.), catalysts for polyurethane foam and room temperature vulcanization of silicone rubber, as well as biocidal applications which include antifouling marine paints, wood preservatives, fungicides, and acaricides. Organotin compounds have also found useful applications in veterinary products, e.g., anthelmintics and coccidiostats for poultry.

It is estimated that about 10,000 tons of organotin compounds are released, primarily to landfills, in the U.S. every year and that 60,000 tons per year of triorganotin biocides are in use worldwide.[209] Inasmuch as these figures have been increasing annually they probably underestimate current usage.

D. HEALTH RISK PERSPECTIVES

The toxicity and health effects of organotin compounds have been reviewed previously.[199] Mono-organotins do not appear to pose an important toxic risk on mammals.[347] The trialkyltins, and to a much lesser extent, the dialkyltins have important toxic actions in mammals which pose significant health risks depending upon chain length and composition. The longer or more complex the chain, the less toxic the tri- and dialkyltin compounds appear to be. The tetraorganotins cause toxicity by virtue of rapid metabolism in liver to triorganotin derivatives. Since the triorganotin compounds are most widely used in commerce, a potential health risk is present.

Aldridge et al.[7] warned of brain damage due to TMT exposure in manufacturing processes and chemical plants. Indeed, Rey et al.[299] reported on the toxicological and clinical aspects of six workers with TMT and methyltin intoxication. One worker died and two remained neurologically disabled. Furthermore, Ross et al.[316] reported the neurotoxic effects of TMT and dimethyltin on 22 chemical workers exposed in a 1978 accident. Therefore, the health risk at these industrial sites appears to be significant.

Adverse tissue reactions to indwelling organotin-stabilized PVC catheters have been reported, verifying that organotin is leachable and may pose a problem for certain medical applications using PVC tubing and plastic implants or inserts.[161,267] Migration of alkyltin compounds from PVC implants causes local irritation and inflammation, but is not considered systemically toxic.[267] Evidence exists that organotin heat stabilizers in plastic food containers may leach into foodstuffs.[241] Migration studies suggest a need for further investigation as to the cumulative effects.[241] The levels ingested, however, are estimated to be well below the concentrations required to produce neurotoxicity.

Trialkyltin exposure alters drug metabolizing enzymes in the liver, thus presenting the potential of pharmacologically altering the action of therapeutic agents.[409] This potential complication has not apparently been examined during the therapeutic management of organotin-exposed patients. The widespread use of organotin compounds may represent another cancer risk factor.[325] Di-n-octyltin dichloride has been reported to be a potential genotoxicant as demonstrated by interaction with the DNA of cultured V79 Chinese hamster cells and induction of mutagenesis in these cells using the hypoxanthine guanine phosphoribosyl transferase assay.[406] Diethyltin (DET), a metabolite of TET, has been shown to stimulate lipid peroxidation in isolated hepatocytes. Thus, these *in vitro* studies raise concern regarding the carcinogenic potential of long-term organotin exposure.

Organotin compounds in aquatic systems may alter standard microbiological assays used to establish water quality. Injury of indicator organisms may give false readings, thus presenting a potential public health threat.[282] There are still many unknowns about the public health risk of the widespread use of organotin compounds. It is fair to say that neurotoxic health risks appear to be overrated, but that immune system perturbations may not.[238] It is noteworthy, however, that the only beneficial biological activity suggested so far for the organotin compounds has been their potential use as antitumor agents.[188]

E. ENVIRONMENTAL CONCERNS

The increasing use of organotin compounds throughout the world raises a valid concern for their environmental impact.[209,234] The increased pollution of the aquatic environment by anthropogenic organotin from industry and shipping has been recently reviewed.[54,96,209] Antifouling paints, which inhibit barnacle growth on the hulls of ships, have been a major concern in recent years due to their toxicity to algae, copepods, clams, crab larvae, oysters, lobster larvae, rainbow trout, brittle stars, and other aquatic and marine invertebrates.[50,96,165,176,185,209,234,284,387] For instance, regeneration of arms of brittle stars (*Ophiuroids*) is inhibited by tributyltin (TBT) oxide and triphenyltin (TPT) oxide (0.1 μg/l seawater).[387] Since regeneration in this organism is mediated by radial nerves, it is suggested that the primary effect of these compounds is their neurotoxicity.

The common use of antifouling compounds on commercial and private ships and boats since World War II has brought significant residual contamination to most salt-water harbors. Recent studies now show contamination of inland waterways as well. Ten percent of water and sediment samples from 265 locations across Canada yielded highly toxic TBT species at concentrations which would cause growth retardation or damage to fish eggs.[234] The high levels of TBT were found mostly in boating and shipping areas, thus providing further evidence for contamination by antifouling paints.

Surprisingly, the world's major source of organotin is from the biomethylation of elemental tin by marine life[64] rather than anthropogenic distribution. Methylation of tin by estuarine and marine microorganisms has been clearly demonstrated.[54,64,164] Whether there is a biological purpose for this biomethylation is not known. Mixed inoculums of microorganisms collected from Chesapeake Bay sediment are able to convert inorganic tin to organotin compounds including TMT and dimethyltin.[164] A strain of *Pseudomonas* isolated from Chesapeake Bay sediments is suspected of producing methylstannanes.[185] The distribution of TMT is widespread in the aquatic environment involving various rivers in the U.S. and Canada, as well as the world's oceans.

The rapidly increasing use of organotins as pesticides in agriculture and in such processes as pressure treatment of wood for preservation raises concern for increased human exposure. The biodegradation and chemical breakdown properties of many of the commercially available products suggest the potential for accumulation in soil and foodstuffs. Increased awareness of the immunotoxicological, if not the neurotoxicological, risks should evoke more scrutiny in the future.

In terms of mammalian neurotoxicology, the general environment does not appear to pose a significant threat at the present time. The greatest risk of organotin neurotoxicity appears to be in manufacturing processes, research laboratories, in the transportation of concentrated organotins and in the applications of biocidal compounds in agriculture, the lumber industry and marine industries. Unfortunately, the aquatic and marine ecosystems are at significant risk as a result of organotin contamination. The shell fishing, crabbing, and fishing industries in the Chesapeake Bay, for example, have suffered in the last several decades by increased pollutants such as the organotin compounds. Environmental concern regarding antifouling paint contamination has led to the curtailment of use by law in California and some other states.

II. PHARMACODYNAMICS

A. ABSORPTION

TBT is not generally well absorbed orally in animals although Guinea pigs appear to be an exception.[140,357] Diethyltin (DET) is not well absorbed orally in the rat.[53] TMT and TET are well-absorbed orally and percutaneously.[55,90,91] Di-n-octyltin is reasonably well-absorbed orally.[279] Lipid solubility is one factor responsible for increased cutaneous and G.I. absorption.[6] All medium trialkyltin compounds enter the systemic circulation by i.p. injection. However, mesenteric and organ adhesions may occur following administration by this route. Administration by the oral route appears to be tolerated well, but has the disadvantage of poor dosage control as related to taste aversion.

B. DISTRIBUTION

Diorganotins and triorganotins are highly lipid soluble compounds.[48] Trialkyltins are more lipophilic than dialkyltins of the same species. Aldridge[6] has shown that the higher molecular weight organotins are more lipophilic. When introduced into the systemic circulation, however, they do not follow a purely lipophilic distribution pattern indicating (as discussed below) that specific binding sites exist for these compounds.[255] Triorganotin hydroxides and chlorides are lipophilic and to varying degrees are soluble in water. Therefore, they circulate in dilute solutions at physiological pH and readily penetrate membranes, rapidly reaching equilibrium in tissues. The distribution kinetics are dominated by the affinity of organotins for hydroxyl and chloride ions in addition to tissue and blood binding sites.[312]

Administration of TMT by the oral route is acceptable for experimentation due to good absorption. Brown et al.[55] and others have studied the uptake of TMT by rat brain. A peak of approximately 9 nmol/g (wet weight) is present 24 to 48 h post-treatment. In another study, the level of TMT in the brain actually increased over 48 h from 1.38 to 1.70 µg/g wet weight following a single 10 mg/kg dose.[310,312] Two doses of 4 mg/kg TMT one week apart gave a total brain content of 2.8 µg/g wet weight or 1.4 µg/g of brain tissue only excluding blood content. This brain concentration was necessary to see the minimal neurotoxic effects of TMT. An LD_{50} dose of 12.5 mg/kg TMT gave 136 mg/g in blood and 3.44 µg/g in brain (i.e., highest value achieved). TMT binds to rat hemoglobin with a high affinity ($K_D = 3$ µM), but has less affinity for the hemoglobin of many other animal species. Seventy percent of injected TMT is present in blood 48 h post-administration.

Hasan et al.[172] demonstrated that it took more than 48 h for a single oral dose of TMT (7.5 mg/kg, p.o.) to produce peak levels in rat brain, perhaps due in part to hemoglobin's high affinity for TMT. Elevated concentrations persisted in blood and brain up to 20 d post-treatment. In another study involving male mice, Doctor et al.[121] treated animals with a single dose of TMT (4.26 mg/kg, i.p.) and observed a rapid distribution of tin. The highest concentrations were in liver, testes, kidney, and lung. TMT accumulated in these tissues for up to 16 h after administration. At 10 h, concentrations of TMT were 1.6, 3.0, 5.7, 1.8, 1.6, 0.13, 2.0, and 2.9 µg TMT/g wet weight in brain, kidney, liver, skeletal muscle, blood, adipose tissue, lung, and testes, respectively.[121] Administration of TMT (3.0 mg/kg) to marmosets results in a brain concentration of 6 to 17 nmol/g wet weight after 24 to 48 h. Thus, TMT is cumulative and persistent, especially in the rat where there is a high binding affinity for hemoglobin. The half-life of TMT is estimated to be 8.5 d.

TET is also well-absorbed orally. In a comparative study, Cook and co-workers investigated the distribution of TET and TMT in both neonatal and adult rats from 1 h to 10 d following treatment.[90-92] In adults, the liver had the highest concentration of tin, peaking at 4 h, whereas peak concentrations occurred at 24 h in kidney and brain.[90,91] No regional differences were detected in the brain; i.e., even distribution was evident. Neonates exhibited peak concentrations at 8 h in brain after high doses (6 mg/kg, i.p.) where levels reached 9.9 ng Sn/mg protein. Moreover, in another study by these authors, a single dose of TMT or TET given to adult Long-Evans rats showed higher levels for TET in whole brain than for TMT even though both exhibited maximum accumulation by 12 h. Tin was eliminated more slowly from rats treated with TMT. In mitochondrial subfractions of whole adult rat brain, peak concentrations of tin (16.6 ng Sn/mg protein) were not reached until 5 d after treatment with TMT (9 mg/kg, i.p. injection). It was concluded that TMT was more persistent than TET and that the rates of elimination of the two compounds differ.[90]

Stoner et al.[358] reported that acute TET was distributed through brain, lungs, liver, kidney, spleen, muscle, heart, pancreas, fat, and blood in a fairly uniform manner in rat. Cremer[98] reported 17.6, 2.9, 5.2, 5.5, 19.8, 6.4 and 1.46 µg/g tissue wet weight for blood, brain, heart, spleen, liver, kidney, and skeletal muscle, respectively, in rats (230 g) treated with 11 mg of total dietary TET over 13 weeks.[98] Rose and Aldridge[312] studied the distribution of radiolabeled trialkyltin ($[^{113}Sn]$-TET) in rats, Guinea pigs, and hamsters. The highest concentrations were found in rat blood and the livers of all 3 species.[312] Compared with rat and cat, blood of Guinea pigs and hamsters does not bind TET to an appreciable extent.[367] These and other studies illustrate that the distribution and metabolism of trialkyltin may be influenced by the ability of hemoglobin to bind these compounds.[7,121,312] In neonatal rats, 1 mg of TET sulfate per kilogram per day administered by gavage for 2 weeks gave tissue concentrations of 0.5, 4.1, 6.3, and 9.5 µg Sn/g tissue in cerebrum, kidney, liver, and blood, respectively.[46] Accumulation of tin in whole brain minus cerebellum and brain stem was studied during chronic exposure of 30 mg TET per l in drinking water.[381] The results show concentrations of 52 ng tin per gram wet weight, 80 ng/g, 930 ng/g, and 1590 ng/g for 0, 1, 2, and 3 weeks respectively. In moles of TET per gram wet weight, the values were 1.75×10^{-10}, 6.3×10^{-10}, 30×10^{-10}, and 60×10^{-10}, respectively. These results represent values on the order of one tenth of that found at autopsy on the Stalinon victims in France.[3]

Systemic administration of tetraalkytin compounds produces toxicity because of rapid conversion to the trialkyltin species. For instance, tetraethyltin is dealkylated in the liver to TET. Dealkylation of tetraalkyltin decreases as the size and stability of the ligand increases.[28] Unconverted tetraethyltin has low toxicity. Arakawa et al.,[28] in studying the distribution of organotin compounds in the rabbit by i.v. injection, found tetraethyltin to be rapidly distributed to liver, kidney and brain; whereas, tetrapropyltin and tetrabutyltin more slowly distributed to these tissues. Nonetheless, 2 h post-injection of TBT and tri-n-propyltin (TPrT), concentrations appeared in liver > kidney > brain and blood; and after 3 h in brain > liver > blood > kidney.[28] Most trialkyltin compounds tested reached peak levels in liver by 3 h.[28] TET has a half-life of 7.3 d.[90]

Evans and co-workers[140] monitored the distribution of radiolabeled bis[tri-n-butyltin]oxide (TBTO) in mice when the compound was administered in drinking water. Higher concentrations were achieved in kidney, liver, spleen, and fat than in muscle, lung, and brain tissues. Blood levels were extremely low in mice chronically exposed to TBT.

Mushak et al.[255] published an extensive study on the distribution of total tin in neonatal rats treated with TMT, TET, TBT, TPT, TPrT, and tricyclohexyltin (TcHT). The highest concentration of tin was detected in liver, kidney, and brain with blood from TMT and TET treated animals also showing significant levels of organotin. Importantly, the neurotoxicity of these trialkyltin compounds did not correlate with the amount of total tin found in the brain. Additional observations suggested that TET, not a metabolite, was the active neurotoxic species in the brain. Administration of 1 mg/kg TET to neonatal rats resulted in tissue concentrations of 102, 114, 492, and 144 ng Sn/g tissue for brain, kidney, liver, and blood, respectively. TMT administration yielded 72, 141, 400, and 3150 ng Sn/g tissue for the same tissues.[255] These data reflect the result of high affinity binding of TMT to hemoglobin as discussed above.

In other studies, various organotin species verified the tendency for high concentrations of the compounds to appear in the liver and kidneys. Administration of 100 ppm tricyclohexyltin in the diet for 90 d revealed a preferential distribution to kidney, liver, and heart with the lowest level in blood.[44] The highest concentrations of di-n-octyltin dichloride are found in liver and kidney after i.v. or oral administration to rats. Low concentrations were found in brain and blood.[279] TBT is not readily absorbed orally and is largely excreted via the feces avoiding systemic distribution.[140] Evans and Cardarelli[140] have suggested that some TBT is distributed to adipose tissue via the chylomicron.

C. METABOLISM AND ELIMINATION

The metabolism of organotin compounds has been reviewed by Aldridge[6] and others.[389] Alkyltins are metabolized *in vitro* by a cytochrome P450-dependent microsomal monooxygenase system in mammalian liver.[8,28,35,99,145-147,201,202,287] The metabolites are generally less toxic. For example, TET is slowly oxidized *in vitro* to yield diethyltin.[72] Further, TBT is oxidized *in vitro* in rat liver by monooxygenase to carbon hydroxylated compounds (i.e., alpha, beta, gamma, and delta-hydroxybutyltin derivatives).[146] Kimmel[202] demonstrated that TBT and dibutyltin are metabolized *in vivo* by carbon hydroxylation of the alkyl groups. The hydroxylation profile of TBT and the lack of oxidative cleavage of tin-carbon bonds yields a free radical rather than an oxenoid.[144,147,148] Aldridge et al.[8] showed that tetrabutyltin derivatives are also metabolized by the microsomal P-450-dependent monooxygenase system of rat liver to yield carbon hydroxylated metabolites identified as alpha, beta, gamma and delta-hydroxybutyltin derivatives, and a gamma-keto compound. Induction of the P-450 microsomal enzyme system by TET has been demonstrated.[173,287,314] TET salts appear to induce heme oxygenase, an effect which parallels cytochrome P-450, its flavoprotein reductase and glutathione-S-aryltransferase activity.[173,287,314]

Phenobarbital, an inducer of the hepatic microsomal monooxygenase system which dealkylates organometals, did not apparently increase the metabolism of TMT to dimethyltin derivatives in an acute study conducted with female SD rats.[417] However, in isolated rat hepatocytes, TET has been shown to break down to ethane whereas metabolism of TET following pretreatment of cells with phenobarbital produced more ethylene (also a volatile product).[409] DET dichloride was not metabolized in this system. TET, and to a lesser extent DET, reduced oxygen consumption and ATP levels in the cells. TET bromide inhibits both oxidation (Phase I) and conjugation (Phase II) metabolism of diphenyl in cells from phenobarbital-treated rats. Diethyltin stimulates lipid peroxidation in isolated rat hepatocytes especially after phenobarbital or 5,6 benzoflavone-treatment of rats before cell harvest.[409]

In agreement with these results are those of Prough,[287] who demonstrated that ethyl substituted organotin derivatives were subject to cytochrome P-450-dependent dealkylation to yield ethane and ethylene in addition to aldehyde formation. The tetra-, tri-, and diethyl-derivatives of tin can be successively dealkylated to yield the respective dealkylated products and the hydrocarbons, both under aerobic and anaerobic conditions. Ethane formation in this case can occur by a reductive as well as an oxidative process. Neuman[265] suggests a free radical intermediate via the reductive process, which could be linked to lipid peroxidation.

Cremer[99] has demonstrated that tetraethyltin is dealkylated to TET in rats and rabbits, both *in vivo* and *in vitro*. Although some conversion occurs *in vitro* in brain slices, the liver is the predominant organ of metabolism *in vivo*.[72,99] In general, tetraalkyltins are converted to highly toxic trialkyltins, which are further metabolized to the less toxic di- and monoalkyltins.[28,72,202] The rate of formation of trialkyltins from tetraalkyltins is dependent on the rate of distribution of tetraalkyltins to the liver. With increased length of the alkyl chain, the rate of formation of trialkyltins decreases. The extent of metabolism seems directly related to the carbon hydroxylation of the alkyl chains and the instability of the hydroxylated metabolite in liver microsomal monooxygenase systems.[28] Iwai and Wada[184] reported conversion of tetralkyltins to trialkyltins in the intestinal mucosa of the rabbit before the compound reached the liver. Tetraethyltin was not particularly toxic in the rat hepatocyte cell culture system.[409]

Byington and Hansbrough[62] suggest that organotin compounds inhibit the glutathione-S-transferase activity of ligandin and may affect the biliary excretion of substances excreted by this mechanism. Doctor et al.[120] reported that TMT administered to mice lowered nonprotein sulfhydryl levels in a tissue-specific and dose-dependent manner. Incubation of TMT with glutathione and mouse hepatic cytosol resulted in no conjugation, confirmed by a lack of reduction of glutathione levels. Further, no increase in oxidized glutathione (GSSG) was observed with *in vivo* TMT. The authors reported no alteration of hepatic ATP levels. Thus, the results suggest that TMT decreases nonprotein sulfhydryl levels *in vivo* either by directly inhibiting GSH synthesis or by inhibiting uptake of precursor amino acids. This will undoubtedly interfere with the disposition or elimination of other drugs, toxicants, and endogenous substances.

Thus, the cumulative evidence to date suggests several pathways for trialkyltin metabolism. They are monooxygenase metabolism through a hydroxylated intermediate with further dealkylation[8,28,53,99,202] and conversion to a free radical intermediate via a reductive process.[265,287] Excretion may be via the urine, bile and/or feces depending on the metabolite formed.

III. GENERAL TOXICOLOGY

The general toxicology of organotin compounds has been the subject of several recent reviews[6,206,348] and will therefore only receive succinct coverage in this review. An insightful and particularly enjoyable review was published less than a decade ago by Duncan. This review includes valuable information about the biocidal uses of organotins.[122] It is noteworthy that the toxicological compromise of other systems (e.g., kidney, blood, and liver) may have a profound effect on the function of the nervous system. In this regard, few comprehensive studies since the report of Stoner et al.[358] have been performed which correlate intersystem toxicity of alkyltin compounds.

Barnes and Stoner[34] tested many dialkyl- and trialkyltin compounds in rats, mice, Guinea pigs, and rabbits. This is a good starting point for an earlier descriptive overview of the toxic properties of organotin compounds. This report demonstrated that dialkyltin compounds produced a generalized illness with damage to the biliary tract being the principal pathology. In contrast, trialkyltin compounds targeted the CNS primarily, producing interstitial edema.[143,204] A year later, Barnes and Stoner[35] provided a more extensive review on the mammalian toxicology of organotin compounds, but without any substantive discussion of the mechanism(s) of toxicity of the chemicals.

According to a World Health Organization report published in 1980,[408] the acute oral toxicity of triorganotin compounds in rats is ethyl > methyl > n-propyl > n-butyl, triphenyl > n-hexyl, n-octyl. Snoeij and colleagues[350] and others[34,82,122,357] concluded that the shorter chain trialkyltins (i.e., TMT and TET) are the most neurotoxic, the intermediate trialkyltins (i.e., tripropyltin and tributyltin) and TPT, a triaryltin, are primarily immunotoxic, and the longer chain length triorganotins (trihexyltin and trioctyltin) have limited toxicity in male rats. Unconverted tetraorganotins have very low toxicity in mammals.

Most studies on the organotin compounds report that the neurotoxic effects are limited to TMT and TET. No signs of neuronal damage or edema have been observed in rats treated with dimethyltin, diethyltin, or any of the higher trialkyltin homologs.[49,255,350]

Subchronic TET exposure by injection (1.5 mg/kg every 3 to 4 d) resulted in a decrease in food and water intake, and a below normal body weight growth curve in adult LE rats.[112] Adult mice exposed daily to TET show a slight decrease in body weight.[153] Most other subchronic or chronic studies report weight loss with TET intoxication.[233] Animals chronically deprived of food and water prior to TET exposure exhibited only slight, transient curtailment of eating and drinking behavior with less effect on body weight.[112] Almost every investigation of chronic TMT exposure reports loss of body weight or growth retardation.[391]

Severe TET toxicity in rats was characterized by "decreased activity, paralysis of hindlimbs, and increased muscular rigidity."[376] The toxicity of TPT in rats, mice, rabbits, Guinea pigs, and hens resulted in CNS toxicity and muscular weakness without CNS edema.[357]

A study of strain and gender differences in susceptibility of rats to dioctyltin dichloride revealed that males are more susceptible than females and that Brown Norway rats are more vulnerable than Lewis rats.[51]

The structure-activity relationship for the cytotoxicity of organotin compounds show the more lipophilic compounds to be more toxic to neuroblastoma cells.[48] The rank order of cytotoxicity in a fibroblastic cell line derived from young bluegill fish (BF-2 cells) was reported to be: dicyclohexyltin > dibutyltin > diphenyltin, dipropyltin > diethyltin > dimethyltin, whereas for butylated tins the rank order was TBT > DBT > tetrabutyltin > butyltin trichloride > tin (IV) chloride.[31]

The human toxicity of trialkyltins corroborates the animal exposure studies. Aldridge and colleagues[7] reported brain damage as a result of TMT exposure citing several human cases and made a comparison of research involving marmosets. Ross et al.[316] reported on the neurological examination of 22 chemical workers who were exposed to TMT and DMT in a 1978 accident. The CNS symptoms included cycles of depression and destructive rage, each lasting a few hours. Other symptoms included forgetfulness, fatigue, loss of libido, loss of motivation, headaches, and sleep disturbances. Some EEG abnormalities were noted in several workers. The CNS effects in the "highly exposed" group persisted beyond 1 year.[316]

In 1981, six patients who had suffered acute TMT intoxication were examined by Besser et al.[40] The industrial incident in Germany involved the cleaning of a tank used in the manufacture of dimethyltin (DMT). Intoxication occurred through inhalation of vapors within the tank. The physicians described an acute limbic-cerebellar syndrome. One patient died and two were seriously disabled. The workers exhibited many signs and symptoms such as hearing loss, disorientation, confabulation, amnesia, aggressiveness, hyperphagia, metabolic acidosis, disturbed sexual behavior, complex partial and/or tonic-clonic seizures, nystagmus, ataxia, and mild sensory neuropathy. The patient who died had severe pulmonary edema, tracheobronchitis, and kidney failure.

A. BODY TEMPERATURE

Acute intoxication with trialkyltin compounds has been shown to decrease core body temperature in experimental animals.[37,103,125,154,172,217,358,366] Acute TET exposure reduced core body temperatures in rats by as much as 6°C depending on the room temperature; however, the effects of acute administration disappeared after 24 h.[125]

TET affects the thermoregulatory mechanisms in mice, inducing a hypothermia which increases in magnitude with decreasing ambient temperature.[154] Following TET (6 and 8 mg/kg, i.p) administration, the mean metabolic rate of the subjects was inhibited by 23 and 66%, respectively, within the first 60 min at 20°C, but not at 35°C. It is remarkable that corresponding colonic temperatures decreased 6.1 and 11.2°C, respectively, at an ambient temperature of 20°C (i.e., normal room temperature). Body temperatures did not significantly decrease at ambient temperatures of 30 and 35°C. Unfortunately, the study was not extended beyond 60 min. Curiously, when given a choice TET-treated mice selected the coolest environment rather than one of higher temperature.[154] Although the mechanism of TET-induced hypothermia is unknown, the authors suggest that inhibition of muscular thermogenesis by prevention of the shiver reflex at lower ambient temperatures could account for the hypothermia when coupled with an inhibition of metabolism. However, a direct effect on central thermoregulatory centers cannot be ruled out in this acute study.

B. LETHALITY

A 1978 compendium of toxicological data including acute oral and dermal LD_{50}s and vapor and aerosol data on numerous organotin compounds has been provided by the International Tin Research Institute.[347] Most of the information was obtained from rodent models. Appendix D^{228} of the I.T.R.I. Publication presents a translation from German of an excellent toxicological assessment report on organotin compounds by Luijten and Klimmer (I.T.R.I. Publication No. 501D). Their report[228] covers pre-1977 references to the toxicological effects of a wide variety of organotin compounds and includes acute LD_{50} and LD_{100} values as well.

Stoner, Barnes, and Duff[358] reported that the acute fatal dose (LD_{100}) of TET for all species including rats, rabbits, guinea pigs, cats, and domestic fowl falls between 3 and 10 mg/kg of body weight independent of the route of administration. TMT is acutely fatal to rats at 30 mg/kg orally and 16 mg/kg i.p. The acute LD_{50} range for TMT in mice is reported to be 2.9 mg/kg, i.p.[94,114] The LD_{50} for TMT in hamsters, gerbils, and marmosets was 3 mg/kg given orally as compared to 12.6 mg/kg in the rat in the same study.[57] Low binding of TMT to rat hemoglobin as compared to the other species was thought to reflect the difference. Since TMT has little binding affinity for human hemoglobin, the human LD_{50} has been estimated to be similar to that determined for the marmoset, i.e., 3 mg/kg.[57] The acute LD_{50} range for TET in mice is 12.6 mg/kg administered by gavage.[55] The LD$_{50}$ for acute administration of TET in rats was 4.0 to 5.7 mg/kg either orally or by i.p. injection.[232] The LD$_{50}$ of triphenyltin acetate was found to be 8.5 mg/kg, i.p. in rat, 3.74 mg/kg in Guinea pig, and 7.9 mg/kg in mouse.[357]

C. BLOOD

Dibutyltin and TBT chloride bind specifically to two sites and one site, respectively, on human erythrocytes.[19] TMT and TET do not bind to human hemoglobin.[6,57,134,312,337] However, TMT and TET bind to rat and cat hemoglobin.[134,310,337] No other hemoglobin has shown comparable affinity for organotin compounds.[312] Diethyltin competes with TET binding only at very high concentrations.[135]

Rose[310] has shown evidence for a histidine residue in the TET binding site on rat hemoglobin. One molecule of rat hemoglobin binds two molecules of TET; each binding site appeared to contain two histidine residues.[310] Siebenlist and Taketa[337] studied organotin-protein interactions in cat hemoglobin. TET binds 2:1 to R-state, but not T-state cat hemoglobin with an affinity constant of approximately 10 µM. A specific, three dimensional binding site consisting of cysteine and histidine residues was proposed for TET. An allosteric effect of TET was suggested for the increase in oxygen affinity and other alterations in the function of hemoglobin. TET is bound at sites on alpha-globin distant from the heme groups. Interestingly, triorganotins also bind to heme oxygenase in the liver.[314]

Other studies reveal that the organotin binding to hemoglobin is likely to be pentacoordinate.[135] The differential binding of TET to cat but not human hemoglobin has facilitated the search for the amino acid residues involved in binding and the verification of pentacoordinate binding. Chemical alteration of histidine residues[134,310] and sulfhydryl groups[135] has been shown to decrease TET binding to hemoglobin, suggesting the involvement of histidine nitrogen and cysteine sulfhydryl groups in the binding sites. Pentacoordinate triorganotin compounds were used by Aldridge et al.[15] to study binding of TET to cysteine and histidine residues in cat and rat hemoglobin. Pentacoordinate binding was evident. Taketa and coworkers[367] confirmed this observation using cat hemoglobin. The target residues which serve as axial ligands in the formation of a pentacoordinate binding complex were identified as cysteine 13 and histidine 20 on the alpha subunit. TET binding results in an increase in oxygen affinity which is similar to the release of protons during the alkaline Bohr effect. TET binding causes a conformational change towards the R-type quaternary structure. Hybrids prepared from the alpha and beta subunits from cat and human hemoglobins established that both the ligands involved in TET binding are in the cat alpha subunit[367] The 2:1 binding ratio of TET:hemoglobin (rat or cat)[134,310] is in agreement with this finding. The interaction of dimethyltin derivatives with rat hemoglobin has been studied with Mossbauer spectroscopy.[33] Pentacoordinated binding in a configuration of trigonal bipyramidal structure was shown with thiol and peptide nitrogen group binding, i.e., cysteine and histidine side chains.[6]

Based on experimentation with erythrocyte membranes, Selwyn et al.[331] suggest that the triorganotin compounds may interfere with the Hamburger effect and thus alter the pH and oxygen and carbon dioxide carrying capacity of the blood. The order of potency for this effect was TBT, TPrT > TPT > TMT.[331] This agrees with studies cited above in which the organotin compounds altered hemoglobin's affinity for oxygen.[314,367]

Dietary triorganotin compounds alter blood composition in mice treated for 7 d.[183] Lymphocytes and total leukocytes decreased, whereas, erythrocytes, hemoglobin levels and hematocrits increased. Elferink et al.[133] reported inhibition of phagocytosis and exocytosis in polymorphonuclear leukocytes by TBT and TPT. Interference with vulnerable sulfhydryl groups related to cell activation was presented as the mechanism of the inhibition.[133] Knowles and Johnson[205] have demonstrated that some triorganotin compounds (e.g., TBT but not TPT) at relatively high concentrations induce aggregation of platelets. TPT was without significant effect in this study.

TET is known to cause hemolysis at higher concentrations *in vitro*.[312,334] One possible mechanism involves the inhibition of erythrocyte hexokinase activity by TET, an action prevented by D-glucose.[277] Erythrocyte hexokinase is inhibited before hemolysis. A correlation exists between the degree of hexokinase inhibition and the extent of hemolysis.[312,334] Byington et al.[63] showed that TPT compounds, but not TBT compounds, have hemolytic activity in dog, hog, rabbit and rat blood. TBT can produce hemolysis in human blood. The concentrations required are consistent with *in vivo* toxicity levels. Our own laboratory has observed mild hemolysis and reduced hematocrits in rats when they are exposed to TET for 3 weeks with 30 mg/l of TET in their drinking water. No effect is seen when rats are given 10 mg/l of TET in drinking water (Bierkamper, unpublished results). Finally, the effects of hemolysis and altered oxygen carrying capacity of the blood have not been adequately addressed as they relate to the neurotoxic syndromes produced by TMT and TET. It is reasonable to suggest, however, that with respect to neurotoxicity, the reduced delivery of O_2 to the brain is not the sole effect of the organotins.

Exposure to organotin compounds alters blood chemistry. Benedek et al.[36] reported hyperglycemia following acute TET administration to rabbits and cats. Paradoxically, the hyperglycemia was inhibited by beta blockers which might be expected to decrease insulin release. Increases in ketone bodies in the circulation of TMT-treated mice has been demonstrated in association with alterations in systemic glucose metabolism.[25] Circulating catecholamine levels increase as intoxication progresses and cause a profound hypoglycemia secondary to increased $beta_2$-adrenergic receptor mediated insulin release. The role of glucose availability in the neurotoxic action of the organotins is uncertain but is likely to contribute, along with O_2 availability, to this central toxicity.[25]

Inorganic phosphate levels rise in plasma after fatal doses of TET sulfate in rats. This rise was attributed to a compromise in renal function. Phosphate levels did not change in the CNS except as modified by interstitial edema.[359] BUN (blood urea nitrogen) levels progressively increased after single dose exposure of rats to TMT (12.25 mg/kg) reaching a group mean of 80 mg/dl by day 14 of post-treatment. Wilson et al.[411] reported elevations of plasma ammonia and serum urea nitrogen 7 d after administration of TMT (5.25 mg/kg, s.c.) to rats; serum alkaline phosphatase was elevated between 7 and 28 d post-treatment. TET ($^1/_4$ LD_{50} dose) did not cause these effects. In addition, creatinine levels have been reported to increase moderately following TMT administration.[305]

Acute TET (9 mg/kg) increased arterial PCO_2 levels and increased cerebral venous blood PCO_2 values within 12 h post-treatment in rats.[286] Cerebral blood flow was decreased by subchronic TET treatment in rats.[212] It was presumed that an alteration in energy metabolism coupled to progressive brain edema modified the flow.[212] Cerebral blood flow decreased dramatically 12 to 24 h after administration of TET (9 mg/kg) in rats, while cerebral blood flow increased slightly 48 h post-treatment. Furthermore, blood pressure is known to increase and then decrease as a result of TET exposure (0.12 to 0.57 mg/kg, i.v.) in the cat. In most species however, blood pressure falls or is not changed in most animals following TET administration and the effect is transient.[358]

D. LIVER

In mammals, dialkyltin compounds are extremely hepatotoxic, whereas trialkyltins are extremely neurotoxic.[35] Dibutyltin diacetate in rodents, for example, causes hepatotoxicity.[68] Electron microscopic evidence shows early mitochondrial swelling and vacuolation of smooth and rough endoplasmic reticulum, leading to progressive depletion of ATP along with congestion of bile canaliculi.

Trialkyltins also show dose-dependent hepatotoxicity.[113] TET binds to high affinity binding sites in Guinea pig liver homogenates.[313] In rats acutely or chronically intoxicated with TBT, Bouldin et al.[49] found bile duct ectasia and portal tract fibrosis in the liver.[34] TPT is hepatotoxic in rats.[113] Reduced conjugation with glutathione as a result of glutathione-S-transferase inhibition may play a role in biliary excretion problems.[120] Reuhl and colleagues reported significant hepatoxicity in monkeys exposed to TMT.[297] Blood chemistry revealed elevated concentrations of alkaline phosphatase and lactic dehydrogenase, consistent with acute, mild liver damage. Hepatocytes were swollen with dilated Golgi and endoplasmic reticulum as well as edematous mitochondria. There was also an increase in lysosomes containing lipoid debris. No abnormalities in blood vessels or biliary apparatus were observed. Damage as a result of ischemia was evident. Since the liver is the primary site of metabolism of the organotin compounds, hepatoxicity may delay detoxification. This in turn would be expected to enhance the neurotoxicity of certain trialkyltins.

E. KIDNEY

Early studies by Stoner and Threlfall[359] suggested a compromise in renal function following acute administration of fatal doses of the trialkyltin, TET, to rats. Most work has dealt with TMT nephrotoxicity since it appears to be more directly nephrotoxic than TET. Rey et al.[299] found renal failure to have occurred in a human TMT fatality. Wilson et al.[411] proposed that renal damage was responsible for TMT-induced hyperammonemia. BUN levels have also been observed to be significantly elevated after TMT exposure.[303]

The pathogenesis of TMT-induced nephrotoxicity in adult rats has recently been investigated by Robertson and colleagues.[305] Single TMT (12.25 mg/kg, p.o.) exposures resulted in significant damage to the renal tubules with lesions detected as early as 2 d post-treatment. The morphological observation of the overloading of hyaline droplets in tubular epithelial cells may have implications regarding the neurotoxicity of TMT in the limbic system (see below). The hyaline droplet inclusions suggest that tubular dysfunction was due to altered intracellular processes of metabolism and transport which rely on energy production. Therefore, TMT may suppress oxidative phosphorylation[13] in high energy regions of the tubular system such as the brush border of the proximal tubule and cause direct cellular damage as has been observed with heavy metals. Decreased cellularity of the brush border, basolateral vacuolation and eosinophilic granular casts in the proximal tubule cells support the theory of energy compromise in rats as well as primates.[272,297,303,305] In an earlier study, TMT treated rats showed no nephropathology 35 d post-treatment, indicating reversibility of TMT nephrotoxicity. Despite the apparent reversibility of TMT nephrotoxicity, it is likely that this toxicity can increase the degree of neurotoxicity seen with TMT intoxication.

F. IMMUNE SYSTEM

Aside from the nervous system, the immune system may represent one of the targets of greatest risk with regard to organotin toxicity. In fact, subtle changes in the immune system may represent the first changes detectable as a result of low level environmental exposure. Suppression of immune responses may have a deleterious effect on the organism's ability to react against foreign antigens. Whether this enhances the neurotoxicity of the trialkyltin compounds remains to be tested.

Weekly oral administration of dioctyltin has been shown to modify mouse immunocompetence.[247] Immune system dysfunction has been shown with di-n-propyltin, dibutyltin,[329] and di-n-octyltin[139,247,248,280,281,328,330] in rats and chickens,[295] but not in Guinea pigs or mice. Males were more susceptible than females to immunotoxicity of dioctyltin.[51] Subchronic exposure to TBT, dibutyltin, and TPT suppressed the immune system in chickens by interfering with thymus function.[162,163] Dietary triorganotin modifies lymphatic tissues in mice by decreasing spleen tissue and lymphocyte count.[183] Snoeij and coworkers[350] have demonstrated that triorganotins of intermediate chain length (i.e., TPrT, TBT, TPT) present the greatest risk to the rat immune system as compared to TMT and TET.

In vitro cytotoxicity of thymocytes and bone marrow cells to TBT has been demonstrated, while erythrocytes were relatively resistant. Direct membrane damage was detected at higher concentrations of TPT. Vacuolation within thymic cells was a common histopathological feature.[139] Dioctyltin dichloride suppresses thymocyte proliferation within 24 to 72 h of exposure.[248] The anti-proliferative effect on thymic reticuloepithelial cells[348] may relate to thymic atrophy and anti-cancer effects. It has been shown that dialkyltin impairs energy metabolism in rat thymocytes.[280] Oxidation of substrate during intermediary metabolism was inhibited, perhaps as a result of inhibition of pyruvate dehydrogenase. This action may account for the thymolytic effects of dialkyltins *in vivo*.[280]

G. ENZYMES

TET is selective in its interaction with proteins, especially enzymes that use ATP as a substrate.[105,333,336] Indeed, in terms of protein interactions trialkyltins are considered to be extremely unreactive chemically.[105,310] Hexokinase, pyruvate kinase, and phospho-fructokinase are examples of enzymes which reveal the selectivity of TET binding.[336] The selective action of trialkyltins on various ATPases is often cited as a primary mechanism of neurotoxicity including CNS edema, muscular weakness, and energy deficits.

The short-chain trialkyltins appear to have the same property as N,N'-dicyclohexylcarbodiimide (DCCD) since they exhibit the ability to inhibit a non-oligomycin sensitive site on mitochondrial and synaptic vesicle ATPase.[264] For example, TET inhibits the oligomycin-sensitive ATP synthase complex of mitochondria by binding to the membrane-bound F_o component.[65] This binding effectively blocks the synthesis of ATP by this enzyme in mitochondria. Dibutylchloromethyltin (DBCMT) chloride *in vitro* has been shown to inhibit the ATP synthase complex of mitochondria. Covalent inhibition by DBCMT and the inhibition by TET appear to be associated with distinctively different, though interactive sites on the oligomycin-sensitive ATPase complex. DBCMT, unlike TET, binds to the ATPase complex in an irreversible manner and can be utilized as a specific affinity label.[67]

Enzymes that require ATP as substrate, e.g., the various cation-dependent ATPases in excitable membranes, are generally inhibited by trialkyltin compounds. Trimethyltin (125 µg/ml) has been shown to inhibit *in vitro* the Ca^{++},Mg^{++}-ATPase associated with cholinergic synaptic vesicles isolated from the electric organ of *Torpedo californica*.[318] Ca^{++} and Mg^{++}-dependent and -independent ATPase activities have been assayed in human neuroblastoma cells (GM3320) growing in medium containing TMT (25 to 100 µM).[74] A dose-dependent effect was measured on three apparently different ATPases. At low concentrations (25 µM) TMT stimulated the Ca^{++}-dependent, Mg^{++}-independent ATPase and inhibited by 70% the Ca^{++}, Mg^{++}-ATPase activity. At higher concentrations, TMT inhibited all ATPase activities measured. Only the low concentration results would correlate with *in vivo* TMT levels resulting from very high dose exposures. Nonetheless, this study represents important evidence that TMT may have an effect on membrane ATPase activity resulting in the ionic imbalances suggested by numerous morphological studies in the brain.[49,87] Furthermore, the synaptic location of the Ca^{++}- and Mg^{++}-dependent enzymes has implications regarding participation in the release of neurotransmitter; that is, inhibition of the enzyme by TMT may interfere with neurotransmission.[74,275]

Bentue-Ferrer et al.[39] proposed that TET inhibits Na^+,K^+-ATPase ("sodium pump") activity. Rabbit brain mitochondrial Mg^{++}-ATPase was found to be more strongly inhibited by TET *in vitro* than Na^+,K^+-ATPase.[394] The IC_{50} of TET for mitochondrial Mg^{++}-ATPase was 1 µM, whereas membrane bound Na^+,K^+-ATPase was less susceptible to TET (IC_{50} = 100 µM). The IC_{50} of the extramitochondrial (microsomal; pellet fraction) Mg^{++}-ATPase for TET was 10 µM, while the IC_{50} of brain cyclic AMP phosphodiesterase for TET was approximately 100 µM. TET, DET, and TMT all inhibited Na^+,K^+-ATPase, cyclic AMP phosphodiesterase, and Mg^{++}-ATPase, with TET being the most potent inhibitor.[394] TET content in brain was dose-dependent, with the administration of 1.5 mg TET/kg/d (i.p.) for 5 d resulting in 2.3 µg TET/g (wet weight) of gray matter and 1.5 µg TET/g of white matter. These organotin levels reflect an approximate tissue concentration of TET of 6 µM for gray and 4 µM for white matter. Thus, one might predict inhibition of the mitochondrial Mg^{++}-ATPase *in vivo*. The failure to find a significant change in assayed enzymes from TET-treated animals (fossil inhibition) may be due to the dilution and removal of TET from particular fractions during sample preparation and does not exclude the possibility that TET has an *in vivo* effect on these enzymes.[394]

Tributyltin (TBT) *in vitro* decreased GABA uptake and release from rat brain synaptosomes by interfering with energy utilization and/or a vesicle associated Ca^{++}, Mg^{++}-ATPase.[264] According to this study, TBT also disrupts loading of norepinephrine into synaptic vesicles by inhibition of a Mg^{++}-ATPase.[374] Neuronal dysfunction *in vivo* would be predicted if trialkyltin concentrations reach the required inhibitory level.

Glutathione S-transferases are a family of enzymes responsible for conjugation of many electrophiles with glutathione. TET has binding sites in liver cytosol[312,313] with glutathione S-transferases as the dominant target sites of TET.[173,372] TET (10^{-7} to $10^{-5}M$) and TPT apparently form a ternary or quaternary complex with the enzyme(s) causing selective inhibition. TET interacts with a number of glutathione S-transferase isozymes in rat liver.[372] Trialkyltin apparently inhibits glutathione-S-transferase activity of ligandin, interfering with the biliary excretion of other substances eliminated by this route.[62]

TET intoxication in rats lowers the activity of particulate 3′,5′-cyclic AMP phosphodiesterase levels in brain.[231] In addition, administration of TET decreased cyclic AMP content in rat cerebrum.[216] Further, 2′,3′-cyclic AMP phosphodiesterases were not affected by TET, TEL, TMT, or DET.[394] When added to the *in vitro* assay, TET and TPT inhibited basal adenylate cyclase activity in brain homogenates. A role for altered cyclic nucleotide metabolism in the etiology of brain edema has not been established.[216]

Relatively high concentrations of TET (25 to 500 μM) have been shown to increase the activity of cyclic AMP-dependent protein kinase assayed *in vitro* from human erythrocytes and bovine brain,[336,368] and in subcellular fractions of rabbit and rat brain.[266] The increased activity appears to be due to the release of the catalytic subunit from the holoenzyme (R_2C_2). However, the interaction of TET with the catalytic and regulatory subunits eventually causes a decrease in enzyme activity. Interestingly, dithiothreitol delayed the loss of activity suggesting sulfhydryl target sites for TET.[368] Noncompetitive inhibition of cyclic AMP binding to the regulatory subunit by TET was associated with increased catalytic activity. The later inhibitory effect of TET on the catalytic activity was noncompetitive with respect to the histone protein substrate for phosphorylation employed in the assay. Since increased phosphorylation of membrane proteins was observed when intact red cells were exposed to TET it is possible that this enhanced phosphorylation of certain substrate proteins is relevant to trialkyltin toxicity.[336] Tributyltin and (2-[(dimethylamino)methyl]phenyl)diethyltin bromide (SnC) were as effective as TET on protein kinase activity, diethyltin was less effective, and TMT was without effect in this assay system.[336] Trialkyltins appear to age in their effects on (activation and then subsequent inhibition) cyclic AMP-dependent protein kinase. The pattern of stimulation and inhibition could provide for a unique mechanism of cellular toxicity and warrants future study.

Hexokinase is inhibited by TET.[333,334] At low temperature, 2 mol of TET bind to 1 mol of yeast enzyme and this binding is prevented and/or reversed by glucose, dithiothreitol and d-mannose.[333,335] The rank order of inhibitory potency is TBT < TET < DET < TMT.[6] Erythrocyte hexokinase is inhibited without hemolysis or inhibition of glycolytic enzymes.[333] TET apparently interacts with sulfhydryl groups on the enzyme, but not with histidine or cysteine groups.[335] Dithiothreitol binds TET and appears to participate in removing this trialkyltin from the enzyme, thus promoting recovery.

TET reportedly activates rabbit skeletal muscle pyruvate kinase *in vitro* when used at low concentrations (<1 μM) and inhibits it at higher concentrations (1 mM).[105] Direct high affinity binding of TET to the enzyme has been demonstrated.[310]

The effects of TET (1 to 50 μM) on protein phosphorylation *in vitro* were investigated in subcellular fractions from rat and rabbit brain.[266] Specific phosphorylation of two proteins, identified as synapsin and the alpha subunit of pyruvate dehydrogenase, was detected. Decreased phosphorylation was detected in a 52,000 mol wt protein tentatively identified as the regulatory subunit of cyclic AMP-dependent protein kinase. Phosphorylation of this protein is not sensitive to TMT. TET-induced phosphorylation of rabbit brain mitochondrial pyruvate dehydrogenase results in a partial inactivation of this enzyme complex. TBT, but not TMT, also affects pyruvate dehydrogenase in a similar manner. TET will therefore interfere with pyruvate utilization through the tricarboxylic acid cycle. The authors contend that since brain tissue contains only a minimal excess of pyruvate dehydrogenase, the inhibition of this sensitive enzyme complex could evoke a cascading negative effect on the bioenergetics of excitable cells which rely almost entirely on aerobic oxidation of glucose for their energy needs. Phospho-dephospho regulation of protein function is a central mechanism in regulatory biology. The fact that the organotins can stimulate specific protein phosphorylation is an intriguing finding that may explain, in part, the neurotoxicity exhibited by these compounds.

Ali et al.[21] examined the effects of TMT on ornithine decarboxylase activity in various regions of mouse brain. Ornithine decarboxylase is the rate limiting enzyme of polyamine synthesis and is capable of responding quickly in damaged cerebral tissue. As a consequence, this may be a good marker for the targets of neurotoxic insult. At a low dose TMT (1 mg/kg) produced a decrease in ornithine decarboxylase activity in caudate nucleus and hippocampus at all times studied (1,2, and 7 d post-treatment). No changes were observed in hypothalamus, cerebellum, and brain stem. At a high dose of TMT (3 mg/kg) ornithine decarboxylase activity in hippocampus, cerebellum and brain stem was increased at 1 and 2 d post-treatment, while other regions were not significantly altered. Increased ornithine decarboxylase levels returned to control within 7 d. Thus, TMT caused selective dose-dependent changes in this marker enzyme.

A number of studies have examined the activity of cholinergic enzyme systems during trialkyltin intoxication. Choline acetyltransferase (ChAT) activity in the rat hippocampus was not changed by several TMT exposure regimens, despite significant changes in morphology and high affinity glutamate uptake.[256] However, TET (1 mM) has been shown to inhibit pseudo-cholinesterase activity *in vitro* in horse serum.[358] Wender and colleagues[398] detected no changes histochemically in acetylcholinesterase, pseudo-cholinesterase and non-specific esterase in the CNS of rats intoxicated with TET with the exception of individual oligodendroglial cells adjacent to edematous areas.[398] Acetylcholinesterase staining was significantly reduced in brain areas of rodents exposed to TMT.[304] Histochemically stained sections of posterior cortex from TET-exposed rats showed modified acetylcholinesterase staining in the hippocampus with enhanced reaction in the outer rim of the molecular layer of the dentate gyrus.[386] Whether acetylcholinesterase activity or just the distribution of the enzyme is changed after triorganotin exposure is not known.

Although the trialkyltins at concentration ranges obtained *in vivo* seem to selectively inhibit only certain enzymes, there exist many scattered reports of interference with other enzymatic activities, sometimes as a result of high concentrations in *in vitro* assays. Triorganotins bind to rat hepatic heme oxygenase causing prolonged induction of activity in a dose-dependent manner.[314] In addition, a depletion of cellular hemeprotein was found. Fluctuations have been observed in glutamate decarboxylase in the rat hippocampal formation 21 d after TMT administration (3 mg/kg, i.p.).[256] Glutamate decarboxylase activity was used as a marker for GABA neurons in this study. Exposure of rats to TMT resulted in a 264% increase in hippocampal beta-glucuronidase activity and an 87% reduction in beta-galactosidase activity.[309] The lysosomal enzyme changes were attributed to neuronal cell loss and to the processes of astrocytic and microglial hypertrophy.[309] Serum alkaline phosphatase and aspartate aminotransferase activities were elevated and depressed, respectively, 7 d after a single dose of TMT (5.25 mg/kg, s.c.) administered to rats.[411] In particulate fractions of rabbit gray and white matter, 5′-nucleotidase activity was stimulated by TET, triethyl lead and TMT, but not DET.[394] The level of serum creatine phosphokinase (CPK), a muscle-associated enzyme, was unaltered by 20 d of TET (20 mg/l in drinking water) exposure in rats.[155] In addition, TET has been shown to inhibit erythrocyte glucose-6-phosphate dehydrogenase, adenylate kinase and hypoxanthine-guanine phophoribosyltransferase,[277] while DET dichloride has been shown to inhibit alpha-keto acid oxidase.[9]

Rocco and coworkers[307,308] examined enzymes linked to myelin lipid synthesis in developing rats exposed postnatally to TET. Three intraperitoneal injections of TET on days 5,6, and 7 after birth resulted in a 15% decrease in body weight and a 50% decrease in forebrain weight at 20 d of age, TET inhibits transketolase, a rate limiting step in the synthesis of glyceraldehyde-3-phosphate, which leads to increased formation of dihydroxyacetone phosphate. Reduction in the capacity to incorporate the glycerol moiety into myelin phospholipids supports the suggestion that TET interferes with the synthesis of membrane lipids at the level of dihydroxyacetone phosphate in the developing rat.[307]

H. OTHER SYSTEMS

Administration of TMT (2.75 mg/kg, i.p.) to mice decreases adrenal epinephrine and norepinephrine levels as early as 8 h post-treatment. This effect preceded symptomatology associated with hippocampal changes. Evidence was also presented for disruption of glucose metabolism.[25] Other investigations report evidence of increased epinephrine and norepinephrine release from the adrenals during acute trialkyltin intoxication. TET has been shown to induce the release of catecholamines from the adrenal gland by central nervous system mediation.[251] Cardiac content of epinephrine has been shown to increase after TET as a result of marked catecholamine release from the adrenals.[251]

Organ atrophy and growth retardation have been reported as a result of trialkyltin intoxication. Testicular atrophy has been reported with chronic TET exposure to rats.[233].Moreover, growth in cell culture of rabbit articular and growth-plate chondrocytes is altered by TMT, TET, dibutyltin, and dioctyltin.[396] These results from cells in culture may explain, at least in part, why other investigators often report stunting of growth and dwarfing in some animals treated with organotins.[75]

IV. NERVOUS SYSTEM TOXICITY

Various reviews of organotin neurotoxicity have been presented over the last 15 years.[149,243,271] The most profound, direct neurotoxicity of organotin compounds appears to be restricted to the trialkyltin compounds.[348] Most excitable tissues, i.e., neurons and their neuroeffector cells, appear to be affected in a dose-dependent manner by either TMT or TET or both. The investigations of other organotin species rarely report frank nervous system damage. For instance, careful study of diethyltin intoxication by Magee et al.[233] failed to reveal a neuropathological profile similar to that caused by TET, despite the fact the animal became ill and died of non-specific, unidentified causes.

A. SYMPTOMATOLOGY

Ross and colleagues[316] reported forgetfulness, headaches, sleep disturbances, depression, and rage attacks among other neurological symptoms in humans exposed to TMT. Many of these symptoms corroborate the limbic system damage observed in experimental animals. LE rats exposed to TMT (7 mg/kg, p.o.) show markedly increased open field activity.[364] Brown et al.[55] described signs of TMT poisoning in rats, i.e., tremors, hyperexcitability, aggressive behavior, weight loss and convulsions with brain tissue levels above 1 µg/g (wet weight) produced by a single dose of TMT near the LD_{50} of 12.6 mg/kg of body weight. TET exposure in mice (2 mg of TET/kg per d, i.p.) resulted in a steady decrease in locomotor activity over 9 d of treatment.[153]

B. EEG ALTERATIONS AND SEIZURES

Conflicting data has been published regarding the effect of TET on electroencephalographic (EEG) activity. Smith et al.[345] did not observe EEG changes after TET administration to dogs given phenobarbital anesthesia. The electrocorticograph (ECoG) of rats exposed to TET (2 mg/ kg) for 5 d was unmodified as determined by spectral analysis.[212] On the other hand, Benedek and coworkers[36] demonstrated EEG changes in rabbits and cats acutely intoxicated with TET. There was an increase in slow waves, with subsequent synchronization of fast beta on slow delta. Moreover, EEG frequencies were shifted to slower values in mice acutely treated with TET. Normal frequency spectra were observed 24 h post-treatment.[153]

Benedek and coworkers[36] have also described a slowing of the EEG in cats exposed to TET (5 to 10 mg/kg, i.p.) within hours of administration. Early synchronization of the EEG in the 3 to 4 per s spectrum yielded two predominant theta-delta waves and progressed to slower beta waves 3 h post-treatment. The EEG changes preceded the CNS edema characteristic of TET poisoning.

Dyer et al.[127] studied neonatal TET exposure on seizure kindling in CD rats. TET was injected on postnatal day 5; kindling consisted of 17 consecutive days of stimulation via implanted hippocampal electrodes. TET altered the after-discharge (AD) duration in cortex only; picro-toxin and pentylenetetrazol activities were not affected by neonatal TET treatment. Gender differences were detected, with females showing slightly more profound effects than the male subjects. Thus, the seizure testing paradigms show little, if any, modification by TET.

Bierkamper (unpublished data) conducted a study on the effects of late gestational/early neonatal exposure of TET (30 mg/l in drinking water) on Fluorothyl ether (bis[trifluoroethyl ether]) seizure thresholds. Pregnant LE rats were exposed during the last 3 d of gestation. Neonates were exposed through their mother's milk for the 21 d postpartum (i.e., until weaning). The exposure regimen consisted of 3 d TET water, then 1 d tap water; then 3 d and so on. Control rats were pair-watered since water consumption decreased by about 15 to 20% in the TET-treated dams. The rat pups were tested as adults when they reached approximately 120 g body weight. The tonic-clonic endpoint seizure of the Fluorothyl test was 225 ± 32 s ($n = 6$) in controls and 382 ± 70 s ($n = 6$) in TET-treated subjects ($P(t) < 0.03$). Similar dosing with cessation of TET at postnatal day 4 did not result in a significant increase in adult Fluorothyl seizure thresholds. A subsequent study with exposure only during the last 3 d of gestation showed no effect on seizure thresholds in adult offspring. These results were surprising inasmuch as increased CNS excitability had been predicted, rather than the marked decrease observed. These results are important since they demonstrate that neonatal exposure to TET has long term effects on CNS excitability, perhaps through permanent changes in myelin-nerve relationships.[36] These experi-ments contradict the results of Doctor and Fox[119] who described an increased severity of maximal electroshock seizures in adult mice treated with TET as neonates. On acute exposure, however, Doctor and Fox found that TET exhibited anticonvulsant properties.[117]

TMT exposure induces seizure activity in rats, mice and other species in a dose-dependent manner. Seizure activity is usually delayed for several days. The lag time may result from the delay in reaching peak TMT brain levels. The seizure activity is associated with extensive, permanent damage to the hippocampal formation. Little substantive evidence exists to explain the mechanism by which TMT induces seizure activity.

Dyer et al.[131] also studied seizure susceptibility of TMT-treated adult rats to pharmacologic convulsants and to amygdala and hippocampal kindling. Administration of TMT (5 to 7 mg/kg, p.o.) lowered the pentylenetetrazol seizure thresholds 5 d post-treatment, indicating an in-creased general susceptibility to seizure activity. This is not unexpected since at 5 d post-treatment, subjects are beginning to show spontaneous seizure activity. Treated rats kindled more rapidly than controls. An increase in hippocampal after discharge thresholds was unanticipated and weakens the argument that the selective effect of TMT on the hippocampus is the origin of all seizure activity in this model. However, hippocampal cell loss induced by

TMT intoxication may account for the increased after discharge thresholds. Furthermore, hippocampal inhibitory circuit failure may account for enhanced kindling.

Sloviter et al.[344] examined the role of seizure activity on hippocampal damage induced by acute administration of TMT (7 to 9 mg/kg, i.p.) in SD rats. Unlike kainate induction of seizures over minutes, TMT seizure production had a latency of 2 to 7 d. TMT seizures were characterized by a clonic movement of the forepaws while rearing, or more severe clonic seizures without tonic convulsions. The seizures were usually brief (5 to 20 s) and occurred between days 2 and 7 post-treatment. Rats with TMT-induced seizures exhibited variable patterns of granule and pyramidal cell necrosis in the hippocampus along with acute dendritic swelling in two limbic regions: one from CA3 to CA1-CA3, and the other from the hilus to the proximal dendrites of dentate granule cells. Rats which did not have seizures by 1 week following TMT treatment showed minimal pathology. The authors suggested that neuronal damage in the hippocampus is therefore mediated by seizure activity.[344]

Ray[289] reported that TMT administration to rats (8 to 10 mg/kg, orally) modified the "EEG" coincident with damage to the hippocampus when followed over an 11 d period. Spontaneous electroencephalographic-like recordings revealed an increasing amplitude with a preponderance of theta rhythm between days 2 and 4. Apparently, the increased theta amplitude was sufficient to increase the entire "EEG". Moreover, the theta rhythm and movement always seemed to be linked in control, but was unlinked in the TMT subjects and even appeared during immobile periods. By post-treatment day 8, the theta rhythm amplitude had fallen below control values coincident with a fall in hippocampal evoked responses; some TMT subjects began to show mild seizure activity at this time point.

Neuropathologic study of human epileptic tissue indicates significant loss of pyramidal neurons in the H1 field.[361] Frequent tonic-clonic seizures were associated with loss of neurons in H1 as well as H1-2 and the CA3 equivalent. Granule cell loss was also observed in the fascia dentata. This study concludes that hippocampus is particularly vulnerable to seizure activity and that neuronal cell loss is an ongoing process in patients with epilepsy involving tonic-clonic seizures.[361]

In a series of papers, Doctor and Fox[115-119,151] have studied the "anticonvulsant" effects of a number of trialkyltin compounds. TMT (4.26 mg/kg, i.p.) administration to mice transiently increased seizure threshold and protected against chemically-induced seizures from pentylenetetrazol, bicuculline and isonicotinic acid hydrazide. These seizure-inducers have all been linked to GABAergic mechanisms. Protection was lost by 14 h and by 16 h spontaneous tremors and convulsions appeared. TMT gave little protection against strychnine induced seizures which involve a glycine-mediated inhibitory neuronal pathway.

Administration of TMT, TET, TPrT, and TBT to mice all provided some transient protection against maximal electroshock seizure in a dose-dependent, graded and structure-related manner. The order of anti-seizure potency was TMT > TET > TPrT > TBT.[117] An extended study on tricyclohexyltin and TPT[116] failed to find similar maximal electroshock seizure protection; rather, the mice exhibited dose-dependent increases in maximal electroshock seizure severity. TPT decreased maximal electroshock seizure severity transiently between 4 to 24 h post-treatment.[116] The anticonvulsant effect of TET was shown not to involve an inhibition of carbonic anhydrase, an enzyme reported to promote seizure discharge propagation.[118]

Fox[151] tested the acute maximal electroshock seizure protective effects of TET in mice in combination with a number of pharmacologic agents known to affect the GABAergic system. It is noteworthy that these tests were done within 1 h of TET administration (1 or 5 mg/kg, i.p.), prior to the development of gross CNS edema and myelin splitting. TET had a dose-dependent, graded effect in protecting against electroshock-induced seizures. Reserpine and yohimbine, but not propranolol, haloperidol, or metergoline, blocked the "anticonvulsant" action of TET. TET inhibited the convulsant effects of two GABAergic blockers, bicuculline and picrotoxin. TET *in vitro* (in the μM to mM range) did not interfere with specific receptor binding of [^{3}H]-GABA or [³H]-WB4101 (an alpha-adrenergic radioligand). In agreement with other studies, the results suggest that the anti-maximal electroshock seizure properties of acute TET involve the GABA neurotransmitter system and perhaps the adrenergic system.[151] Fox has suggested that acute TET exposure may block the re-uptake of GABA and/or norepinephrine, thus prolonging their physiological (anticonvulsant) effect. Unfortunately, it is not known if TET entered the brain within the short 30-min period prior to testing since measurements were not made. Thus, it is difficult to determine whether the effects of TET are direct or indirect.

Neonatal exposure of mice to TET bromide (0.25 to 5.0 mg/kg, s.c.) for up to 20 d, delayed in a dose-dependent manner the onset of the normal maximal electroshock seizure responses (i.e., clonic and tonic seizures).[119] TET administration retarded growth and decreased the maximal electroshock seizure severity during development. Adult mice exposed as neonates exhibited a long-term increase in maximal electroshock seizure severity. A TET (1 mg/kg) challenge at 75 d in these pre-exposed animals did not alter maximal electroshock seizure responses as compared to control. One would expect complete reversibility of any non-permanent TET-induced lesions by postnatal day 75 and beyond. Thus, the increased maximal electroshock seizure severity in the adult mice could be due to residual myelin reduction and/or other structural deficiencies; the presence of TET and intramyelinic edema can be ruled out.

Prull and Rompel[288] reported EEG changes as a result of organotin intoxication in humans. These investigators suggested seizure generation as a result of chemical hypoxia.

C. BLOOD BRAIN BARRIER

With few exceptions, studies find no direct modification of the blood brain barrier (BBB) by organotin compounds. Rossner and Nusstein[317] claim that TET (1 to 10 mg/kg) significantly increases the permeability of the BBB in mice. Jo'o and colleagues[191] demonstrated an increased permeability of the BBB as measured by Evans blue fluorescent microscopy 48 h after i.v. TET (5.0 or 7.5 mg/kg) given to rats. Electron microscopy revealed astrocytic swelling and signs of leakage of electron dense material from the capillary lumen into intracytoplasmic vacuoles in the vascular endothelial cells. Accumulation of electron dense particles (i.e., Thorotrast) was also found in basement membrane. These effects were not observed with a lower dose of TET (2.5 mg/kg).[191] The increased appearance of serum albumin and transferrin in soluble brain protein fractions following higher doses of TET also suggests a breakdown in the BBB.[190] Furthermore, Susuki et al.[363] have suggested an alteration in the BBB in neonatal rats during TET exposure. This may be due to the immaturity of the BBB in neonates.

In general, however, TET exposure does not appear to alter significantly the blood-brain barrier[32,57,233,360,375,377] when dye injection of trypan blue or pontamine sky blue is used and CNS tissue is examined for dye penetration of the BBB 14 min to 48 h later.[233] Torack et al.[377] found no change in BBB permeability to trypan blue injected into mice. Aleu et al.[18] showed no breakdown in BBB to trypan blue in adult rabbits subchronically treated with TET (1 mg/kg/d). The use of radiolabeled albumin has also failed to show a change in BBB permeability with TET.[197] Furthermore, the rate of [²⁴Na]-influx into brain tissue was the same in control and TET-treated animals, suggesting an intact BBB.[197]

Another complex diffusion barrier, the placenta, is readily penetrated by TMT during prenatal exposure.[278] Acutely administered [¹⁴C]-TMT has been shown to cross the placental barrier in significant amounts, with fetal blood concentrations of the radioisotope reaching 50% of maternal blood concentrations, and fetal brain levels of TMT approximating those of maternal brain.[219] Similarly, triphenyltin acetate has been shown to cross the placental barrier and the BBB during gestational exposure in rats,[213] inducing behavioral changes in the offspring as well as a high mortality rate in the nursing neonates.

D. CNS EDEMA AND MYELIN INJURY

Cerebral edema was a prominent finding in the victims of the Stalinon incident in France in the mid-1950s. The edema was confined to white matter, i.e., heavily myelinated nerve tracts.[6] In 1955, Stoner et al.[358] demonstrated in experimental animals that trialkyltins are toxic to CNS white matter. Dietary administration of TET (20 mg/kg) to rats induced interstitial edema of white matter of brain and spinal cord without obvious neuronal damage.[233] The edema consisted of progressive increases in water, sodium, and chloride contents of the CNS and resulted in an increase in CSF pressure.[214,233] Myelin appeared to be the target and source of the edematous change.

In 1957, Magee et al.[233] presented the first thorough experimental study describing TET-induced edema in the CNS. Acute and chronic experiments were performed on albino rats in order to examine the neuropathology with histological and biochemical methods. The prominent lesion associated with TET administration was the formation of interstitial spaces within the heavily myelinated white matter, without apparent loss of myelinated fibers. The interstitial edema of myelin at approximately 2 weeks was accompanied by increases in brain water content, brain weight, and sodium content without a change in potassium content. The changes in the spinal cord were usually more profound than the brain. The optic nerve was involved, but no ophthalmoscopic damage was observed in the retina or the optic disc. The CNS edema persisted as long as the organotin was present. Neuronal damage was not detected with any exposure regimen. The CNS lesions produced by TET were completely reversible with no observable residual effects.[233] These authors concluded that the accumulating fluid had the composition of an ultrafiltrate of plasma.[233]

In a subsequent study, Torack and coworkers compared the cerebral edema of TET poisoning in mice to human conditions of increased intracranial fluid.[377] The most prominent finding of TET intoxication was enlargement of clear glial cell processes. Mitochondria became enlarged only in the advanced stages of poisoning when focal rupturing of clear glial cell membranes was detected. Fluid accumulation did not occur in the extracellular compartment, but appeared to be largely limited to glial cell enlargement.

Smith et al.[345] have compared lead encephalopathy in children with TET-induced cerebral edema in laboratory animals. Lead encephalopathy was found to produce vascular damage with patchy exudative edema (i.e., the presence of a protein-containing fluid in the extracellular space). TET intoxication in dogs and rats, on the other hand, produced a striking interstitial edema in the white matter only, with no evidence of vascular damage.

In 1963, Aleu et al.[18] studied TET-induced edema (1 mg/kg/d for 5 to 7 d) by light and electron microscopy in rabbits. As in earlier studies, severe edema of the white matter within myelin sheaths was confirmed. Swelling of clear glial cells within the myelin sheath was noted. Oligodendrocytes and neurons were not affected. Vacuoles were filled with "plasma filtrate". Sodium content of white matter increased while the Na^+ to K^+ ratio changed from 0.64 in control white matter to 1.41 in TET treated animals. Chloride content expressed as mEq Cl/100 g dry weight increased 100% in white matter as a result of TET exposure.[18]

Lee and Bakay[211] also presented an ultrastructural study on TET-induced edema in the CNS of rats. Young rats were fed TET hydroxide (40 to 50 ppm in drinking water and food) until various stages of symptomatology and were then removed from TET exposure in order to study their recovery. Myelin splitting was observed in white, but not gray matter. The splitting preceded any neurological symptoms. The splitting always occurred at the less dense lamellae resulting in vacuole formation. The vacuoles were clear until demyelination commenced at which time they became more electron dense. Demyelination, phagocytosis and remyelination were observed. Curiously, the "stainability" of the myelin decreased with advancing toxicity. The ultrastructure of the CNS returned to normal 45 d following removal from TET exposure.

The ultrastructural picture of TET-induced CNS edema was nearly completed by the study of Hirano et al.[174,175] in which the intramyelinic and extracellular spaces were examined in rats exposed to TET acetate (5 mg/kg). Typical intramyelinic edema was observed following injection of TET and compared to local intracerebral implantation of TET plugs. The implants resulted in a regional, not generalized, myelin splitting, and exhibited a severe inflammatory edema with involvement of the extracellular space. When injections and implants of TET were made simultaneously, the results suggested that there is no direct communication between the extracellular spaces and the interiors of the myelin splits or between individual adjacent splits within a single myelin sheath. The implants further suggested that TET has a direct toxic action on the surrounding tissue, rather than an indirect metabolic derangement in the liver or elsewhere.[175] Edema fluid in splits appears clear, like an ultrafiltrate of plasma. Only in severe inflammation, cellular necrosis and demyelination does the fluid become more electron dense. Horseradish peroxidase did not penetrate myelin vacuoles which remained clear.[174] These authors conclude that the intramyelinic splits are effectively sealed from the extracellular spaces.

Jacobs et al.[186] examined the acute effects of TET (10 mg/kg, i.v.) in rats. Intramyelinic vacuoles occurred within hours. Since no effects were found on retinal rod cells or peripheral myelin, it was concluded that TET has a direct effect only on the membrane of the oligodendroglial myelin.[186] Squibb et al.[355] demonstrated a dose-dependent intramyelinic edema of major CNS white matter tracts in rats exposed to TET for 2 weeks. Severe edema was observed in the cerebellum of rats exposed to three doses of 3 mg/kg over two weeks without neuronal necrosis. Schwann cells were unaffected. Complete reversibility of the lesions was reported after 4 weeks recovery in most animals. The TET myelin target is therefore highly specific with surprisingly little involvement of, or damage to surrounding tissues.

TET-induced edema of the optic nerve has been studied by a number of investigators. TET (1 mg/kg, i.p. for 6 d) intoxication caused the rabbit optic nerve to become edematous.[306] The water content and Na^+ to K^+ ratio of the optic nerve increased. Electron microscopy revealed intramyelinic vacuolation of the optic nerve.[326] Unlike myelin in the peripheral nervous system (e.g., sciatic nerve), this central myelin is derived embryologically from oligodendrocytes and is the doubly convoluted plasma membrane of oligodendroglial processes. Wender et al.[402] isolated optic nerve myelin proteins from control and TET-intoxicated rats and found no specific changes in individual myelin proteins related to TET exposure.

The structure-activity relationships involved in the intramyelinic edema are interesting. TMT[55] and DET[233] are inactive; and in decreasing order of activity, TET, tri-n-propyltin, and tri-n-butyltin are active.[34] Further, Aldridge et al.[17] have examined the structure-activity relationships of analogs of TET and TMT. Dimethylethyltin (14 mg/kg, p.o., oral LD_{50}) given to rats resulted in toxicity similar to that seen with TMT, i.e., damaged neurons especially in the hippocampus, and production of moderate intramyelinic vacuolation. Methyldiethyltin (7.5 to 10 mg/kg, oral LD_{50}) administration produced results that are similar to those produced by TET (i.e., it resulted in marked vacuolization of myelin sheaths). Methylethyl-n-propyltin bromide acted essentially like methyldiethyltin and TPrT which also cause the water content of brain and spinal cord to increase.[34] It is concluded that one or more ethyl groups are required for myelin toxicity and edema, while direct neuronal toxicity requires one or more methyl groups.[17]

In producing CNS edema, TET also lowers oxygen consumption, alters blood glucose levels and decreases colonic temperature. Ambient temperature influences the rate of CNS edema formation by TET, i.e., if hypothermia is allowed to occur by not raising room temperature to 34°C, edema formation is delayed.[218] EEG activity is depressed and hyperglycemia is observed.[366] The hyperglycemia and slowing of the EEG precede CNS edema during TET poisoning in rabbits.[36] Interestingly, the beta receptor antagonist pindolol prevents the blood glucose elevating effect of TET and of epinephrine, suggesting a stress-like mechanism of sympathetic origin is responsible for the hyperglycemia.[36] Indeed, adrenalectomy prevents the hyperglycemic effect of TET.[358]

The effect of TET intoxication on cerebral edema in neonatal rats has been examined by Wender and coworkers.[399] In contrast to other studies, administration of TET (10 mg/kg, i.p.) on postnatal day 6 did not cause cerebral edema. However, decreased brain and body weights reflect an inhibition of overall growth and development by TET intoxication. Furthermore, the authors provided evidence of delayed myelinogenesis in the TET-treated pups. Apparently, mature myelin must be present for TET to induce the typical intramyelinic edema. Other studies support this speculation. Veronesi et al.[386] noted delayed myelination although an abundance of oligodendroglia were present. Exposure of neonatal rats to TET (6 mg/kg, i.p. at postnatal day 5) caused permanent deficits in neuronal as well as glial development. Gliotypic proteins were reduced in hippocampus. Neonatal TET exposure caused a reduction in brain weight and a disruption of myelinogenesis, but cerebral edema was not observed.[270,293,399]

It is interesting that long-term exposure of neonatal rats to TET is reported to induce some CNS edema with myelin splitting and vacuolation with normal neurons and axons.[363] Severe, diffuse status spongiosis was observed in CNS white matter of asymptomatic neonatal rats exposed chronically to TET.[363] The intramyelinic vacuoles in the white matter were comparable to those seen in mature rats exposed to TET. Acute administration of an LD_{50} dose of TET to these neonatal rats caused death within 3 d and extensive, diffuse hemorrhagic damage, with only slight spongiosis in the CNS.

In seeking the mechanism of TET-induced intramyelinic edema, researchers have turned to a comparison of other toxic agents or metabolic uncouplers which cause an identical lesion.[70,200,221,224] Hexachlorophene is an example. This bacteriocidal agent was the active ingredient of a widely used hospital soap, Phisohex, whose use was curtailed out of concern for neurotoxicity especially in neonates and children.[378] Hexachlorophene [2-2′ methylene bis (3,4,6 trichlorophenol)] and TET *in vivo* both cause an increase in rat CNS water content, $[Na^+]$, brain glucose, brain to blood glucose ratios, and $[Cl^-]$.[221] No change in brain $[K^+]$ was detected in rat tissue 24 h after a single injection of hexachlorophene. Hyperglycemia was found without a change in hematocrit. Cerebral glutamate levels were significantly lowered after TET exposure,[100] but not following hexachlorophene administration.[221] Acetylethyltetramethyl tetralin (AETT) also uncouples oxidative phosphorylation and induces disruption of myelin reminiscent of the action of TET, hexachlorophene,[70] and halogenated salicylanilides.[69] In another study, the administration of 2-acetyoxy-3:5-diiodo-4′-chlorobenzanilide to rats produces a reversible CNS edema characterized by myelin splitting.[223] Like TET, this compound uncouples oxidative phosphorylation in isolated rat brain and liver mitochondria. ATP synthesis was reduced while ATP hydrolysis and pyruvate oxidation were stimulated.[223]

Lock and Aldridge[225] have reported a high affinity binding site for TET *in vitro* in rat brain myelin with an affinity of approximately 60 μM at pH 7.5. TMT which does not cause CNS edema has an affinity 30 times less than that of TET. Although hexachlorophene and 3,5-diiodo-4′-chlorosalicylanilide produce CNS edematous lesions similar to those seen with TET, they do not interfere with the high affinity binding of TET to myelin in brain. This binding affinity for TET is similar to that of rat hemoglobin[310] and rat liver mitochondria.[12] The authors suggest that TET binds to a protein moiety within myelin, but that the site may or may not be involved directly in the production of edema since hexachlorophene also causes edema but does not interfere with high affinity binding of TET.[221,225] The striking similarity among chemicals which induce TET-like intramyelinic edema is their ability to uncouple oxidative phosphorylation, thus establishing a common link for further investigation.

A number of studies have focused on possible changes in membrane lipids induced by TET which might explain TET's effect on CNS myelin. Smith[346] conducted a biochemical study of demyelination in rats chronically treated with TET (10 mg/l drinking water for up to 4 weeks). Spinal cords were more edematous than the brain, sometimes weighing twice as much as normal due to accumulation of water. Myelin loss was not extensive. *In vitro* uptake studies in slices of brain or spinal cord from TET-treated rats showed [^{14}C]- acetate uptake into myelin lipids depressed, and [^{14}C]-leucine uptake into myelin proteins increased. Additional experiments verified faster turnover in myelin proteins as compared to myelin lipids in response to subchronic TET. The enhanced protein synthesis undoubtedly marks a stimulation of repair or remyelination processes. A lecithin-rich, partially deproteinated "floating fraction" of lipids with more high-molecular-weight proteins was thought to be from split-off lamellae from disintegrating myelin.[346] The floating fraction may therefore represent lost myelin.

Eto and co-investigators[137] studied the lipid composition of rat brain myelin in TET-induced edema in rats by oral feeding of the intoxicant. Abnormalities in chemical composition (e.g., increased cholesterol esters and cholesterol and decreased galactolipids) were attributed to non-specific secondary myelin degeneration subsequent to the edematous condition. Interestingly, the yield of myelin from intoxicated rats was decreased to almost half normal.[137] Owsianowski[274] reported significant and longlasting inhibition of labeled mevalonic acid incorporation into cholesterol of the cerebral white matter in rats exposed to TET (4 mg/kg). The defect in cholesterol biosynthesis may contribute to the degeneration of myelin.[274] Total content of cholesterol esters (CE) of rat white matter decreased as much as 50% within 5 h of TET exposure (4 mg/kg, i.p.) although edematous dilution was not taken into consideration adequately in this study.[397] Free cholesterol also decreased. A decreased nonesterified or free fatty acid content of 25% suggests another mechanism in addition to simple edematous dilution.[397] The authors did not, however, take into account the rapid post-mortem changes in free fatty acids in brain tissues.[73] Normally, CE content in brain is low and only increases as a result of brain injury.[42] Thus, the initial decrease in CE by TET exposure may be related to edematous dilution while the observed accumulation of CE by 28 d may signal TET-induced white matter (myelin) degeneration and/or repair.

TET *in vivo* inhibits the incorporation of $[^{32}P]P_i$ into rat brain phospholipids when body temperature is allowed to drop, but not if body temperature is maintained.[311] Yamada and co-workers[412] have reported an interaction of trialkyltins *in vitro* with phospholipids and inorganic phosphates in bacteria. TET-induced demyelination is somewhat unique since macrophages do not appear to be involved in myelin removal and there is a lack of sudanophilic staining and cholesterol esters are elevated.[157,346] It is concluded, therefore, that TET does not produce any selective lipid alterations except those that would be found during secondary demyelination.

Demyelination with chronic TET exposure may have subtle effects on the CNS. For instance, Amochaev et al.[26] reported a slowing of auditory evoked potentials as a result of TET-induced myelin dysfunction in rats exposed via drinking water (20 mg TET per liter) for 2 weeks (see below). Brain weights and water content were increased, but myelin yield (mg/g brain) was apparently decreased. The apparent loss of myelin and the resulting demyelination responses of the auditory evoked potentials were reversible by 2 weeks of recovery.[26] Therefore, CNS demyelination would, in general, be expected to slow nerve conduction velocities in the myelinated tracts of the CNS, interfering with the temporal aspects of signal processing.

The source of the edematous fluid of the intramyelinic vacuolation has been the subject of several studies which were discussed above as related to the BBB. A recent study, however, is worth reiterating as it examined vascular permeability with higher resolution than earlier work. Hulstrom and colleagues[180] studied the vascular permeability in acute TET-induced brain edema in the hamster utilizing tracer molecules. CNS fluid began to accumulate within 4 h of TET exposure (5 to 10 mg/kg, i.v.) and progressed to the typical condition of intramyelinic edema during the next 20 h. After systemic injection, neither horseradish peroxidase, nor fluorescene isothyocyanate (FITC) labeled dextrans were present in the accumulated fluid in myelin vacuoles. Furthermore, no evidence of vascular leakage of the tracers was found in the cerebral vessels of the TET-treated animals. The authors conclude that the edema must be formed intracellularly and is not due to vascular leakage or a breakdown in the BBB.

In the absence of selective membrane lipid changes, some investigators have explored the role of alterations in energy-dependent ionic flux imbalances in myelin to explain CNS edema. Bakay[32] presented the typical description of ionic imbalance in a study of the morphology and biochemistry of chronic TET-induced CNS edema in rats. Water content of brain increased 28.5%. Increased [Na⁺] and [Cl⁻] in white tissue represented a 60 to 70% increase in electrolytes;

K⁺ concentrations decreased. Exchange of [²⁴Na] between plasma and edematous white matter was delayed compared to control. As other investigators have concluded, the edema fluid appears to be plasma ultrafiltrate.[32] In another investigation, Na⁺ to K⁺ ratios were doubled in white matter of rabbits subchronically exposed to TET (1.5 mg/kg/d, i.p., for 5 d), reflecting an increase in [Na⁺] and a decrease in [K⁺] coincident with increased water content. Gray matter was unaffected.[394] Since carbonic anhydrase and Na⁺,K⁺ ATPase are associated with rat CNS myelin,[291] maintenance of normal myelin may require a significant amount of ion transport. Under normal conditions, these enzymes may be important for maintaining the compact structure by actively extruding salt and water.

Katzman and colleagues[197] studied ionic balance in the white matter of rabbits after TET treatment. As in other studies, [Na⁺] was up in white matter, while [K⁺] was down significantly. The Na⁺ to K⁺ ratio nearly doubled. Water content in rabbit brain was dramatically increased by TET (1 mg/kg/d for 5 to 7 d). No alteration in sodium-dependent ATPase activity in crude homogenates of white matter from rabbits intoxicated with TET was observed.[197] The tissues of edematous subjects contained approximately $10 \mu M$ TET. TET employed *in vitro* $(5 \times 10^{-4}M)$ inhibited both Na⁺,K⁺-ATPase and Mg⁺⁺-ATPase enzymes, whereas only the Mg⁺⁺-dependent enzyme was moderately inhibited with $10 \mu M$ TET *in vitro*. Extracellular space changed very little during TET exposure. The exchange of radiolabeled sodium was slower in TET-treated subjects. Sodium uptake and efflux were decreased during TET intoxication when intramyelinic edema was present. This may be due to inhibition of the sodium pump secondary to blockade of oxidative phosphorylation. The results of this study provide evidence that the fluid accumulation is localized within the intramyelinic vacuoles and is not from an extracellular source.

That TET induces an increase in cerebral fluid by inhibition of Na⁺,K⁺-ATPase has been an obvious suggestion presented in numerous studies. Indeed, this ATPase is inhibited by TET (100 μM) *in vitro*.[197,218] However, it is unlikely that such concentrations could be achieved in brain with doses of TET that produce central edema.[218] Evidence that tissue dilution during homogenization results in the loss of TET from the preparation (by as much as 100-fold) suggests that ATPase assays may not be done in the presence of the TET concentration which is normally present *in vivo*.[197] Torack[376] supported this concept of *in vivo* ATPase inhibition, demonstrating that TET intoxication caused the inhibition of a Mg⁺⁺-ATPase system in white matter of exposed rats. Impaired ATP utilization was proposed to induce edema.[375]

The effect of TET-induced edema on water content, electrolyte distribution and the kinetics of [²²Na⁺], [³⁶Cl⁻] and [¹⁴C]-sucrose uptake in cerebral cortex was reported by Reed and coworkers in 1964.[290] Brain water content, and Na⁺ and Cl⁻ concentrations increased with chronic TET exposure. Potassium levels did not change. An increase in the permeability and volume of glial cells was suggested by the results and agrees with the results of Torack et al.[377] who reported that TET increases glial volume. Changes in the contribution of fast and slow compartments to [³⁶Cl⁻] uptake in brain tissue suggests a decrease in the extracellular compartment and an increase in the intracellular (slow) water component. This shift in [Cl⁻] is likely to affect the membrane potential. It is estimated that membrane potential could depolarize as much as 15 mV. It is not known whether this effect would also be evident in neurons.

The potential interactions of TET with other preexisting neurological syndromes has been probed. Subchronic TET treatment in rats did not alter an experimental allergic encephalomyelitis reaction despite widespread vacuolation of CNS myelin.[158] Apparently, TET and the experimental allergic encephalomyelitis reaction involve different sites on central myelin. Another neurological abnormality in mice, termed the "quaking" syndrome, has been examined in the presence of TET. Oddly, the quaking mouse is resistant to TET-induced cerebral edema.[261] An absence of intramyelinic edema was correlated with an absence of changes in CNS water content without a shift of edema to other compartments. The spinal cord water content of untreated quaking mice was already higher than controls, however, TET treatment did not alter this. In control C57BL/6J mice and phenotypically normal littermates, CNS water content increased and typical intramyelinic edema was induced by TET (5 or 10 mg/kg, i.p.). The resistant quaking mice still exhibited the typical non-specific responses to TET such as lethargy, but were often slower to recover. This suggests that TET's central effects are not simply due to CNS edema alone. Furthermore, this intriguing report suggests that a specific myelin abnormality in the quaking mice interferes with TET's ability to induce intramyelinic edema.[261]

Suzuki and colleagues expanded the previous short communication to a full publication in 1981.[262] TET (5 or 10 mg/kg) given by intraperitoneal injection to 13 day old pups failed to produce intramyelinic edema in the spinal cord of quaking mice from two different genetic backgrounds (B6C3H-qk and BTBRTF/Nev-qk). No edema was present in the brains or spinal cords of the TET-treated quaking mice. The presence of large spherical vacuoles in the oligodendroglia (often containing moderately electron-dense floccular materials) was the only unique morphological feature found in the TET-treated quaking mouse. The authors postulated that the extra interlamellar tight junctions found in the quaking mouse myelin might mechanically restrict the spread of intramyelinic vacuoles. Abnormalities in the oligodendrocytes in the quaking mice suggest a more fundamental metabolic abnormality which makes them more resistant to TET-induced CNS edema.[262]

One of the most elegant studies on TET-induced intramyelinic edema was published by Kirschner and Sapirstein in 1982.[203] X-ray diffraction was used to probe the *in vitro* and *in vivo* effects of TET on the amount of membrane disorientation and its periodicity in the lamellar myelin. In essence, the results indicate that TET *in vitro* induces an ordered swelling of CNS myelin. Rather than a vacuolar, vesicular disruption of the myelin lamellae, the X-ray diffraction revealed ordered swelling with the maintenance of a planar, concentric geometry. Later phases of the process may lead to gross swelling and vacuole formation as observed by electron microscopy.[174,175] Interestingly, TET modifies the myelin periodicity in the CNS but not in sciatic nerve. The repeating unit in myelin (i.e., a pair of membranes with their cytoplasmic surfaces apposed) gives predominant reflections at about 80 and 40 Å since the individual membranes are almost equally separated by their cytoplasmic and extracellular boundaries. TET causes a change in spacing which is reflected as new X-ray reflections. The patterns with TET were different than the disorderly and disoriented pattern caused by hypotonic treatment. Inhibition of carbonic anhydrase, a myelin-associated enzyme involved in ion and fluid balance, did not change the native structure of myelin, nor did it alter the TET-induced structural changes. This study detected ordered swelling of CNS myelin, wherein the swollen structure coexists with membrane arrays which have normal periodicity. Carbonic anhydrase does not appear to be linked to the intramyelinic edema, since swelling still occurred in the presence of the carbonic anhydrase inhibitor methazolamide. However, TET-induced myelin swelling was prevented by replacing mobile ions in the medium with isotonic sucrose. Consequently, Kirschner and Sapirstein ruled out non-specific disruption of myelin integrity in favor of a theory of specific ionic transport followed by obligatory fluid movement to explain the intermediate swelling state induced by TET.[203]

E. VASCULAR EFFECTS

Earlier investigations on CNS vascular permeability in TET-treated animals used macromolecular tracers such as trypan blue, Evans blue, or [^{131}I]-labeled albumin, and reported no change in vascular permeability.[18,32,197,233,375,377] The ultrastructure of blood vessels in almost all studies appears normal.[375]

Hultstrom and co-workers[180] studied vascular permeability in acute TET-induced brain edema using tracers of smaller molecular size than previously examined. Hamsters were injected i.v. with 5 to 10 mg of TET per kilogram to induce intramyelinic edema within 20 h. Horseradish peroxidase did not escape from cerebral vessels during edema development. Moreover, no change in vascular permeability to molecules larger than 3000 mol wt was detected utilizing FITC-dextrans from 3000 to 70,000. Sodium fluorescein, a green fluorescent tracer (376 mol wt) showed no sign of extravasation as a result of the TET-induced edema. All the tracers were therefore confined to the lumen of parenchymatous vessels, except for the occasional presence of smaller weight FITC-dextran in the edematous optic nerves and corpus callosum in a few TET-treated animals. Hence, vascular permeability appears normal in acutely TET-treated brains of hamsters.[180]

Although significant vascular permeability changes are doubtful with TET administration, all vascular effects cannot be ruled out. Arakawa et al.[28] have observed vasodilation in the ears of rabbits immediately after i.v. injection of tetraalkyltin compounds. Moreover, neonatal rats that died within 3 d of a single injection of TET (5 mg/kg, i.p.) have diffuse hemorrhagic encephalopathy with edema.[363]

Pluta and Ostrowska[286] reported changes in the cerebral micro circulation after administration of TET (9 mg/kg) to rats. The vascular network was reduced as a result of edema-induced ischemia.[286] This was associated with a decrease in cerebral blood flow and a compensatory increase in O_2 extraction. The authors suggested two stages of brain edema correlated with changes in cerebral blood flow and changes in vascular permeability.[286]

F. NEUROTRANSMITTER SYSTEMS

A brief overview of the effects of organotin compounds on central neurotransmitter systems was presented by Hanin and collaborators in 1984.[167] In essence, the trialkyltin compounds have been shown to affect most major neurotransmitter systems. Effects of the organatins on ATP-cotransmission in both adrenergic and cholinergic nerves are quire possible given the action of these compounds on ATP formation and utilization. No studies to date have addressed this intriguing possibility.

1. Acetylcholine

Mailman et al.[235] found acetylcholine (ACh) and choline (Ch) concentrations in hippocampus and other brain regions unaffected by postnatal TMT and TET treatment of rats. Hanin et al. reported no change in ACh or Ch concentrations in brain tissues obtained from rats exposed neonatally to TMT (1 mg/kg/d) or TET (0.3 mg/kg/d) for approximately 29 d with intermittent changes in dosing schedule.[167] These authors, having conducted a relatively extensive analytical study on the TET and TMT toxicity in the developing rat, warn of relying on measurements of neurotransmitters and metabolites as the sole index of neurotoxicity. Nonetheless, the majority of published studies rely on concentration measurements of neurotransmitters in various brain regions, without actually assessing the *functional* status of the transmitter system. Thus, a prudent view should be taken so as not to misinterpret the data reviewed below.

The nicotinic side of the cholinergic system is compromised under certain conditions by the presence of trialkyltins. Intracellular recordings from mouse sternomastoid muscle revealed a progressive depression in evoked ACh release in the presence of 10 μM TET *in vitro*.[24] Spontaneous release of ACh, as reflected in miniature endplate potentials (mepps) was unaffected during 70 min of TET exposure. Endogenous ACh release from the alpha-motor neurons of rat phrenic nerve (hemidiaphragm) is compromised by subchronic TET exposure in LE rats.[43] Administration of TET bromide (30 mg/l of drinking water) for up to 3 weeks induced decreases in evoked ACh release, especially at higher stimulation rates (20 Hz). Post-stimulation basal release was markedly elevated in preparations dissected from TET-treated animals. ACh release returned to normal values after 3 weeks of recovery. The mechanism of neurotoxicity may involve mitochondrial injury and/or a direct disturbance of the excitable membrane in the region of the nerve terminal. The authors suggest that, based on the unusual post-stimulation elevation in basal release, ionic imbalances (e.g., Ca^{++}) within the nerve terminal may prevent the uncoupling of the ACh release mechanism, allowing excessive release to continue despite cessation of nerve stimulation. A decreased capacity of cholinergic nerve terminals to respond to moderately demanding stimulation may explain the symptom of muscular weakness following TET exposure in laboratory animals and humans.[43]

Recent studies from this laboratory (Bierkamper and Lewis, unpublished data) have demonstrated an enhanced facilitation of ACh release in phrenic nerve-hemidiaphragm preparations as a result of 3 weeks TET exposure (10 mg/l of drinking water *ad libitum* or 6 mg/kg per week) in rats. The data were generated by paired-pulse testing during intracellular recordings. A significant depolarization of the resting membrane potential was discovered in the myofibers from rats exposed to 3 weeks of TET. Moreover, miniature endplate potential frequency was consistently elevated as a result of TET exposure. These and other studies strongly suggest Ca^{++} mismanagement by cholinergic nerve terminals during frequent neuronal firing.

Organotin induced changes in the muscarinic side of the cholinergic pathway in the CNS have also been reported in many studies. Muscarinic receptor binding is studied conventionally with the radiolabeled muscarinic antagonist quinuclidinylbenzilate ([^{3}H]-QNB). The dissociation constant, (K_D), of muscarinic receptors and their number (B_{max}) can be determined in equilibrium binding studies in membranes. Sumner and Hirsch[362] reported preliminarily that TMT (2.5 mg/kg) significantly depressed [^{3}H]-QNB binding in mouse frontal cortex and amygdala 19 d post-treatment. A decrease in QNB binding due to acute TMT exposure in mice has also been reported in other studies,[20,22] but only when tissue sampling was delayed until after 24 to 48 h post-treatment when TMT concentrations begin rising in the brain.[22,94,340] Scatchard analysis of saturation binding of radioligand indicated that reduced binding is the result of decreased receptor affinity (an increased K_D) and not a reduction in the B_{max}.[22]

Administration of TMT hydrochloride (3.5 mg/kg, p.o.) to Wistar rats caused a 21% decrease in muscarinic receptor density (B_{max}) as measured by [^{3}H]-QNB binding in the hippocampus 2 weeks post-treatment.[226] The K_D of the receptor for the antagonist radioligand was not changed. The reduced receptor density was correlated with a loss of hippocampal pyramidal cells and deficits in behavioral measurements including retention of passive avoidance and extinction of active avoidance tasks.[226] In another study, a single injection of TMT hydroxide (7 mg/kg, i.p.) in rats induced a 15 and 18% decrease in [^{3}H]-QNB binding in the amygdala and the striatum, respectively.[362]

Slikker et al.[340] followed [^{3}H]-QNB binding to a membrane fraction derived from frontal cortex and hippocampus of TMT-treated mice 2, 7, or 14 d after a single dose (1 or 3 mg/kg, p.o.) of the organotin. Binding increased within the first 24 h following treatment, and then decreased in hippocampus. The onset of tremor in the mice began approximately 24 h after exposure, at a time when receptor binding was decreased.[340]

In addition to causing receptor modifications, TMT may interfere with other signal-transduction mechanisms in the muscarinic system. For example, TMT was used *in vitro* to inhibit the ATPase activity thought responsible for loading synaptic vesicles with ACh.[318] Moreover, TMT inhibits ACh release from *Torpedo* synaptosomes, although only at high concentrations *in vitro* (0.5 to 3.5 m*M*). Morot-Gaudry[252] considers TMT an anion-hydroxide ionophore which inhibits oxidative phosphorylation and H^+-ATPase. By analogy with probenecid (organic anion transport inhibitor) and phenylglyoxal (arginine-specific reagent which inhibits chloride exchange in erythrocytes), TMT is postulated to interfere with calcium-induced release of ACh via a modification of anion (e.g., chloride) movement in the nerve terminal. It is interesting that none of these inhibitory compounds blocked ACh release from the synaptosomes when evoked by the calcium ionophore A23187, suggesting that influx of calcium can still evoke neurotransmitter release. Nonetheless, TMT may interfere with the voltage-dependent calcium flux normally required for ACh release by interfering with ionic gradients across the nerve terminal membrane.[252]

It has been suggested that changes in the CNS cholinergic system might correlate with the clinical symptoms of memory deficits, sleep disturbances and depression in humans, and seizure activity in rodents after organotin exposure.[167]

2. Norepinephrine

Mailman et al.[235,236] reported no alteration of norepinephrine (NE) levels in the cerebellum and brain stem of rats treated chronically with TMT (1 mg/kg every other day) or TET (0.3 mg/kg/d). Valdes and co-workers failed to find changes in the NE system when conducting uptake studies in hippocampal synaptosomes from rats chronically treated with TMT (2 to 4 mg/kg per week, i.p. for 3 weeks).[382] Neonatal exposure to low, variably scheduled doses of TMT and TET for 29 d produced no change in the concentration of NE in the brain stem and cerebellum. Nonetheless, other studies in male mice find NE concentrations in whole brain to be decreased 48 h after an acute oral dose of TMT.[20]

Messing and Sparber[245] found increased radioligand binding to beta-adrenergic receptors in rat forebrain after TMT exposure, suggesting a change in some of the effector sites of NE neurotransmission. Moreover, uptake of NE assayed in mouse forebrain synaptosomes appeared to be reduced 2 and 14 h after acute administration of TMT (4.26 mg/kg, i.p.); however, synaptosomal NE uptake was significantly inhibited when TMT was added to the assay *in vitro* yielding a IC_{50} of 43 μM. It is possible that synaptosomes were prepared too soon after *in vivo* exposure to allow TMT concentrations to accumulate in the brain. Another trialkyltin compound, tributyltin, has been shown to inhibit the loading of NE into synaptic vesicles through inhibition of a vesicle associated ATPase.[374]

Bentue-Ferrer et al.[39] reported a decrease in NE content of hypothalamus, mesencephalon, and cerebellum in Wistar rats treated with TET chloride (2 mg/kg/d, p.o. for 5 d).[39] In another study that supports these findings, TET acetate (10 mg/kg) reduced rat brain NE concentrations within hours of injection and sustained the reduction for 48 h in surviving animals.[251] At 5 mg/kg TET, the reduction in NE brought levels to 50% of control between 2 and 4 d. The NE level gradually increased to within 20% of control by day 21 post-treatment.[251] The reduced tissue content of NE was associated with enhanced NE release.

Acute TET, TBT, and dibutyltin intoxication in rats lowered brain NE levels between 2 and 48 h post-treatment;[306] this effect paralleled the release of catecholamines from the adrenals. Tetraethyltin (40 mg/kg) decreased brain NE content in rats within 24 h of injection. Levels of the neurotransmitter remained decreased for 48 h.[88] These changes in the CNS adrenergic system could correlate with the clinical symptoms of rage, depression, headaches, and mood changes seen in man following organotin poisoning.[167]

3. Dopamine

Ali et al.[22] reported dopaminergic alterations with acute TMT (1 to 3 mg/kg) exposure in mice. Neurochemical analyses in the frontal cortex, hippocampus, and caudate nucleus were performed by HPLC at 48 h, 1 and 2 weeks post-treatment. The concentration of homovanillic acid (HVA, a dopamine metabolite) decreased significantly in caudate nucleus by 2 weeks, while dopamine (DA) and 3,4-dihydroxyphenylacetic acid (DOPAC) levels were unaltered. In support of previous work[110] no change in dopaminergic receptors was detected by [³H]-spiroperidol binding assays in the caudate nucleus and frontal cortex. Pargyline was used to inhibit monoamine oxidase in order to determine if decreased HVA concentrations were due to a decrease in DA turnover or an increase in HVA elimination. The results suggested a decrease in DA turnover in caudate nucleus even though DA concentrations and receptor binding parameters were unaltered. Taken together these studies demonstrate that the dopaminergic system is less vulnerable to TMT perturbation than most other major CNS neurotransmitter systems. Nevertheless, other studies are in variance with this conclusion. For instance, dopamine receptor binding increased in whole brain 7 to 14 d after TMT administration (3 mg/kg) to mice; HVA levels increased within 48 h post-treatment.[20]

DeHaven and co-workers[110] studied the effects of TMT chloride (3 or 7 mg/kg, p.o.) on the dopaminergic and serotonergic (see below) systems in the CNS of the LE rat. The concentrations of neurotransmitters and their metabolites were measured in six major brain regions. Decreases in DA and DOPAC were detected in the nucleus accumbens (7 d post-treatment with 7 mg/kg TMT), but not in the striatum, olfactory tubercle, septum or amygdala/pyriform cortex. DOPAC to DA ratios did not change in any brain area. Preliminary binding studies with [³H]spiperone failed to reveal alterations in dopamine receptor affinity or density. Extending this study from 14 to 28 d using the 7 mg/kg dose, these investigators[109] provided additional evidence that TMT decreased DA concentrations in the nucleus accumbens up to 21 d and transiently increased DOPAC and HVA concentrations in the frontal cortex only at 14 d post-treatment. Tissue concentrations of dopaminergic markers were normal in all brain regions by day 28. In another study, chronic TMT administration (1 mg/kg every other day for 28 d) to neonatal rats caused a decrease in DA in striatum, but did not alter the dihydroxy-phenylacetic acid or HVA metabolites.[235] In addition, administration of TMT to rat pups provoked a significant decrease in DA in the striatum, but not the brain stem at the end of the first month;[167,235] DOPAC and HVA levels were normal.[167]

The longterm consequences of neonatal exposure to TET as addressed by apomorphine responsiveness in rats have been postulated to involve the dopaminergic system.[169] Dopamine concentrations decreased only in the hypothalamus of rats treated with TET chloride (2 mg/kg/ d for 5 d); HVA levels increased in striatum.[39] Chronic treatment of neonatal rats with TET[167] did not alter DA, HVA, or DOPAC levels in various brain regions when sampled at postnatal day 29. Tetraethyltin decreased DA levels in rat brain 24 h after a single injection of 40 mg/kg.[88] This action is likely due to the rapid metabolism of tetraethyltin to triethyltin. No changes in tyrosine hydroxylase, dopamine-beta-hydroxylase or monoamine oxidase were found in rat brain after TET administration.[97] Finally, central changes in the dopaminergic system could relate to the observed clinical symptoms of rage, mood changes.[167]

4. Serotonin

Serotonin (5-HT) and its metabolite, 5-hydroxyindolacetic acid (5-HIAA) were altered in brain tissue from LE rats exposed acutely to TMT.[110] Concentrations of 5-HT decreased in striatum and nucleus accumbens while 5-HIAA increased in the hippocampus 1 week after 3 mg/ kg TMT exposure. Concentrations of 5-HIAA were increased in the striatum, nucleus accumbens, septum, amygdala/pyriform cortex and hippocampus 7 d post-treatment with 7 mg of TET per kilogram administration of TMT. In agreement with these findings, Wilson et al. reported a significant elevation in brain 5-HIAA concentrations 7 days after an LD_{40} dose of TMT administered to rats. This dosage regimen decreased the concentration of 5-HT only in the amygdala/pyriform cortex. Indirectly estimating the turnover of 5-HT by calculating the 5-HIAA to 5-HT ratio, an increased turnover was predicted in all brain regions measured at the higher dose of TMT. An extended study showed maintenance of serotonergic changes through 28 d post-treatment.[109] These experiments demonstrate that the serotonergic system of the rat is extremely sensitive to moderate acute exposure to TMT. The mechanism(s) behind the perturbation of this neurotransmitter system is unknown.

The effects of TMT on the serotonergic system in mouse brain[22] appear to differ from those determined in the rat.[110] In mice, 5-HIAA concentrations were significantly decreased in the caudate nucleus 2 weeks after TMT administration (3 mg/kg); 5-HT levels were unchanged. No significant changes in either 5-HT or 5-HIAA were observed in the frontal cortex, hippocampus, or brain stem over a 2 week period post-treatment. Since the rat and mouse studies did not compare identical brain regions, a species difference in the effects of TMT on the serotonergic system cannot be concluded, but this is likely. Doctor and co-workers[114] demonstrated decreased 5-HT uptake into forebrain synaptosomes of mice when assayed 2 and 14 h after TMT administration (4.26 mg/kg, i.p.). When added to the assay *in vitro*, TMT at a concentration of 24 μ*M* produced a 50% inhibition of 5-HT uptake.

In 1961, Moore and Brody demonstrated that TET treatment decreased the concentration of serotonin in the brains of rats exposed to TET by i.p. injection of 5 to 10 mg/kg. Serotonin content decreased to 75% of control values within hours of TET administration and remained at these low levels for 4 h. The 5-HT levels were still reduced by 15% at 14 d post-treatment.[251] In a more recent study, 5-HT was found to be decreased in striatum, hypothalamus and mesencephalon of rats treated with TET chloride (2 mg/kg/d for 5 d); 5-HIAA levels were increased in these regions suggesting increased metabolism of 5-HT.[39] In other studies, brain concentrations of 5-HT were shown to be decreased in adult rats 24 to 48 h after administration of several organotins: TET, TBT, and DBT.[306] Serotonin concentrations have also been shown to decrease in rat brain 8 h after a single injection of tetraethyltin (40 mg/kg).[88]

As with other amine neurotransmitters, disturbances in the central serotonergic systems could be related to the clinical symptoms of organotin exposure. For the serotonergic system, it is possible that depression and sleep disturbances are secondary to organotin exposure.[167,316]

5. GABA

Gamma-aminobutyric acid (GABA) concentrations in mouse forebrain, mid-brain and hippocampus did not change 2 and 14 h after an injection of TMT (4.26 mg/kg, i.p.); however, assay of active uptake of GABA into synaptosomes from forebrain was significantly inhibited.[114] Naalsund and Fonnum[257] reported that TMT's effect on homogenates of hippocampal tissue was minimal with regard to GABA release, although GABA receptor binding was decreased.[257] Wilson et al.[411] observed a decrease in GABA concentrations in hippocampus and frontal cortex, 7 d following administration of a dose of TMT equivalent to $0.75 \times LD_{50}$ in rats. Valdes et al.[382] reported that chronic TMT administration to LE hooded rats caused an increase in the high affinity component of GABA binding, but a decrease in the total number of binding sites in hippocampal synaptosomes. These findings correlated with morphological damage to the hippocampal pyramidal cells and other neurons.

Administration of TMT to neonatal rats caused a significant decrease in GABA levels in the striatum.[235] Chronic treatment of neonatal rats with TMT (1 mg/kg every other day for 27 d) decreased GABA concentrations in hippocampus as a result of generalized neuronal cell loss, but did not alter GABA levels in cortex or other brain regions studied.[235] Similarly, chronic exposure of developing rat pups to TMT, but not to TET, induced a significant reduction in the concentration of GABA in the hippocampus without affecting the cerebellum, brain stem, hypothalamus, striatum or frontal cortex.[167] Loss of neuronal tissue is the likely explanation for the selective reduction of GABA by TMT exposure.

Veronesi et al.[386] treated neonatal LE-rats with TET (6 mg/kg, i.p.) on postnatal day 5 and detected neuronal damage in the CNS on postnatal day 20 as assessed by benzodiazepine binding. Benzodiazepine binding to the GABA-receptor complex was severely and selectively depressed in the hippocampus of TET-treated rats, suggesting an actual loss of granule cell population or a degeneration of its dendritic arborization.

Cremer[100] discovered that TET treatment of rats lowered GABA concentrations in the brain. Hippocampal benzodiazepine binding was severely depressed in neonatal rats treated on postnatal day 5 with TET (6 mg/kg, i.p.) and sampled on postnatal day 20.[386] This may signal granule cell loss inasmuch as the benzodiazepine receptor complex is thought to be located post-synaptically on the soma and dendritic arborizations of the granule cells.

Triphenyltin (TPT) decreased GABA uptake in rat brain synaptosomes when added to the *in vitro* assay. This action of TPT is thought to be due to inhibition of a high affinity, energy-dependent transport mechanism present in the synaptosome.[195] Indeed, Kanner[195] utilized TPT as a tool to provide direct evidence for an active uptake process for GABA into synaptic vesicles. Further, the uptake as well as efflux of GABA has been found to decrease in rat brain synaptosomes exposed to tributyltin.[264] The decreased efflux may be secondary to reduced uptake and storage of GABA into the vesicles and may therefore involve inhibition of the ATPase-associated transport process.[195,264,374]

Slight changes in glutamate decarboxylase in the hippocampus of rats exposed to single or multiple doses of TMT suggest minor, if any, changes in GABAergic neurons to the extent that glutamate decarboxylase can be used as a marker for these specific neurons.[256]

6. Glutamate

TMT treatment has been shown to decrease glutamate levels in rat hippocampus 7 d post-treatment.[411] Glutamate uptake studies in tissues from rats chronically treated with TMT revealed an increased affinity and a decreased capacity in hippocampal synaptosomes, corroborating extensive morphological damage to this structure.[382] TET intoxication in rats also results in a lowering of glutamic acid in the brain.[100]

High affinity glutamate uptake (HA-Glu) can be used as a marker for glutamergic neurons. Naalsund and collaborators[256] studied glutamine uptake in adult male Wistar rats who received a single injection of TMT (8 mg/kg, i.p.) or a multiple injection schedule of TMT (3 mg/kg, i.p. at 1 week intervals). Glutamine uptake was significantly reduced by as much as 42% (single TMT dose) and 68% (multi-dose schedule) in hippocampus at 21 d post-first injection, indicating a loss of functional glutamergic neurons. Pyramidal cell projections known as Schaffers collaterals are believed to employ glutamate as a neurotransmitter.[256] Naalsund and Fonnum[257] extended their investigations to the *in vitro* effects of TMT on hippocampal homogenates. TMT (50 μM) enhanced the release of [³H]L-glutamic acid and [³H]D-aspartic acid from synaptosomes; [³H]-GABA release was not altered. Confirming the *in vivo* exposure results, glutamine uptake was inhibited by TMT (50 μM to 5 mM) in the *in vitro* assay. GABA uptake was also inhibited at higher concentrations (500 μM to 5 mM). The calcium dependence of enhanced glutamate release would suggest synaptic release; in contrast, inhibition of sodium-independent binding of glutamate and GABA would suggest an interference at receptor binding sites. Indeed, the presence of TMT decreased the number of receptor binding sites (B_{max}), while the affinity constant (K_D) remained unchanged.[257] Since glutamate is an "excitotoxic" amino acid, its enhanced release, perhaps by anionic exchange, could lead to neuronal damage in the hippocampal formation. This is only speculation at present. The concentrations of TMT required to induce these acute effects *in vitro* were about five times higher than the estimated concentrations achieved *in vivo*.[256,257]

Glutamate releasing neurons appear to be extremely sensitive to external chloride concentrations, enhancing release as the chloride concentration falls.[168,257,258] Since GABAergic cells do not show the same sensitivity to changes in the external chloride concentration and in view of trialkyltin-associated changes in ionic balance, one might predict a perturbation in extracellular chloride as the trigger for the selective glutamate efflux.[257]

7. Other Neurotransmitter Substances

Wilson et al.[411] reported elevated levels of glutamine in all brain regions measured 7 d after TMT treatment of rats. Glutamine synthetase was not altered in brains of rats treated with doses of TMT equivalent to 0.75 × the LD_{50}.[411] Coulon and colleagues[97] showed that TET alters octopamine levels in rat hypothalamus and brain stem. The effect of this alteration on neuronal function was not addressed. Taurine levels in hippocampus and frontal cortex decreased 7 d after treatment of rats with TMT (5.25 mg/kg, s.c.).[411]

In summary, TET causes few major alterations in neurotransmitter concentrations despite extensive CNS edema. In fact, Mailman et al.[235] indicated that based on their neurotransmitter measurements alone, one would conclude that TET is not particularly neurotoxic. Nonetheless, taken together the many published studies reviewed here provide strong evidence that neurotransmitter systems are perturbed by trimethyltin and triethyltin. The dopaminergic, cholinergic, and serotonergic systems seem to be particularly vulnerable. Although regional brain studies have been performed, the changes in neurotransmitters are generally only grossly described and have not been related to physiological or behavioral changes. The role of these neurochemical changes in the neuropathological manifestations of the trialkyltin compounds remains open to further investigation.

G. HIPPOCAMPUS AS TARGET OF TRIALKYLTIN TOXICITY

The hippocampus is a complex intracerebral structure whose functional significance and physiology is only superficially understood.[47] A general understanding of the circuitry of the limbic system is helpful in interpreting the effects of the organotin compounds on the hippocampus. Loullis and co-workers[226] and Veronies and Bondy[386] have provided a worthwhile illustrated introduction to hippocampal circuitry. Walsh and Emerich[390] have recently offered a mini-review on the hippocampus as a common target of neurotoxic agents which may be useful to the reader embarking on an understanding of hippocampal circuitry susceptible to toxicological insult. Briefly, the hippocampus contains a complex signal flow pathway that runs from the pyriform/entorhinal cortex through the dentate granule cells to the CA3 pyramidal neurons (mossy fibers in the CA3 region receive excitatory input from the granule cells) and the CA1, 2 pyramidal neurons. The entorhinal cortex receives inhibitory input from brain stem neurons (e.g., medial raphe and locus coerulus nuclei). The hippocampal basket cells provide inhibitory modulation of the signal traffic between the fascia dentata (granule cells) and the pyramidal neurons of Ammon's horn. If the basket cells become damaged, granule cell hyperstimulation may overwhelm the CA3 region causing neuronal cell damage.

Veronesi et al.[386] provided a good discussion of hippocampal development which is useful in understanding the regions final circuitry. In normal hippocampal development, major afferents arrive in the dentate gyrus during the first week of life and quickly partition to their appropriate levels on the granule cell dendrites. These fibers originate from the ipsilateral entorhinal cortex and the pyramidal cells in both the ipsi- and contralateral Ammon's horn. They terminate in the molecular layer on granule cell dendrites in a strictly ordered pattern, with the entorhinal afferents making synapse on the distal two-thirds of the dendrites and the commissural (contralateral) and associational (ipsilateral) fibers terminating on the proximal one-third. A third afferent fiber system, the cholinergic septohippocampal tract, terminates in a stratum between the granule cell bodies and the zone of the commissural fiber synapses. In this region, synaptogenesis is most active between postnatal days 4 to 11, when the number of synapses double every day. Synaptogenesis might be the crucial target event in the neonatal model of TET-encephalopathy.

1. Effects of TMT on Hippocampus

A considerable body of evidence suggests that the hippocampus is a primary target of TMT toxicty. The majority of investigations have utilized rodent models to study the histopathological, neurochemical, physiological, and behavioral aspects of TMT neurotoxicity. Since the hippocampal neuropathology of mice and rats has some distinct differences,[87] these species will be discussed separately. Moreover, the developmental studies have been separated from the adult studies due to different responses to TMT in the neonate.[296]

In general, mice appear to be more sensitive than rats to the CNS effects of TMT. Dyer and co-workers[87] found that mice showed the neurological signs of tremor and aggression earlier than rats. Two strains of mice (BALB/c and C57BL/6) developed symptoms at 2 d (TMT chloride 3.0 mg/kg) whereas adult LE and SD rats showed signs at 3 d and 5 d, respectively (7.5 mg of TMT per kilogram). The neuropathology was also more prominent in mice than rats. LE rats were more sensitive to TMT than SD rats. Mice had more lesioning to the hippocampal fascia dentata and brain stem than rats. Rats showed more prominent neuronal damage in the olfactory cortices (pyriform/entorhinal and olfactory tubercle) and the pyramidal cell region of Ammon's horn in hippocampus. Although complete dose-response relationships were not established, these apparent strain and species differences must be taken into account when discussing the CNS toxicity of TMT especially in the hippocampal formation.[87]

a. Mouse

The neuropathology of mouse hippocampus during acute TMT intoxication has been explored by light and electron microscopy.[83-85,87,298] Chang and co-workers[83-85] injected adult BALB/c mice with a single dose of TMT (3 mg/kg, i.p.) and examined brain tissue 48 h later. Scattered neuronal degeneration was observed in the neocortex, pyriform cortex, amygdaloid nucleus and brain stem. Extensive neuronal necrosis was observed in the granule cells of the fascia dentata of the hippocampus. The shrunken necrotic neurons exhibited clumped nuclear chromatin (pyknotic nuclei), some multiple clumping of nuclear chromatin fragments (karyorrhexis) and highly eosinophilic cytoplasm. Although morphologically normal at the light microscopic level, the pyramidal neurons of Ammon's horn showed early signs of cell injury (e.g., vacuolation of endoplasmic reticulum and increased numbers of lysosomes) under the electron microscope. Intracellular edema with severe distention of the cytoplasmic membrane, observed by electron microscopy, was a prominent feature of granule and pyramidal neurons. The intraneuronal edema with subsequent vacuolation was postulated to be the result of suppression of ATP-driven pumps in the plasma membrane which would lead to rapid fluid and electrolyte imbalances in the neuron.[84] Mitochondria, however, were unchanged morphologically at this 48 h time point. The onset of pyramidal cell damage was different than that observed in the rat.[49]

In adult mice, Ali and co-workers[21,22] reported that a single dose of TMT (3 mg/kg) increased the activity of ornithine decarboxylase, a marker of neurotoxic insult, and modified muscarinic receptor affinity in the hippocampus within 2 d of exposure. No studies have been conducted that correlate TMT-induced changes in morphology with the observed increase in ornithine decarboxylase activity.

Mouse hippocampal slices have been the subject of several electrophysiological studies on TMT *in vitro*.[29,30] Changes in spontaneous activity were observed by extracellular recording when Armstrong et al.[29] applied TMT chloride ionophoretically to transverse hippocampal slices. Increased spontaneous firing was observed in 68% of the granule cells in the dentate gyrus, whereas a decrease in activity was found in all other regions including CA1, CA2, and CA3. The changes in firing rates were reversible except after repeated doses, when irregularities in rate signaled irreversible toxic effects. These results agree with observations by Allen and Fonnum[23] who showed that TMT decreased the amplitude of CA1 population spikes evoked in rat hippocampal slices. Moreover, histopathological studies in the mouse have demonstrated that the granule cells of the fascia dentata have a profound sensitivity to the toxic insult of TMT when compared to pyramidal cell changes.[83,87] Maintenance of tissue slices in low Ca^{++}, high Co^{++} containing buffers to inhibit synaptic activity revealed a blockade of increased granule cell firing. Thus, the increased firing may be due to increased neurotransmitter release or other calcium-dependent mechanisms related to synaptic transmission.

The effects of TMT on evoked responses in the CA1 pyramidal cell layer and dentate gyrus have also been explored by Armstrong and co-workers[30] in mouse hippocampal slices. Superfusion of TMT (10 μM) produced an 80 to 90% reduction in population spike amplitude induced by orthodromic stimulation in both the CA1 pyramidal cell and dentate gyrus regions of the hippocampus. Dose-dependency, reversibility and an increased latency of effect were established with lower concentrations of TMT (1 to 5μM). This study demonstrates that the orthodromic excitability of both pyramidal and granule cells is significantly decreased by TMT *in vitro*, thus supporting earlier observations in hippocampal slices from rat brain.[23] Preliminary intracellular studies from this group revealed TMT-induced membrane depolarization, which when prolonged may account for the loss of excitability in response to orthodromic stimulation.[30]

b. Rat

In 1979, Brown et al.[55] published an important study on the neurotoxicity of TMT in adult Wistar rats. The behavioral and neuropathologic sequelae of intoxication were followed after single (10 mg/kg) or repeated weekly doses (4 mg/kg for $\leq$ 4 weeks) of TMT by gavage. The symptoms of toxicity included tremors, aggressive behavior, hyperexcitability, weight loss, and seizures. Light microscopy revealed bilateral and symmetrical neuronal alterations involving hippocampus, pyriform cortex, amygdaloid nucleus, and neocortex. The most vulnerable regions were the large pyramidal cell layer (CA2 and CA3) of the hippocampus, the pyriform cortex, and the amygdaloid nucleus. Neuronal degeneration was characterized by loss or dispersal of Nissl substance, followed by clumping of nuclear chromatin with subsequent shrinkage and fragmentation of the nucleus within eosinophilic cytoplasm. These changes were observed 1 d after the second dose of 4 mg/kg or 2 d after a single TMT dose of 10 mg/kg and corresponded to brain tissue concentrations of approximately 1.4 μg of TMT per gram wet weight. In single, high-dose experiments, early and consistent involvement of the fascia dentata was noted. Unlike TET whose effects appear largely reversible in rat and man, TMT-induced lesions appeared to be permanent with severe damage and hippocampal cell loss observed at 70 d post-treatment. This original study[55] has ignited widespread interest in the effect of TMT on the limbic system and set the foundation for a flood of subsequent papers.

The ultrastructural changes due to TMT intoxication in Long Evans rats were examined in an extensive investigation by Bouldin and colleagues.[49] Adult rats were exposed to TMT hydroxide acutely (5 mg/kg/d for up to 5 d) or chronically (1 mg/kg/d for 14 to 16 d) by gavage and sacrificed 1 d after the last dose. Neonatal rats were treated with 1 mg/kg on alternate days from postnatal day 3 through 29 and sacrificed on postnatal day 30. This maximally tolerable dosage regimen resulted in a 20 to 30% loss in body weight and induced tremors, hyperexcitability, and seizures. Pathological lesions were limited to the CNS. Necrotic neurons were observed in the neocortex, pyriform cortex, hippocampus, basal ganglia, brain stem, spinal cord, and dorsal root ganglia. Preferential involvement of the hippocampus and pyriform cortex was evident, supporting Brown and co-workers' previous study[55] and observations by Ray.[289] Cellular changes were first detected by light microscopy after 3 to 4 d of TMT treatment. Progressive granulation of neuronal cytoplasm was most prominent in the pyramidal cells of the CA1-CA4 regions, but was also evident in the fascia dentata. Electron microscopic evaluation of tissue sections revealed multifocal collections of dense-cored vesicles and tubules, and membrane-delimited vacuoles in cytoplasm and proximal dendrites of degenerating neurons. The vesicles and tubules had acid phosphatase activity similar to Golgi-associated endoplasmic reticulum (GERL). The emergence of these unique GERL-like membranes in TMT-intoxicated neurons may indicate enhanced synthesis of lysosomes and autophagic vacuoles. This seems likely since autophagic vacuoles and polymorphic dense bodies accumulated in the neuronal cytoplasm subsequent to the appearance of GERL-like structures. Thus, TMT intoxication appears to trigger increased autophagic activity as the cell attempts to manage cellular injury. Cytoplasmic vacuoles, often the size of mitochondria, were found only in neurons which also contained cytoplasmic accumulations of dense-cored vesicles and tubules. Neuronal necrosis was characterized by condensed electron-dense cytoplasm, pyknotic nuclei, dense-body accumulation in the cytoplasm and proximal dentrites (the ultrastructural correlate of the granules found by light microscopy) and autophagic vacuoles. Ultrastructural changes differed very little between acutely and chronically treated adults and neonatal rats. However, acutely intoxicated rats (short-term, high-dose) had greater neuronal necrosis in the fascia dentata, while chronically intoxicated rats (long-term, low dose) had neuronal necrosis principally localized in Ammon's horn. The mechanism for this differential vulnerability is uncertain.

Brown et al.[56] observed neuropathic changes in hippocampus of rats treated with TMT (10 mg/kg) as early as 12 h following treatment. By 24 h a general vacuolation of Golgi cisterns and smooth endoplasmic reticular membranes were found. Within 48 h post-treatment vacuolation of Golgi cisterns decreased and accumulation of dense bodies and secondary lysosomes increased, coincident with progressive toxicity and subsequent death of pyramidal and granule cells.[56]

One of the most exciting findings of Bouldin et al.[49] comes from the testing of a series of other alkyltin compounds. Neonatal rats were exposed to TET (1 mg/kg/d), TPrT (3 mg/kg), TBT (10 mg/kg), tricyclohexyltin (30 mg/kg), or TPT (30 mg/kg) from postnatal day 3 through 30. No neuropathology was observed in tissues sampled 24 h after the last dose by any of the compounds except TET, which induced the typical neurological abnormalities and myelin splitting.[46] TBT produced bile duct ectasia and portal tract fibrosis in the liver. Additional experimentation was provoked by suspicion that the biotransformed product of TMT, DMT, might produce neurotoxicity in the TMT models. However, neither DMT (35 mg/kg/d) nor diethyltin (10 mg/kg/d) produced the paresis, tremors, convulsions, fatalities or neuropathologic changes characteristic of TMT treatment of developing rats. This series of experiments emphasizes the exquisite structural specificity of TMT in producing selective limbic system neurotoxicity.

Dyer, DeShields and Wonderlin[124] focused on the hippocampus in TMT-intoxicated rats to determine if gross morphology could be used as a correlate in other physiological and behavioral studies of TMT toxicity.[130] A single dose of TMT (5, 6, or 7 mg/kg) was administered by gavage to adult LE hooded rats. Hippocampal sections for light microscopy were obtained within 11 d, at 30 d, or at 105 d post-treatment. TMT administration caused pyramidal cell loss within the hippocampal formation, primarily in the CA3 region. Following a single administration of 6 mg/kg TMT, loss of pyramidal cells appeared first in CA3c, then CA3b and CA3a regions. Even after the longest post-treatment period, damage was extensive, but cells remained intact in the CA1 region. These results differ from those of Brown et al.[55] who observed significant thinning and cell loss in the CA1 region. TMT treatment shortened the length of the line of pyramidal cells from CA1 through CA3c in a dosage and time-dependent manner; loss of cells appeared to begin in CA3c and progress through CA3b and CA3a with dosage and time following treatment. The authors claim that the pyramidal cell line length can be used as a simple dosage-dependent correlate of hippocampal damage induced by TMT. The degenerative changes observed in this single dose study did not differ significantly from the multi-dose study of Bouldin et al.,[49] thus raising questions about the dose-dependency of the proposed morphological marker or correlate.

A time-course study of TMT-induced neuropathy in adult LE rats was published in 1983 by Chang and Dyer.[78] This valuable study explored the progressive neurotoxicity of a single dose of TMT (6 mg/kg by gavage) from 8 h to 60 d post-treatment. Histopathology was examined in the hippocampal formation (i.e., CA1,2; CA3a,b; CA3c, and fascia dentata), pyriform and entorhinal cortices, olfactory tubercle, and brain stem by light microscopy. Although not morphometric, this study provided the first long-range, controlled comparison of the major targets of TMT neurotoxicity in the CNS. Interestingly, the fascia dentata and brain stem showed neuropathic changes by post-treatment day 3 (pyknotic nuclei and scattered necrosis of granule cells, and extensive chromatolysis of large neurons) which subsided by the 3rd week with limited damage. Extensive cell loss was observed in the pyriform and entorhinal cortices, and in CA1-3c regions of Ammon's horn from weeks 2 to 3 through day 60. The order of sensitivity of the pyramidal cells of Ammon's horn to TMT exposure was CA3c > CA3a,b ≥ CA1,2. Uniform susceptibility was not displayed by these cell fields along the septo-temporal axis (see below). Significant cell loss was also observed in the olfactory tubercle through day 60. Extensive cell destruction was always accompanied by proliferation of astroglia cells (gliosis) except in the CA3a,b regions. No explanation was provided for this unusual finding.

Discrepancies in the apparent vulnerability of neurons to TMT intoxication in various hippocampal regions may have been due to histological sectioning techniques.[80] Studies showing CA1 and CA2 sensitivity and CA3a resistance were based on coronal sections;[55,382] whereas those demonstrating CA3a and CA3b sensitivity were based on sagittal sections.[80,124,130] Chang and Dyer[80] addressed these interstudy discrepancies in TMT-induced hippocampal lesions by a careful examination of the full septotemporal axis of the hippocampal formation. LE hooded rats received a single dose of TMT (6 mg/kg p.o.) for these comparative studies; tissue was collected at 3, 15, 30, and 60 d post-treatment. Histopathological examination of serial sections in the coronal and sagittal planes revealed region-specific and cell field-specific damage. The granule cells of the fascia dentata were most involved ventrally and dorsolaterally (temporally). The CA3a and CA3b pyramidal cells were most sensitive dorsomedially (septal pole), and least sensitive at the temporal pole. Neurons in CA3c were affected along the entire extent of the dorsal hippocampus, but showed more severe involvement at the septal pole. The CA1 and CA2 neurons were principally affected at the mid-septotemporal levels. An inverse relationship between regional lesion severity of the granule cells and the pyramidal neurons in Ammon's horn was discovered. Sparing of CA3a and CA3b pyramidal neurons was apparent in portions of the hippocampus which showed extensive granule cell destruction. Conversely, when pyramidal cells exhibited significant damage, granule cells in the corresponding fascia dentata appeared to be intact. The authors suggested that functional granule cells are required for pyramidal cell destruction (in CA3a and CA3b). Indeed, hyperactivity of the granule cells may induce CA3 degeneration as has been shown with kainic acid.[343] Moreover, granule cell destruction prevents kainic acid-induced CA3 destruction.[259]

Additional studies[256,344,382] have included histopathology to complement further exploration of the neurochemical and neurophysiological effects of TMT in adult rats. Valdes et al.[382] described extensive CA3c damage with sparing of the CA3b and CA3a in adult LE rats chronically exposed to TMT. Since hippocampal damage can be caused by convulsant chemicals[260] and recurrent seizure activity,[319,361] Sloviter and his collaborators[344] examined the histopathological role of TMT-induced seizures as compared to kainic acid lesions[414] by light and electron microscopy. Adult SD rats received a single injection of TMT hydroxide (7 to 9 mg/kg, i.p.) or kainic acid (12 mg/kg, i.p.). Major motor seizures commenced 3 d after TMT exposure; in contrast, kainate-induced seizures began within minutes after injection. Less than half of the TMT-treated rats displayed seizures; all kainic acid-treated rats had clonic-tonic seizures. TMT-treated rats which did not have seizures showed minimal pathology after 1 week. Sharp differences and some similarities were observed in the two "seizure" models. Acute dendritic swellings in the distal granule cells, hilus, and mossy fiber region of the hippocampus was noted within 4 h of kainic acid treatment. These authors[341,342] have previously shown that electrical stimulation of the perforant path induces epileptiform activity in the granule cells of the dentate gyrus, thus causing damage to the hilus of the fascia dentata and to the mossy fibers in the CA3 region which receive excitatory input from the granule cells. The granule cells must be present for kainic acid to produce this pattern of neuropathology and to induce seizures.[259] TMT damage, although delayed, was more extensive and variable, involving granule and pyramidal cell necrosis and acute dendritic swelling in two limbic regions, the CA3 (and/or CA1) and from the hilus to the proximal dendrites of dentate granule cells. Seizure-associated dendritic swelling induced by TMT was noted in CA1-CA3 and in the inner dentate molecular layer. Lesions to the CA1-CA3 pyramidal cells were apparently similar in both models in later post-treatment periods.

These results provide evidence that TMT may damage the pyramidal cells of CA3 and granule cells of the dentate gyrus via seizure activity. It is generally accepted that the events of seizure activity are neurotoxic to the cells involved.[259,260,344,361] Thus, this hypothesis appears to have merit. The authors offer the interesting speculation that TMT may provoke the release of an "endogenous excitotoxin" (e.g., glutamate or aspartate) in pathways in which seizure activity is induced. However, Zimmer et al.[417] have offered evidence that TMT does not behave like a kainic acid-like excitotoxin. Phenobarbital failed to prevent TMT-induced neuropathy in the rat limbic system, while it did block kainic acid-induced damage. Moreover, Zimmer and his colleagues[417] question the need for sustained electrical "seizure" activity to produce the TMT lesions. These two groups[344,417] agree that TMT and kainic acid operate via different mechanisms but share similar endpoint neuropathic changes in parts of the limbic system.

Following neuronal degeneration by Fink-Heimer staining, Naalsund et al.[256] studied hippocampal damage 3, 7, 14, and 21 d after a single dose of TMT (8 mg/kg, i.p.) administered to adult Wistar rats. The CA4 and ventral CA3 pyramidal cells were most sensitive to TMT toxicity and were virtually abolished by day 21 following a single dose of TMT or by day 35 following multiple injections (3 mg/kg/week for 3 weeks). Pyramidal cells in CA1 were less sensitive to the effects of TMT and dorsal CA3 neurons were spared. This pattern of degeneration was in agreement with previous studies.[49,55,124,255,382] Changes in hippocampal electrical activity measured with implanted electrodes were correlated with the first appearance of neuronal degeneration at 3 d, even though biochemical changes (e.g., decreased high affinity glutamate uptake) were delayed until day 7 post-treatment. Nerve terminals in the highly sensitive CA3 and CA4 regions are thought to release glutamate as their neurotransmitter; thus, it appears that decreased high affinity glutamate uptake is secondary to the loss of pyramidal cell neurons.[382]

TMT has been shown to induce zinc depletion in rat hippocampus.[79] A single injection of TMT (6 mg/kg, i.p.) resulted in a progressive loss of heavy metal staining by Timm's method in the hippocampus formation which was qualitatively correlated with a loss of zinc, especially from the mossy fibers and the granule cell layer.

A series of behavioral studies[321,365,393] on TMT neurotoxicity in the rat also included a structural correlation. Light microscopy revealed extensive damage to the hippocampal pyramidal cells especially in the CA3 region. The morphological damage was related to changes in behavioral performance and was comparable to that seen in previous studies.[124,130]

The latest report in this long run of morphological studies on TMT-induced limbic system damage offers a detailed morphometric analysis of the hippocampus with light microscopy.[302] Adult male LE rats were exposed to 3 mg/kg TMT for three consecutive days; tissue was collected on days 4, 8, 14, and 28 post-treatment. The neurotoxicity of TMT is delayed both symptomatically and histologically. The quantitative results are consistent with previous qualitative studies and demonstrate a rank order of cell loss as CA4 > CA1 > CA3 granule cells. The volume of the degenerating hippocampus decreased within 4 d post-treatment and was offset by ventricular expansion. The greatest cell loss occurred between 14 and 28 days post-treatment. Cell loss in CA3 was primarily located in the CA3b sub-region. Decreased nuclear volume was a distinguishing feature preceding cell loss in the CA3 and CA4 regions; however, cell loss in CA1 occurred without loss in nuclear volume, suggesting two different mechanisms for TMT-induced pyramidal cell destruction. Overall, the dorsal hippocampus appeared most affected. These investigators presented a dramatic visualization of the TMT-induced degenerative changes using a three dimensional comparative reconstruction of control and treated hippocampi by computer-generated overlays of microscopic sections.

c. Prenatal Exposure

Prenatal exposure to TMT also appears to have an adverse effect on the limbic system.[278] Pregnant SD rats were given a single injection of TMT (5, 7, or 9 mg/kg, i.p.) on gestational days 7, 12, and 17. Maternal and pup body weights were decreased by TMT. However, the protein content of brain was unaltered. The neuropathology suggests that TMT crosses the placental barrier. *In utero* TMT exposure between gestational days 12 and 17 represented the most sensitive time window for neuropathic changes in the hippocampus. Although degeneration was not extensive, neuronal alterations were observed in the CA3 and CA4 regions of Ammon's horn. Scattered neurons appeared swollen, with granular, brightly eosinophilc cytoplasm. Rarely, neurons with markedly granular cytoplasm and/or pyknotic nuclei were found in Ammons' horn and the developing fascia dentata, but were observed more frequently in the entorhinal and superior parietal cortex. Gliosis was not found. The time-course of TMT sensitivity suggests that differentiated neurons must be present in developing rat brain for TMT to induce certain neuropathic changes. The presence of maternal toxicity seemed to be a prerequisite for prenatal TMT toxicity as observed postnatally. However, the contribution of other adverse effects of TMT (e.g., kidney and/or liver damage) on maternal health as they relate to gestational toxicity were not addressed.

d. Neonatal Exposure

Exposure of neonatal rats to TMT induces extensive neuropathic damage to the hippocampal structures.[75,298] Histopathological examination by light and electron microscopy was performed on postnatal days 15, 21, and 24 following TMT administration (6 mg/kg, i.p.) to neonatal SD rats on postnatal day 11. Treated pups had stunted growth. The results demonstrate a highly sensitive age period between postnatal days 9 and 15. Rapid destruction of the pyramidal cells occurred within days such that Ammon's horn was almost completely destroyed by postnatal day 21. As described previously,[49,55,78] the pyramidal neurons became extremely edematous before disintegration and death. Dystrophic calcification and gliosis were observed in the pyramidal cells. The granule cell layer of the fascia dentata was only mildly affected. Compensatory enlargement of lateral ventricles offset the severe atrophy of the hippocampus. Although the neuropathological findings at the cellular level in the TMT-treated neonate are similar to the adult, it is apparent that neonatal brain is more vulnerable to the neurotoxic effects of TMT than adult brain.

Reuhl and colleagues[298] followed the temporal progression of TMT-induced CNS lesions in neonatal BALB/c mice by light microscopic examination. Tremor and microscopic changes in the hippocampal and pyriform cortical neurons occurred within 16 h of a single i.p. injection of TMT (3 mg/kg) on postnatal day 3. Eosinophilia and cytoplasmic granularity leading to pyknosis and karyorrhexis were characteristic of the degenerating neurons. By 5 to 7 d post-administration, virtually all the neurons of the hippocampal CA3 region of Ammon's horn were involved in degenerative or necrotic changes. Severe damage was also observed in the pyriform cortex. Degenerating neurons were detected in the cerebellum (i.e., Purkinje and granule cells), cerebral cortex and basal ganglia, although damage was relatively less than the pyriform cortex. The pathological changes of the granule cells in the fascia dentata were not as extensive as those reported for adult subjects.[82,83] Damaged neurons in the hippocampus and pyriform cortex exhibited widespread intracellular vacuolation within 1 week post-treatment. The authors suggested two separate, though possibly related origins for the distinctive vacuole formation. Small vacuoles were noted throughout the neuronal cytoplasm and were presumed to coalesce into larger clear vacuoles displacing the nucleus and cytoplasm into the cell periphery. This possibility is supported by histochemical studies in the LE rat where some membrane-delimited cytoplasmic vacuoles demonstrated dinucleosidase activity and other vacuoles did not.[49] Other neurons with a severely distorted or marginated nucleus contained one large vacuole filled with floccular, eosinophilic debris. Neonatal and adult subjects appear to follow a similar morphological pattern when exposed to acute TMT, except for a different regional distribution of the hippocampal lesions.[83,298]

e. Electrophysiology

The electrophysiological manifestations of TMT-induced hippocampal damage in the rat have received limited attention. Ray[289] observed an increase in EEG theta rhythm[47] coincident with the commencement of hyperactivity and tremor in rats given a single dose of TMT (10 mg/kg, p.o.) and implanted with cortical and depth electrodes. Hippocampal evoked responses transiently increased, then declined by 4 d post-treatment. Naalsund and colleagues[256] followed a decline in spike amplitudes from the hippocampus and an altered EEG in rats exposed to TMT. Aberrant EEG patterns of hippocampal origin appeared in TMT treated rats during motor activity.

Hasan, Zimmer, and Woolley[171,172,416] conducted chronic electrophysiological studies to determine the *in vivo* effect of TMT on limbic system function. A single dose of TMT (7.5 mg/kg, p.o.) was administered to adult female SD rats previously implanted with bipolar electrodes in the olfactory or prepyriform cortex, dentate gyrus and the distal CA3 region of the hippocampus. The averaged evoked potential (AvEP) method was used to examine two limbic system pathways: (1) stimulation of the olfactory or prepyriform cortex to evoke potentials in the dentate gyrus (i.e., the PPC-DG path), and (2) stimulation of the hilus of the dentate gyrus to activate the mossy fiber system (whose axons terminate on apical dendrites of the CA3 pyramidal cells) to evoked potentials in the CA3 subfield (i.e., DG-CA3 path). No change in AvEP was observed during the first 24 h post-treatment; however, changes emerged by 48 h. It is noteworthy that corresponding brain tin concentrations were low at 24 h (0.25 µg/g) and doubled by 48 h (0.58 µg/g); the tin concentration doubled again by day 4 and then remained elevated throughout the 20 d examination period. Female subjects did not show tremors and convulsions as severe as male rats.[130] The AvEP amplitude of the PPC-DG path was potentiated between 2 and 7 d with the peak occurring between 4 and 6 d. The magnitude of the increased amplitude was nearly ninefold in some animals; the rise-time of the response was increased by 50%. Spontaneous activity remained normal during this time, with no evidence of seizure patterns. AvEP amplitude of the DG-CA3 path decreased at 3 d post-treatment and was markedly depressed by 3 weeks. Control AvEPs in both paths were remarkably stable over the 20 d recording period. Curiously, the PPC-DG path response was potentiated at a time when the DG-CA3 path response was not. Furthermore, the PPC-DG path and the DG-PPC path responses declined after 6 d. The authors ruled out kindling[130] to explain the potentiation of the PPC-DG path response. Nevertheless, it is clear that the dentate granule cells behaved in a hyperexcitable manner. The potentiation could represent alterations in the bisynaptic pathways from prepyri-form cortex to the dentate gyrus via altered neurotransmission and/or could reflect a decrease in tonic inhibition or an increase in facilitation. A toxic insult to the GABAergic basket cells (which normally exert an inhibitory tone on the dentate granule cells) may affect the regulatory process known to mediate phasic recurrent inhibition in the hippocampus. TMT is known to reduce recurrent inhibition in the dentate gyrus within 24 h.[123] This theory is also supported by neurochemical evidence showing the concentrations of GABA in rat hippocampus to be reduced by TMT.[235] Moreover, GABA uptake into synaptosomes and/or metabolism may be compro-mised by TMT.[114,381,382] Regardless of the precise mechanism, it is clear that inhibitory control was significantly reduced during evoked responses. The pattern of linked responses in the PPC-DG and DG-CA3 pathways is in line with the histologically derived theory that excess stimulation (disinhibition or hyperexcitation) of cells in the dentate gyrus may destroy select CA3 pyramidal cells downstream.[76,289] Since the majority of the previous papers on TMT toxicity have viewed static pathology through the microscope, this study is particularly valuable because it examines the actual functioning *in vivo* of the involved brain pathways.

The *in vitro* effects of TMT on rat hippocampus slices have been reported by Allen and Fonnum.[23] Electrophysiological methods were applied to a brain slice preparation from adult male Wistar rats. CA1 pyramidal cell activity was recorded with extracellular electrodes. Pyramidal cells were synaptically stimulated via electrodes in the Schaffer collaterals or antidromically via an electrode in the alveus. Bath application of TMT (0.6 µM) had a depressive effect on CA1 pyramidal cells. Stimulation via the Schaffer collaterals electrode revealed a reduction in response (i.e., decreased population spike amplitude). Stimulation via the alveus electrode also decreased the population spike amplitude. Antidromic stimulation of the CA1 neurons via the alveus electrode represents excitability of the neurons since it is not dependent on the synaptic release of neurotransmitter. The authors suggest, therefore, that TMT alters synaptic transmission in this hippocampal pathway *in vitro* by depression of the postsynaptic CA1 pyramidal cell. The mechanism of this depression is unknown, but it may play a role in the increased after-discharge threshold reported by Dyer et al.[131] In this regard, Garber et al.[152] recently presented preliminary evidence of increased spike thresholds in CA1 pyramidal cells by TMT *in vitro*. Intracellular recordings from CA1 cells in a rat hippocampal slice preparation revealed a decrease in membrane input resistance, elevated spike thresholds and occasional depolarization in the presence of 50 to 100 µM TMT.[152]

f. Primates and Other Species

In relating the neuropathological studies on TMT toxicity to the human condition, it is important to note that most studies have primarily involved rodents. On the other hand, Reuhl and co-workers[297] have published a study on organotin toxicity in the monkey *(Macaca fascicularis)*. Acute exposure to TMT chloride (0.75 to 4.0 mg/kg, i.v.) resulted in a predictable clinical picture: slow emergence from anesthesia, transient reduction in food and water consumption, excessive salivation initially without other parasympathetic signs, slight hand tremor by 72 h, loss of dexterity, and a hyperreactive startle reflex prior to deterioration into a state of stupor within 5 to 8 d. Light and electron microscopic examination revealed neuronal loss and damage in the CA3 and CA4 regions of Ammon's horn in the hippocampus. The neuropathology did not differ from that described previously.[49] Degenerating neurons exhibited pyknotic nuclei, dense cytoplasm, dilation of the endoplasmic reticulum and Golgi apparatus, and increased numbers of lysosomes. Gliosis and phagocytic activity were evident in the degenerating areas. Little neuronal damage or destruction was noted outside the hippocampus. As with the other species studied, the limbic system of primates appears to be highly vulnerable to TMT, although the spatial and temporal distribution of lesions within the limbic system may differ slightly from other species and with other dosages.

The limbic system toxicity of TMT has also been described for marmosets, hamsters, and gerbils.[57] The clinical progression of TMT neurotoxicity in the marmoset parallels that of the cynomolgus monkey.[7,57,297] The clinical symptoms are not dissimilar to those of TMT exposure in man,[7,150,316] however, the lack of human histopathology prevents further comparisons. The LD$_{50}$ in these three other species was estimated to be 3 mg/kg as compared to 12.6 mg/kg in the rat.[57]

Brown and colleagues[57] measured tissue concentrations of TMT in marmosets following acute exposure. Unlike data obtained in the rat, marmoset hemoglobin and plasma had a relatively high affinity for TMT. TMT was uniformly distributed in different regions of the brain and exceeded the concentration in blood within 24 h (first time point measured) of exposure.[57]

2. Mechanism of Hippocampal Toxicity

In 1986, Chang[76] presented a unified hypothesis for the pathogenic mechanism of TMT neurotoxicity to the mammalian limbic system. Taking into account a large body of morphological,[49,55,75,78,87,124,256,278,344,382] neurochemical,[22,110,114,226,235,381,382,386,417] and physiological data,[23,171,172,289,416] a hyperexcitatory cascade among neurons of the entorhinal cortex, fascia dentata, CA3 and CA1,2 regions has been devised to explain the pathological lesions induced by TMT in the limbic system.

Ray[289] was the first investigator to propose and to present evidence for the hyperexcitability theory of organotin neurotoxicity in a report in 1981. This report coincided with a series of studies on kainic acid[343,414] which suggested that this excitatory neurotoxicant impaired recurrent inhibition and induced severe degeneration in the CA1-CA4 fields and dentate gyrus of the hippocampal formation.

Chang's[76] working hypothesis evolved, in part, from the comparative evaluation of the TMT-induced mouse and rat neuropathies (described above). In mice exposed to TMT, the pyramidal cells of Ammon's horn were spared while the dentate granule neurons were destroyed. Conversely, in rat, the dentate granule cells suffer minimal damage while Ammon's horn was extensively involved. This inverse pathological relationship makes more sense when one considers the experiments of Sloviter and colleagues[343,344] and Nadler and Cuthbertson.[259] Sustained electrical stimulation of the perforant pathway for 24 h abolished recurrent inhibition, evoked granule cell hyperactivity and caused degeneration of the CA3 pyramidal cells and hilar interneurons;[342,343] kainic acid produced identical results.[166,343] Destroying the granule cells prior to kainic acid treatment prevented CA3 damage.[259] Further strengthening the idea that the granule cells are necessary for downstream CA3 destruction, Chang and Dyer[80] have shown in rat hippocampus that temporal pole portions of the hippocampus showed more extensive granule cell damage and marginal CA3 injury. In the septal pole portion of the rat hippocampus, a moderate degree of injury to the dentate granule cells was correlated with extensive CA3 pyramidal cell destruction.[80] Finally, developmental studies in neonatal rats[75] demonstrate that the pyramidal cells of Ammon's horn are not vulnerable to TMT (prior to postnatal day 3) until the mossy fibers (axons of the dentate granule cells which project to the CA3 pyramidal neurons) reach developmental maturity. Taken together, these numerous observations present a compelling argument that the pathogenesis of TMT in the hippocampus involves the functional interaction of these two cell types.

Dentate granule cell hyperactivation is also suggested by other studies. Dyer and co-workers[124] demonstrated that TMT reduced recurrent inhibition in the dentate gyrus, which would increase granule cell activity, perhaps by basket cell inhibition. Sustained perforant pathway (i.e., entorhinal cortex) stimulation of the granule cells caused a loss of zinc content in the mossy fibers,[342] as does TMT.[130] Armstrong et al.[29] demonstrated a consistent increase of spontaneous firing of cells in mouse dentate gyrus exposed to TMT *in vitro*; CA3 cells correspondingly showed a decrease in activity. Zimmer and co-workers[416] showed potentiation of granule cell responses prior to a decline in responsiveness following *in vivo* TMT exposure.

Hyperactivation of granule cells by reduction of recurrent inhibition is supported by several additional studies. TMT's ability to inhibit the uptake of GABA by synaptosomes[115] may transiently enhance inhibition, but will eventually impair recurrent inhibition. TMT-induced alteration in the uptake of glutamic acid and GABA,[382] if due to damage to the GABAergic basket cells, will lead to decreased inhibition in the loop involving the dentate granule cells. The loss of this GABA system would remove an important inhibitory control over granule cell traffic to the CA3 region, thus allowing hyperactivation of this pathway and injury to the CA3 pyramidal neurons.

The extreme vulnerability of CA3 pyramidal cells has been established for other agents or conditions as well, including excess stimulation (kainic acid),[259,343,414] aging,[230] epilepsy,[361] ischemia,[208,319] carbon monoxide hypoxia,[208] antimetabolites (3-acetylpyridine),[249] and excessive stimulation of CA3 pyramidal neurons.[81,259,343,344] While the neuronal pathway involved in the hyperexcitation damage appears clear, the mechanism of the damage remains unclear.

3. Behavioral Studies

Walsh and DeHaven[389] have recently reviewed the behavioral aspects of trialkyltin toxicity. Therefore, only a brief synopsis will be presented in this review. All behavioral studies reported here used the rat as a model unless otherwise indicated and all conclusions are based on the neurotoxic insult of TMT to the limbic system.

Ruppert et al.[321] found that despite morphological damage to the CA1-CA3c regions at all doses, TMT only induced hyperactivity at the highest dose (7 mg/kg), as determined in a figure-eight maze method and in the open field paradigm.[364] Rats also had altered spatial pattern activity. In contrast, mice showed a reduction in spontaneous motor activity after a single dose of TMT (3 mg/kg).[403] Wenger et al.[404] confirmed this observation in two strains of mice (C57BL/6N and BALB/c) demonstrating an early decrease in spontaneous motor activity in doses above 1 mg/kg; increased motor activity was observed within 1 week post-treatment in both mouse strains. Severe problem-solving deficits were discovered by Swartzwelder et al.[365] utilizing the closed-field Hebb-Williams maze. Rats treated with TMT chloride (7 mg/kg, p.o.) exhibited a marked pattern of perseveration. Walsh et al.[393] studied TMT-treated rats (6 mg/kg) subsequent to training in an automated eight arm radial maze to explore retention and performance of maze tasks. A marked and persistent impairment of radial-arm maze performance was observed throughout the period of testing (up to 70 d post-treatment). A TMT-induced hyperactivity was also documented. Evans et al.[138] studied activity following administration of TET (3mg/kg) and TMT (7 mg/kg) to Fisher rats and found that treated animals were much more active during nocturnal dark cycles. Rats on TMT and TET remained sensitive to environmental lighting. TMT disrupted diurnal rhythm so as to produce significant hyperactivity in the light period on day 3.

Increased nocturnal rearing activity was observed in TMT rats, but no differences were found in total 24 h activity. These results differ from those of Wenger et al.[404] who concluded that diurnal patterns were not different during TMT intoxication, perhaps due to a lack of temporal resolution in their monitoring method. Chronic neonatal exposure of rats to TMT also altered radial-arm maze performance in older animals.[246] Forty days after a single dose of TMT (7 mg/ kg, p.o.), an increase in the rate of lever responses was observed throughout an ascending fixed ratio series regardless of the food reinforcement schedule.[364] In addition, rats were unable to perform the short-term working memory components of the task paradigm. Permanent impairment of maze performance was suggested by this study as a result of a single dose of TMT. Rats exposed to a single dose of TMT (7.5 and 10 mg/kg) exhibited a decrease in response to a differential reinforcement schedule that provided a food reward for correctly spaced responses.[405] Impairment of retention of passive avoidance conditioning has been described in rats treated with a single dose of TMT at 5, 6, or 7 mg/kg;[391] all three doses were equally effective in disrupting retention performance. Retention deficits were not due to sensory deficiencies since footshock sensitivity did not change. TMT-induced changes in passive avoidance behavior may be linked to cholinergic deficits due to pyramidal cell loss in the hippocampus.[226]

That TMT disrupts learning and memory is consistent with the observed limbic system damage.[166,391] For example, spatial memory is impaired following kainic acid-induced damage to hippocampal CA3 pyramidal cells.[166] Noland and co-workers[268] have demonstrated that prenatal and neonatal exposure to TMT induces learning deficiencies in weaned rat pups. Flavor aversions have been produced by TMT in LE rats.[210,232,239,301] Nation and Bourgeois[263] have shown that TMT (7 mg/kg) alters performance in appetitive acquisition and extinction trials in adult male rats. Johnson et al.[192] showed that TMT-treated rats persistently made fewer reinforced spontaneous alternations in a T-maze than controls for more than one month after exposure. The TMT-treated rats consistently ran faster through the T-maze on the second trial of reinforced spontaneous alternations than did controls. The rate of learned alternations did not differ between control and TMT-treated rats. The simple T-maze is apparently insensitive to the spatial learning deficits associated with TMT neurotoxicity.[192,393] TMT has been shown to produce a decrease in the percent of reinforced responses and a decrease in the rate of reinforcement in TMT-treated rats maintained under a differential-reinforcement-of-low-rate 15 s schedule of water presentation[240] and in mice responding to a multiple fixed-ratio 30, fixed-interval 600 s schedule of reinforcement.[403] Sparber et al.[354] demonstrated that TMT-induced learning deficits were reversed by a vasopressin analog, desglycinamide-8-arginine vasopressin. The mechanism of action of this peptide to reverse TMT induced toxicity is not known, but the low levels of endogenous vasopressin in the CSF of patients with Karsakoff's psychosis suggests that it is necessary for normal limbic system function.[237]

4. Effects of TET on the Limbic System

Few studies have been published on the effects of TET on the limbic system. TET, unlike TMT, does not appear to target the hippocampus in a selective manner in adult rodents. During development, however, the limbic system shows greater vulnerability to TET. Veronesi et al.[386] have demonstrated a derangement of hippocampal afferent circuitry in rats exposed to a single injection of TET (6 mg/kg) on postnatal day 5. Histopathological examination of the CNS at 20 d of age revealed severe damage to the neuronal populations of the entorhinal cortex and transitional cortex; disruption of synaptogenesis in the afferent layer of the dentate gyrus was also remarkable. The damage was consistent with structures undergoing rapid development at postnatal day 5. Timm's stain for heavy metal deposition showed decreased staining in the CA2, CA3, and CA4 regions of the hippocampus and an absence of stained laminae in the outer molecular layer of the dentate gyrus. The staining pattern seen in the TET treated pups was similar to that seen in neonatal surgical deentorhinalization.[415] The neuropathology of this single dose of TET indicates an absence of perforant synapses on the zones normally occupied by terminals from the medial and lateral perforant paths of the dentate gyrus. Severe neuronal damage in the posterior cortex and an interruption of hippocampal afferent pathways would be expected to have a profound effect on behavior. However, in an electrophysiological study on the effects of TET exposure (single 3 to 9 mg/kg dose on postnatal day 5) in neonatal rats, Dyer and co-workers[127] found no direct functional evidence of hippocampal damage at 60 d of age. Therefore, the extent of residual deficits in limbic system function as a result of neonatal exposure to TET remains uncertain.

H. SENSORY SYSTEMS

Flavor aversions in adult LE rats have been reported following TMT and TET exposure.[210,232] TET induces changes in the consumption of distinctively flavored solutions. It is not known whether this is due to a change in gustatory sensitivity or results from hippocampal damage. Riley and co-workers[239,301] also demonstrated a mild disruption in the acquisition of long-delay conditioned taste aversion learning induced by TMT in male LE rats without apparent changes in sensory responsiveness. The modification of taste aversion learning is likely to involve TMT-induced deficits in working memory.

Pain reception (i.e., nociception) in mice was adversely affected by TMT in a dose-dependent manner in the absence of other overt signs of toxicity.[94] The antinociceptive effects of TMT were investigated pharmacologically to determine whether a specific neurotransmitter system was involved. The preliminary data suggest that the GABAergic system is primarily involved in the antinociceptive response to TMT with the muscarinic side of the cholinergic system providing a secondary influence.

According to Tilson and Burne[371] chronic exposure of rats to TET induces neuromotor deficits, but does not affect reactivity to pain. To the contrary, rats exposed acutely to TMT and TET exhibit dose- and time-dependent antinociceptive effects. Hotplate latencies were generally increased by TET (2.7 mg/kg, s.c.) and TMT (5.25 mg/kg) as monitored over a 21 d post-treatment period.[392] This preliminary study did not delve into the mechanistic differences between the two agents, nor did the report address the influence of peripheral muscular deficits on the TET responses. Harry and Tilson (1981) showed that air puff responses were significantly decreased 55 d after a single neonatal injection of TET (3 mg/kg) administered to rats. The latency in response was presumed to be due to delayed neural development of the response in the treated rats.[170]

Chang and Dyer[77] have studied the histopathology of sensory neurons following TMT treatment (6 mg/kg, p.o.) of LE rats. Changes in the retina by day 3 post-treatment involved swelling of the optic fiber layer, and scattered pyknotic neurons in various layers including the ganglion cell layer. Loss of many ganglion cells and thinning of other neuronal layers was verified by day 30 post-treatment (see Visual System section below). Early changes in the inner ear consisted of swelling and vacuolation of the hair cells which preceded extensive destruction of the hair cells and spiral ganglion cells by 30 d post-treatment. This destruction of structures within the inner ear undoubtedly results in permanent loss of auditory function (see Section IV. J. below). As has been shown previously,[49,55,289] the olfactory tubercle (olfactory cortex) and pyriform cortex are extremely sensitive to TMT damage, exhibiting cytoplasmic vacuolation and lysosomal accumulation. Extensive neuronal loss and necrosis were observed between 15 and 30 d in the cortices. In spinal cord processes, the dorsal root ganglia showed accumulation of lysosomes, formation of myeloid bodies and cytoplasmic vacuolation and dissolution of the Nissl substance in some neurons as examined by electron microscopy. Although permanent damage was less extensive in the dorsal root ganglia, the neuronal components were clearly vulnerable to TMT and showed a pattern of toxicity characteristic of other central neurons. Swelling, budding conformation (suggesting mitochondrial replication), hypertrophy, hyperplasia, and vacuolation of mitochondria were observed in some neurons especially in the later stages of TMT intoxication.

Selected somatosensory functions have been tested following TMT (7 mg/kg) exposure in LE rats.[177] Hot-plate testing which addresses response to thermal pain revealed increased latencies to hind-paw licking. Conduction velocity was unaltered in the tail. Somatosensory evoked responses derived from implanted cortical electrodes and tail stimulation showed an increase in the latencies of all peaks. The authors speculate that somatosensory deficits are of CNS, not peripheral origin.

I. VISUAL SYSTEM

The sensitivity of the visual system to the toxic effects of metals is discussed in the chapter by Fox elsewhere in this volume. Trialkyltin compounds produced numerous visual system effects that vary with the species of organotin. Exophthalmos was observed in rats as a result of long-term TET feeding experiments.[233] Chang and Dyer[77] reported swelling in the optic fiber layer of the retina, and necrotic changes in the ganglion and inner nuclear layer within 72 h of a single dose of TMT (6 mg/kg, p.o.) administered to LE rats. By 30 d post-treatment, the acute toxicity of TMT had disappeared, but evidence remained of ganglion cell loss and thinning of the other retinal layer, indicating permanent damage.[77]

Toews et al.[373] demonstrated that chronic TMT (4 mg/kg, p.o.per week for 4 weeks) administration to LE rats induced increased protein and glycoprotein synthesis as assayed by radiolabeled precursor incorporation into isolated retinas. This general effect on protein synthetic rates was not apparently associated with neural degeneration. TMT did not alter axonal transport of newly synthesized proteins and glycoproteins in optic nerve. Sparse neuronal necrosis and minimal ultrastructural abnormalities were noted in the retina. These studies suggest that the functional impairment of neuronal endoplasmic reticulum and Golgi apparatus is not the earliest biochemical alteration associated with TMT toxicity as has been proposed by others.[49,56]

Dyer et al.[128] found changes in visual evoked potentials (VEP) in LE rats exposed acutely to TMT 4 to 7 mg/kg). Their results demonstrate a reduction in the amplitude of the retinal input into the visual pathways and a delayed onset of response. The various early and late peaks of the VEP had reduced amplitudes.

Wender and co-workers[401] reported intramyelinic edema in the optic nerve from rats chronically treated with TET (5 mg/l drinking water for 16 d; 10 mg/l during next 14 d). Electron microscopic examination at 10, 28, 60, and 74 d revealed myelin fragmentation and disintegration in some fiber bundles with little damage to axons.

TET induced morphological changes in optic nerves of adult New Zealand white rabbits exposed to 1 mg TET/kg/day for 6 d.[326] Marked intramyelinic vacuolation of the optic nerve closest to the optic chiasm was observed. Water and sodium accumulated to a greater extent in the proximal as compared to distal portions of the optic nerve. This may be due to the mechanical constriction and thicker meningeal investment of the proximal region where movement and stretching take place.[326]

A preliminary report by Gerren et al.[153] indicated an increased amplitude in the first component of the visual flash-evoked potential and an increase in the latencies of the responses in TET-treated mice (8 mg TET/kg, i.p.) within the first 24 h as compared to controls.

In a complementary study, exposure of CD rat pups to TET (3, 6, 9 mg/kg, i.p.) on postnatal day 5 altered VEPs recorded after 60 d of age.[127] TET treatment, measured via implanted electrodes, increased the latency of P2, P3 and N3, of the VEP in a dose dependent fashion. Visual function is permanently impaired following a single injection of TET on postnatal day 5. Gender-specific alterations were detected and revealed that females generally had higher N1P2 amplitudes than males; females also had shorter P1 latencies and longer N3 latencies for VEP.[127] The data suggested that neonatal TET toxicity did not affect retina, optic tract, or the lateral geniculate body, but appears to induce damage at the level of the visual cortex.

Dyer and Howell[126] have studied the acute effects of TET exposure on the VEP and hippocampal afterdischarge in adult male LE rats. Subjects received 0.19 to 1.50 mg TET/kg, i.p. daily for 6 d. As with neonatal exposure, TET administered to adult rats increased the latencies of VEP responses (i.e., P1, N1, P2, N2, and N3) consistent with a generalized slowed conduction velocity in myelinated fibers of the optic nerve. Implanted electrodes provided stimulation and recording sites for the hippocampal afterdischarge testing. TET exposure had a biphasic action on the afterdischarge parameters. Low concentrations of TET enhanced initial spike frequency (i.e., enhanced excitability), while high concentrations of TET exhibited a depressive effect postictally. The prominent effect of TET appeared to be anti-excitatory, raising the afterdischarge threshold at moderate dosages and limiting afterdischarge spiking.[126] Caution was raised by these investigators regarding ambient temperatures during testing since relatively minor housing temperature changes may alter the VEP results in rats exhibiting hypothermia from TET exposure.[125]

Boyes and Dyer[52] studied pattern reversal and flash evoked potentials in LE rats exposed acutely to TET (4.5 or 6.0 mg/kg, i.p.).[52] Pattern reversal (i.e., alternating black and white bars or checkerboards on a video monitor) evokes a powerful response in the visual pathways which has been used clinically to assess neurotoxicant exposures. In rats exposed to TET, significant increases in latency and changes in the amplitudes of response peaks were observed at the higher dosage. Flash evoked potentials were unaltered with both dosages of TET. Testing was performed 24 h post-exposure. Unfortunately, testing was done at only one time point which was prior to gross morphological changes in myelin.[175,306,326,402] The changes in pattern reversal potentials could have been distorted by hypothermia and ambient holding temperatures, thus raising questions about the actual sensitivity of this test for neuronal damage.

J. AUDITORY SYSTEM

TMT has been shown to cause extensive and permanent damage to the hair cells and spinal ganglion cells of the Organ of Corti in the inner ear of adult LE rats within 15 to 30 d after a single dose (6 mg/kg).[77] Ray[289] reported TMT-induced (8 to 10 mg/kg, p.o.) changes in auditory evoked potentials (AEP) in rats between 4 and 8 d after a single dose. The overall peak to peak amplitudes of the AEP were elevated. Enhancement of the secondary component in the AEP was not accompanied by changes in latency.[289] The induction of ototoxicity by TMT in albino rats resulted in a frequency-dependent hearing loss as detected by reflex inhibition audiometry.[132] The same results were obtained in the LE rat indicating that the presence of cochlear melanin is not needed for the toxic action of TMT.[132]

Disruption of the acoustic startle pathway by a single injection of TMT (4 to 6 mg/kg) has also been reported in LE rats.[320] TMT exposure increased the startle response latency, while decreasing the number of responses, response amplitude and sensitization by background noise. The auditory impairment was observed within 2 h of treatment and at dosages below those necessary to produce the full TMT syndrome (see below). Residual impairment still existed 4 weeks post-treatment. The decrease in the number of responses to a 120 dB tone stimulus was most remarkable and suggests lesioning of the primary neuronal pathway subserving the auditory startle.

Young and Fechter[413] examined the progression of auditory toxicity in rats treated with a single injection of TMT (2 to 6 mg/kg, i.p.). TMT interfered with the ability of a low-intensity prestimulus to inhibit the acoustic startle response; high-intensity prestimuli evoked a normal inhibitory response. High prestimulus tone frequencies were especially affected. Auditory acuity was altered by TMT in a frequency-specific, dose-dependent manner. These results are consistent with a cochlear lesion as has been observed histopathologically by Chang and Dyer.[77] Interestingly, slow recovery from the TMT-induced auditory impairment was observed in the low dose subjects reminiscent of recovery from intense noise damage. The authors gave no description of the central symptoms of TMT treatment (e.g., tremors or convulsions) which would be expected at the higher dosage range.[130,344]

Fechter and colleagues[142] have extended their studies of TMT ototoxicity to examine a cochlear locus of injury. The results demonstrate a frequency relationship in hearing impairment and recovery. Mid-range frequency loss slowly recovered, while very high frequencies did not. Loss of only high frequencies is indicative of damage to the outer hair cells in a restricted area of the basal cochlea. Additional sites of reversible injury are likely to include structures associated with the hair cells (e.g., stereocilia), the spiral ganglion cells, and associated support cells.

Amochaev et al.[26] reported a slowing of AEP as a result of TET-induced myelin dysfunction in rats exposed via drinking water (20 mg TET/l) for 2 weeks. The apparent loss of myelin and the responses of the AEP to demyelination were reversible by two weeks of recovery.[26] Reiter and co-workers[292] observed deficits in acoustic startle response as a result of TET administration to rats. Squibb et al.[355] reported a reversible, dose-dependent depression in acoustic and air puff startle responses in rats exposed subchronically to TET (1 to 3 mg/kg for 3 doses of 2 weeks). These observations coincided with histopathological evidence of CNS intramyelinic edema in the large white matter tracts. Harry and Tilson[170] demonstrated a decrease in acoustic startle responsiveness at 60 d of age in rats exposed as neonates to a single dose of TET (3 mg/kg, s.c.) on postnatal day 5.

Fechter and Young[141] also reported on the auditory toxicity of chronic TET exposure in rats and provided a valuable method for discriminating the nonauditory component of toxicity using reflex modulation audiometry (i.e., pure tone prestimulus inhibition of the startle reflex).[141] TET treatment (30 mg/l in drinking water *ad libitum*) for 3 weeks resulted in severe neuromuscular weakness and hindlimb paresis. Prior to the overt neuromuscular involvement, however, treated rats had a marked decrease in acoustic startle reflex amplitudes. The depression in startle behavior was detected within 3 d of exposure and worsened over the 3 week course of treatment. Pure tone prestimuli produced normal modulation of reflex behavior even though the startle amplitude baseline dropped more than 80%. The motor deficits could therefore be differentiated from the auditory function. The deficits appeared to be completely reversible in recovered rats. It was concluded that TET predominantly disrupts neuromuscular function, but has little effect on hearing per se.

The primary startle circuitry is neuronally distinct, including the auditory nerve, the cochlear nucleus, the nucleus of the lateral lemniscus and the nucleus reticularis pontis caudalis of the brainstem and, finally, descending neurons to the primary motor neurons in the spinal cord.[106] In other words, the circuit likely involves five synapses plus the neuromuscular junction. The results of Fechter and Young[141] suggest that TET primarily affects the auditory startle response at the level of brainstem and below (i.e., the descending spinal motor pathways).[141] The resolution of the method could not detect TET-induced damage above this level.

Chronic TET intoxication in young rats was accompanied by a significant increase in the peak latencies of all brain stem auditory evoked potentials (waves I, II, III, IV) and the interpeak difference (IV–I).[26] Rats were exposed to TET (20 mg/l) in drinking water for 2 weeks prior to testing. Body weights and brain myelin yields were markedly reduced by TET treatment. The authors[26] attributed the increased AEP latencies to a "demyelination" effect of TET. The modification of the AEP could not have resulted from frank demyelination with myelin loss, inasmuch as AEP latencies returned to normal after 2 weeks recovery.

K. CRANIAL NERVES AND BRAIN STEM

Chang et al.[82,85] reported degeneration of large neurons in the brain stem 48 h after mice were given a single injection of TMT (3 mg/kg). The prominent neuronal degeneration in the brain stem, mesencephalic trigeminal nuclei and median raphe nuclei, was characterized by chromatolysis and hyalinization with eventual disintegration and vacuolation of the neurons. Nissl substance (rough endoplasmic reticulum) was lost. Electron microscopic observations revealed distention in the cytoplasmic membranes of the endoplasmic reticulum and Golgi complex with widespread intraneuronal vacuolation. Moderate mitochondrial damage in the form of vacuolar inclusions were also found, suggesting a source of the smaller cytoplasmic vacuoles seen in the neurons. A disruption in protein synthesis in these cells may be responsible for the subsequent death of the cell. In the adult LE rat, a single exposure to TMT (6.0 mg/kg, p.o.) produced chromatolytic changes in large mesencephalic trigeminal neurons of the brain stem, but without detectable cell loss.[78] TMT exposure in marmosets causes chromatolysis in the motor nuclei of the brainstem.[7] Chromatolysis was also observed in reticular neurons.[271] Correlations between these lesions and their possible physiological effects have not been studied.

Treatment of New Zealand white rabbits with TET (1 mg/kg/d) for 6 d failed to produce gross edema or sodium accumulation in cranial nerve V.[326] Large vacuoles were noted, however, in the part of the nerve which consists of central white matter. This study demonstrates how sharply the central vs. peripheral demarcation divides the effect of TET on central myelin.

L. SPINAL CORD

Subchronic TMT and TET induces chromatolysis in spinal cord neurons (e.g., ventral horn neurons) of the rat.[7,271] TMT treatment of rats (10 mg/kg) induced marked vacuolation of Golgi formations in dorsal root ganglion cells and increased the presence of autophagosomes.[56]

A single dose of TMT (3 mg/kg, i.p.) administered to C57BL/6N mice induced neuronal degeneration in the spinal cord (between C4-S1; L1-L4) within 48 h.[86] Light and electron microscopy revealed the familiar picture of neuronal changes seen in other parts of the CNS. Chromatolysis, cytoplasmic degeneration, and vacuolization in both medial and lateral motor nuclei of the ventral horns was observed. Chromatolysis commenced around the nucleus, then spread to the entire neuron. Many neurons were observed with extensive dilation of the Golgi complex, swelling of mitochondria, increased numbers of lysosomes, and small cytoplasmic vacuoles within the rough endoplasmic reticulum. Large intraneuronal vacuoles appeared to form as a consequence of the extensive intraneuronal edema. Progressive distention of the cytoplasmic membrane systems appeared to induce prominent neuronolysis. These authors suggest what is a recurring theory offered to explain the toxicity of organotins; that mitochondrial damage and/or a failure of a plasma membrane ATP dependent pump may cause a disturbance in fluid balance and result in the widespread intraneuronal edema.[86]

Richman and Bierkamper[300] described severe spinal cord edema and compression in rats exposed to TET (30 mg/l in drinking water) for 2 to 3 weeks and during early recovery periods. Extensive vacuolation of the ventral and dorsal roots was observed. This reversible condition altered nerve conduction velocity at the level of the spinal cord reflexes.[2] This work[300] suggested that spinal signal traffic through the lower motor systems might be interrupted simply by a compression neuropathy resulting in a disuse type paresis.

Explants from fetal mouse spinal cord have been cultured in the presence of TET *in vitro*.[159] Addition of TET (1 μM) was highly toxic to the cells, causing rapid and extensive necrosis and disintegration of cellular elements. Typical intramyelinic vacuolation was observed at lower concentrations of TET (10^{-7} to $10^{-9}M$), but was always accompanied by other significant changes in neurons and glial cells. Aside from a generally more profound insult by TET acting *in vitro* rather than *in vivo*, mitochondrial damage was most remarkable in the cultured cells. Structural abnormalities such as swelling, disruption of cristae, increased electron density of the matrix and elongation were most notable in both glial and neuronal cells at various stages of degeneration after 24 to 72 h of exposure. Increased phagocytic activity was also observed. Although these results are difficult to interpret with regard to the effect of TET on the spinal cord *in vivo*, they support a primary direct action of TET on membranes since the myelinotoxic effects were always associated with parallel changes in neurons and astrocytes.[159]

M. PERIPHERAL NERVES
1. Nerve Conduction Velocity

TMT does not apparently affect peripheral nerve conduction velocity. However, controversy surrounds the effects of TET in this regard. Legrain and MacKenzie[212] detected no modification in nerve conduction velocity in rats treated chronically with TET. Likewise, Aizenman et al.[2] found no change in sciatic nerve conduction velocity after 3 weeks TET exposure (30 mg/l drinking water, *ad libitum*) in rats.[2] Sciatic nerve excitation thresholds and conduction velocities were elevated in adult mice treated with TET for 9 d. Apparently, the stimulating electrodes initiated activity at the level of the spinal cord.[153]

Graham et al.[156] measured a 33% decrease in nerve conduction velocity of sciatic nerve in adult rats exposed to TET sulfate (20 mg/l of drinking water *ad libitum*) for 20 d (67.9 m/s for control vs. 45.4 m/s for TET). The dosage regimen approximated a 20 d LD_{50}. Demyelination was not reported to be a factor in the slowed conduction velocity. TET treated animals exhibited weight loss, hindlimb wasting and paraplegia. These investigators reported myelin splitting at the intraperiod line in peripheral Schwann cells but failed to appraise the extent of myelin vacuolation. Myelin splitting was most prominent at the sciatic notch. Widening of the nodes of Ranvier and changes in the paranodal regions were noted. Paranodal demyelination changes were also noted by Aizenman and Bierkamper (unpublished observations) in sciatic nerves from rats exposed to TET (30 mg/l of drinking water) for 3 weeks. Except for an occasional fiber undergoing Wallerian degeneration, the neurons appeared normal. The TET-induced lesion was reversible upon removal of TET. Thus, the change in nerve conduction velocity in the absence of segmental demyelination may be attributed to a decreased resistance and increased capacitance of the internodal and paranodal myelin. In other words, TET intoxication may cause a "functional" rather than "structural" demyelination.[156] In summary, TET may increase,[153] decrease,[156] or not change[2] nerve conduction velocity in rodent sciatic nerve during prolonged exposure. These differences appear to be methodologically derived.

2. Peripheral Neurons

An early report by Stoner et al.[358] suggested that TET had an effect on peripheral motor nerves, affecting neuromuscular transmission. On the other hand, TMT does not cause significant neurotoxicity to motor neurons in the periphery.

Graham and co-workers[155-157] published several histological studies on peripheral nerves and muscle during subchronic TET exposure in rats. These studies distinguished the peripheral neuropathy of TET from other more widely studied agents (e.g., acrylamide, diisopropylfluorophosphate, and 2,5-hexanedione) which induce degenerative distal axonopathies. These authors exposed rats to 20 mg TET/l of drinking water for 2 to 3 weeks and subsequently examined segments of sciatic nerve by light and electron microscopy. Although this work is widely misquoted as having documented significant peripheral nerve damage and myelin splitting, the results actually demonstrate that less than 4% of sciatic nerve fibers showed intramyelinic vacuole formation. In general, the Schwann cells, capillaries, axons, extracellular space, and tissue appeared normal.[157]

Graham et al.[155] found no significant changes in the intramuscular neurons in soleus and extensor digitorum longus (EDL) muscles following up to 23 d of exposure to TET (20 mg/l in drinking water). While fibers undergoing Wallerian degeneration were observed, sprouting and regeneration were not. Nerve terminals were normal in soleus and EDL muscles as determined by electron microscopic examination. Increased numbers of neurofilaments and neurotubules in sciatic nerve have been observed in TET-intoxicated rats exposed via drinking water for 20 d with no change in the ratio. Minor histopathological damage has been reported in peripheral nerves as a result of subchronic, high dose TET exposure in rats. Axonal degeneration, however, was not thought to be significant.[229,271] Protein composition and content did not appear to differ in sciatic nerves from control and TET-treated mice.[153]

Jacobs et al.[186] found no evidence of ultrastructural damage to peripheral nerves and myelin of rats acutely intoxicated with TET, although intramyelinic vacuole formation and astrocytic swellings were observed in the CNS. Moreover, neither intraneural injection nor direct application of TET to the sciatic nerve *in vivo* caused any ultrastructural changes in tissues examined 24 h later.

In a histopathological study of peripheral nerve, Richman and Bierkamper[300] found minimal degenerative changes in the axons of the sciatic nerve following 3 weeks of TET (30 mg/l in drinking water) administration to rats despite chromatolytic reactions in the cell bodies in the spinal cord. Examination of the soleus muscle revealed signs of denervation and reinnervation based on fiber type changes during a recovery period. Nonetheless, denervation changes were judged to be minimal and could not explain the severe paresis and pronounced muscular weakness after 3 weeks of TET exposure. Aizenman and Bierkamper (unpublished results) observed paranodal retraction of myelin in a significant number of fibers in the sciatic nerve in rats undergoing the same TET treatment schedule.

TET (1 to 100 μM) *in vitro* had no effect on the compound action potential of the sciatic nerve isolated from the toad *Bufo marinus*.[24] However, Allen and co-workers[24] demonstrated that TET (10 μM) *in vitro* depressed evoked transmitter release in the mouse sternomastoid nerve-muscle preparation as measured by intracellular recording. The presence of spontaneous miniature endplate potentials from the preparation indicated that the neuronal release mechanism and postjunctional ACh receptors were intact. Further, the study demonstrated that the action potential mechanisms were unaffected in another species. The failure of neuromuscular transmission was postulated to be associated with an inhibition of, or a decrease in stimulated ACh release.

Tan and Ng[369] studied the *in vitro* effects of TET (100 μM) on phrenic nerve axons and brain tissue. TET inhibited [^{3}H]-colchicine binding to crude as well as purified tubulin preparations derived from brain tissue. TET in high concentration (>100 μM) blocked normal assembly of microtubules from tubulin in phrenic nerve axons as observed by electronmicroscopy. TET did not interfere with preassembled microtubules.[370] It is important to note that these concentrations of alkyltin are not likely to be attained *in vivo* during non-lethal exposure. In a follow-up study, Tan et al.[370] showed that TET, TBT, TPT, and tribenzyltin interfered with tubulin polymerization and [^{3}H]-colchicine binding to purified tubulin from rat brain in a dose-dependent manner. It is clear from this work that trialkyltins can modify tubulin structure, perhaps through an interaction with sulfhydryl groups.

Exposure of LE male rats to TET bromide (30 mg/l in drinking water *ad libitum*) caused reversible alterations in the release of ACh from the isolated vascularly perfused phrenic nerve hemidiaphragm preparation.[43] Bierkamper and Valdes demonstrated a progressive decrease in stimulated (20 Hz) ACh release over 3 weeks of exposure. The most curious effect observed was a post-stimulus rise in spontaneous (unstimulated) release of ACh. These authors attributed the enhanced unstimulated release to possible membrane depolarization and/or calcium misman-agement by the cholinergic nerve terminals. The deficits in stimulated release had a deleterious effect on force of muscular contraction and were attributed to interference with mitochondrial function (i.e., the source of acetyl-CoA for ACh synthesis) and/or a direct membrane effect of TET such that the release mechanism for ACh was disrupted.[43]

3. Peripheral Myelin

Many studies have demonstrated a profound differential effect of TET on central vs. peripheral myelin. Central myelin from oligodendrocytes is highly sensitive to TET and readily exhibits intramyelinic edema (i.e., myelin splitting and vacuolation). By comparison, peripheral myelin is resistant to the effects of TET although slight changes may be noted with longer term, high-dose treatment.

The presence of TET and TMT at high concentrations (>100 μM) *in vitro* caused no myelin damage within the phrenic nerve in isolated rat hemidiaphragm preparations. Intramyelinic edema was not observed in sciatic nerves of rabbits exposed subchronically to TET (1 mg/kg/ d i.p. for 6 d), even though extensive water content changes and vacuolation were noted in the optic nerve.[326] Myelin of the peripheral nerve, unlike central myelin, is derived from the Schwann cell. Numerous differences (e.g., immunological staining, lipid composition, and enzymatic profiles) are apparent between oligodendrocytic and Schwann cell myelin. The relative selectivity of TET for central nervous system myelin suggests that it may provide a useful tool to characterize fundamental differences in the structure and function of myelins.

Peripheral myelin splitting at the intraperiod lines in larger fibers of sciatic nerve has been reported near the sciatic notch in severely intoxicated rats exposed to TET (20 mg/l in drinking water) for 20 d.[156,157] In older rats (12 months), a significant percentage of fibers in the ventral (<15%) and dorsal (<10%) roots showed intramyelinic vacuolation, but less than 3% of sciatic nerve fibers were involved. Younger rats (3 months) exhibited approximately half as much involvement in these three areas, suggesting that younger animals possess greater repair potential. In support of this idea, Blaker et al.[46] reported that chronic TET administration inhibited CNS myelination, but did not alter sciatic nerve myelination in developing rats. No gender differences were found.[157] Motor fiber involvement was consistently greater than sensory fiber involvement in the spinal roots. Distortion of myelin at the paranodal regions and widening of the nodes of Ranvier was also observed.[156] Assays of myelin sulfatide, unesterified cholesterol, protein and collagen showed no differences in sciatic nerve samples between control and TET treatment.[156]

Kirschner and Sapirstein failed to find any X-ray diffraction evidence that TET induced myelin swelling or splitting in rat sciatic nerve, whether from *in vitro* or *in vivo* administration of the alkyltin.[203] These data differ from Graham and Gonatas[157] who reported that up to 4% of fibers (too low for X-ray diffraction resolution) exhibited damage. However, direct application of TET to nerve had no effect on myelin after 24 h.[186] It is reasonable to conclude that myelin of Schwann cell origin is remarkably resistant to the effects of TET when compared to CNS myelin.

N. DEVELOPMENTAL NEUROPATHOLOGY

An overview of the developmental neuropathology of trialkyltin compounds was published by Reuhl and Cranmer in 1984.[296] Exposure of neonatal mammals to TMT or TET induces many of the same symptoms and lesions as adult exposure. Differences are apparent, however, and will be discussed in terms of the developing nervous system.

Paule et al.[278] concluded that prenatal TMT exposure results in postnatal toxicity in pups only in the presence of maternal toxicity. Histopathological changes were noted in hippocampal regions CA3 and CA4 of the offspring. Exposure of the neonatal mouse to TMT produces a different regional pattern of injury than in acutely treated adult mice.[298] Nonetheless, extensive degeneration was observed in the hippocampus and pyriform cortex, suggesting enhanced vulnerability to TMT induced damage during development.[298]

Evidence has been presented by Harry and Tilson[170] that neonatal TET exposure in rats retards normal development as measured by spontaneous motor activity, acoustic startle responses, two-way shuttle box experiments and other neurobehavioral tests. Reiter et al.[293] reported deficiencies in performance in a figure-eight maze and homing orientation test before weaning in TET-exposed pups. When adults, neonatally exposed rats were consistently more active in the maze. Thus, a single neonatal (postnatal day 5) exposure to TET (9 mg/kg, i.e., LD_{50}) in the developing rat appears to produce permanent behavioral alterations.[293]

Blaker et al.[46] studied myelinogenesis in developing rats exposed to TET, concluding that this trialkyltin compound inhibits forebrain growth and CNS myelination, but not sciatic nerve myelination. O'Callaghan and Miller[269] reported decrements in neurotypic and gliotypic proteins as a result of neonatal rat exposure to TET (3 or 6 mg/kg, i.p.). TET also induced a permanent decrease in synapsin I in hippocampus.[269] TET caused permanent deficits in neuronal as well as glial development, especially in the hippocampus. In addition, myelin specific basic protein content was decreased in hippocampus, indicating interference with myelinogenesis.[270] Disruption of myelinogenesis has also been confirmed by other studies.[293,363,386,399]

Rocco and co-workers[308] suggest that TET causes a selective defect in the ability of the developing brain to synthesize membrane lipids. The conversion of dihydroxyacetone phosphate into phospholipids may be the central point of the defect which is postulated to interfere with myelinogenesis.[307]

O. MISCELLANEOUS ORGANOTIN NEUROTOXICITY

Walsh et al.[388] studied the behavioral toxicity of a TBT ester in the rat as part of the United States Navy's interest in antifouling coatings for the hulls of large vessels. Tributyltin[methacrylic-CO-methylmethacrylate] ester administration through drinking water resulted in significant behavioral changes and weight loss after 5 months of exposure. The reversible changes observed utilizing a Sidman avoidance paradigm were increased shock rate, decreased responding, and a shift in the interresponse time distribution. Histopathological changes were not found in brain tissue when examined at the light microscopy level.

Lehotzky and colleagues[213] studied the prenatal neurotoxic effects of the fungicide triphenyltin acetate by exposing pregnant rats on day 7 to 15 of gestation. A high mortality rate was observed in nursing pups. Spontaneous locomotor activity transiently increased around 23 to 36 d and returned to normal levels by 90 d of age. Although the avoidance learning was more rapid in organotin exposed pups, the preservation of the learned motor responses was decreased. In addition to the subtle behavioral changes noted in this investigation, this study provides further evidence that TPT crosses the placenta and the blood brain barrier during gestation.

Exposure of male pups to monomethyltin[268] via the dam's drinking water throughout gestation and postpartum to 21 d resulted in alterations in learning behavior at a dose of $^1/10$ the LD_{50}. In addition, dimethyltin[268] exposure pre- and postnatally resulted in learning deficiencies in young male rats.

Interference with sleep patterns has been established for trialkyltins. TET acetate (5 mg/kg) treatment to adult rats nearly doubled the sleep time induced by hexobarbital and phenobarbital as compared to control rats.[251] The enhanced "sedation" by TET may have resulted from a significant reduction in brain 5-HT concentrations in these rats.[251]

P. THERAPY

Cerebrolysine (Ebeve Co., Austria) has been shown to protect neonatal rats against TET-induced retardation of brain maturation.[400] The mechanism of action of this brain "hydrolysate" is unknown and its therapeutic application in humans is uncertain.

Theophylline is weaker than silymarin (Legalon®) in protecting against TET-induced edema in rat; mannitol is ineffective and promotes cerebral deterioration.[423] Varkonyi et al.[384] reported that while theophylline prevented myelin changes in TET-induced brain edema, it caused severe damage to cerebral capillaries. In control rats, theophylline induced glial edema. Treatment of TET-induced edema with theophylline is therefore counter productive. Therapy with escin (Reparil®) has been reported to prevent the development of severe brain edema as a result of TET exposure in rats,[385] and has a protective effect against TET-induced brain edema in the dog as well.[418-422]

TET-induced cerebral edema in rodents may be reduced by dexamethasone therapy.[360] Concurrent administration of dexamethasone with TET bromide (1.25 mg/kg/d for 7 d) reduces mortality and lessens the severity of brain edema. Oddly, dexamethasone significantly decreased liver, brain, and blood levels of TET during exposure. It is conceivable that dexamethasone increased the excretion or metabolism of TET, but no data were presented. With respect to central edema, the authors speculated that the steroid could have a direct action on brain water-ion transport system countering the action of TET.[360]

Other compounds such as vincamine, *Gingko biloba* extract, dihydroergotoxine, and levoeburnamonine have been reported to counteract the edematous effects and symptoms of TET toxicity.[39] *Ginkgo biloba* extract accelerates the recovery from TET-induced CNS edema in rats.[231] Its mechanism is thought to be due to its modulating influence on cellular cyclic AMP levels by activation of membrane bound phosphodiesterase.[231]

Zimmer et al.[417] have shown that phenobarbital does not protect against TMT-induced neuropathology in the limbic system as has been shown for kainic acid-induced seizure-related damage.[360] Administration of pyruvic or citric acid to rats did not alleviate the symptoms of TET toxicity.[136] One may conclude from the paucity of studies on therapy that the therapeutic management of trialkyltin intoxication has yet to be adequately investigated.

Q. SUMMARY OF TMT SYNDROME

In 1982, Dyer et al.[130] described the TMT syndrome in LE hooded rats from a series of experiments employing single dose exposures close to the LD_{50}.[129] The LD_{50} of TMT was shown to be inversely weight-dependent. In large rats (450 g) the LD_{50} was 7 mg/kg and in small rats (250 g) it was 10 mg/kg body weight. TMT also caused generalized discomfort and adhesions of abdominal viscera, particularly in the liver, stomach, spleen, and small intestine when the alkyltin is injected intraperitoneally. Exposure to TMT causes a decrease in body weight[55] and a mild decrease in core temperature. TMT intoxication also produces an altered response to anesthesia.[297]

The behavioral syndrome resulting from TMT intoxication includes tail chasing and mutilation, anorexia, running fits, hyperactivity, vocalization, seizures (i.e., forelimb clonus and rearing; sometimes tonic-clonic episodes) within the 1st week post-treatment. Subjects react violently to handling and exhibit aggressiveness and hyperexcitability.[55] A dose-dependent fine tremor at rest is a classic sign of TMT intoxication and usually appears within 2 to 3 d after significant exposure.[55] The neurobehavioral toxicity has been reviewed elsewhere.[243]

The hippocampus is decidedly the most vulnerable CNS target for TMT. Dyer et al.[87] in a comparative species/strain study of TMT toxicity reported a different neuropathic profile for rats and mice. For mice, the order of neuronal vulnerability to TMT was observed to be: granule cells of fascia dentata > spinal cord motorneurons > brain stem neurons > olfactory cortical neurons (pyriform/entorhinal cortices and olfactory tubercle) > Ammon's horn pyramidal neurons. For rats, the order of vulnerability appears to be: olfactory cortical neurons > Ammon's horn pyramidal neurons (especially CA3c) ≥ granule cells of fascia dentata > brain stem neurons. Rats did not exhibit spinal cord motorneuron damage. The mechanism(s) of action of TMT has not been completely determined. TMT was considered by Cremer[104] to be a specific excitotoxin acting in a manner that shares some homology with kainic acid. On the other hand, Brown et al.[56] suggest that the Golgi complex may be the site of critical metabolic lesions initiated by TMT whereby disturbances in protein synthesis and transfer result in neuronal cell damage and death.[56] Costa and Sulaiman[95] have found that TMT inhibits protein synthesis in mouse brain shortly after injection (3 mg/kg) coincident with a drop in body temperature. *In vitro* confirmation of decreased protein synthesis has led to the suggestion that this is the principal mechanism of TMT neurotoxicity.[95]

The human neurotoxicity of TMT closely parallels information gathered from animal experimentation. The case involving six workers in 1981[40] who inhaled TMT while cleaning a tank used in the manufacture of dimethyltin provides detailed information on the human TMT syndrome. Urinary organotin levels paralleled the development and severity of symptoms. Clinical signs were consistent with damage to the limbic system and cerebellum. EEG recordings showed bursts of delta activity in both temporal regions of the brain. The neuropathology assessed by light and electron microscopy on tissue from the one fatality showed lesions similar to that found in experimental animals. Cytoplasmic lamellated inclusions (i.e., zebra bodies) were observed in neurons of the amygdala. Seizures, hearing loss, disorientation, confabulation, memory disturbances, aggressiveness, and sensory disturbances corroborate the findings of animal studies. Lung, kidney, and liver damage, along with metabolic acidosis and pulmonary edema were present in the patient that succumbed.[40]

R. SUMMARY OF TET SYNDROME

TET intoxication is characterized by two dominant features, progressive muscular weakness and CNS edema. The pathobiology of acute TET intoxication was reviewed by Torack et al.[375] in 1970 in an attempt to define mechanisms. Recent reviews on the toxicity of TET have been provided by Aldridge,[6] Snoeij et al.,[348] and Walsh and DeHaven.[389] TET exposure produces a decrease in body weight,[112,242,355] a sustained decrease in body temperature,[217,366] chromodacryorrhea,[355] and decreases in grip strength.[355] Food and water intake decrease in a dose-dependent and temporal manner. In the latter stages of chronic exposure, paresis of the hindlimbs and general muscular weakness restrict feeding. There is a progressive decrease in locomotor activity in mice[153] and rats.[153,212] The gross effects of TET intoxication appear to be completely reversible.[355]

Reviews of behavioral studies[243,292,294,311] indicate that TET exposure alters schedules of reinforcement and responses to pharmacologic challenges.[111,169] Schedule-induced and schedule-dependent behaviors are sensitive to chronic TET administration in LE rats.[111] TET administration caused decreased lever pressing, drinking, and running behavior for more than 1 week post-treatment.[242] Neonatal TET exposure has been shown to induce elevated emergence latencies prior to the first avoidance test in adults in a two-way shuttle box design.[170] Radial arm maze performance is altered by chronic neonatal TET exposure.[246] Increases in spontaneous motor activity have also been reported following TET exposure of neonatal rats.[170,246] Motor activity increased between 28 to 60 d of age, then decreased below control by 90 d of age as a consequence of a single injection of TET on postnatal day 5.[170] Flavor aversions are rapidly produced by TET in rats before hindlimb weakness occurs.[210] Deficits in startle responsiveness have been described,[355] and acute TET administration decreases motor activity within hours.[292] TET exposure in rats produces decreased motor activity, grip strength and operant response rate and startle amplitude in parallel to CNS edema and myelin vacuolation.[294] Chronic TET administration (3 weeks) to rats resulted in performance decrements in maze activity, acoustic startle responses, open field behavior, and landing foot spread.[292]

Some authors suggest that TET intoxication could represent a model of brain aging based, in part, on neurotransmitter studies.[39] Gerren and co-workers[153] have argued that TET toxicity may be a model for degenerative disorders. Despite a mixture of neuropathological data supporting such an association it is likely that TET intoxication is in fact a poor or misleading model of CNS degenerative disorders.

V. MECHANISMS OF NEUROTOXICITY

In considering the possible mechanism(s) of organotin-induced neurotoxicity, one must ask why nervous tissue is so sensitive to the effects of these compounds. Nervous tissue is excitable; neurons, their effector cells, and to some extent auxillary supporting cells utilize the excitability of the cell membrane to transmit signals. The membrane is therefore a very active site wherein action potentials and graded potentials occur through carefully maintained transmural ion gradients. The maintenance of these critical ion gradients and the rapid conduction of signals requires energy. The trialkyltin compounds possess properties which perturb excitable tissues in several ways. First, the trialkyltins have a direct effect on biomembranes and lipid bilayers, promoting anionic flux by a non-specific carrier mechanism. This would tend to destroy the critical ion gradient necessary for proper signal transfer and basic cell excitability. Second, the trialkyltin compounds interfere with various ATPases (e.g., the sodium pump) in the membrane which function to maintain ion gradients against the constant perturbations of normal neuronal activity. Third, the trialkyltin compounds have a profound action on the mitochondrion, uncoupling oxidative phosphorylation and robbing the cell of ATP. Any of these three primary actions of the organotin compounds would have a deleterious effect on nervous system function.

It is likely that *in vivo* several of these mechanisms may be working together. For the purpose of the discussions below, however, the mechanisms will be considered separately.

A. METAL INTERACTIONS WITH ZINC

Chang and Dyer[79] reported progressive zinc depletion in rat hippocampus as the result of a single injection of TMT. Zinc depletion in the mossy fibers and terminals appeared to occur before significant neuronal damage. Endogenous zinc may play a regulatory, perhaps inhibitory role in these fibers. These investigators postulate that an imbalance in zinc and/or calcium in the granule cell mossy fibers may lead to the hyperexcitability presumed to destroy the CA3 pyramidal neurons. The relationship of calcium to zinc, as well as zinc to copper or other multivalent metals in these cells remains unclear. It is not unlikely that divalent tin could antagonize the actions of Ca^{++} and Zn^{++}. Sato et al.[324] studied zinc-binding sites in rat hippocampus. Kinetic analysis suggests cytosolic zinc-binding protein(s) may be responsible for sequestration of zinc in the hippocampus. Gestational exposure of rats to lead decreased the amount of zinc associated with a cytosolic zinc-binding protein at postnatal day 30. The authors provide evidence that other environmental agents or metals may interfere with zinc in the hippocampus.[323] Thus, this is an area worthy of future investigation in terms of the selective action of TMT on hippocampal function (see Chapter 4).

B. EFFECTS ON ENERGY METABOLISM

The uncoupling of oxidative phosphorylation by the organotins, even if modest, is likely to have a profound effect on signal transmission and excitation-response coupling. The prevailing question that remains, despite studies demonstrating effects of organotins on ATP generating and utilizing systems, is whether the concentrations of the trialkyltins achieved *in vivo* can reach the levels that are required to demonstrate these effects *in vitro*.

TET interferes with oxidative phosphorylation in rat brain tissue *in vitro*.[98] Evidence includes lower oxygen consumption in brain slices and an increase in the lactic acid to pyruvic acid ratio, an effect similar to that produced by cyanide. Many studies have documented that TBT and TET inhibit oxidative phosphorylation in mitochondria *in vitro*.[108,338,352,356] TET at a concentration of 1 μM inhibits the uptake of pyruvate, malate, citrate, and beta-hydroxybutyrate into rat liver mitochondria *in vitro*,[338] yet this concentration of TET does not block an increase in resting respiration (i.e., increased oxygen uptake). KCl-containing medium is required for TMT and TBT to inhibit pyruvate uptake. TET enhances the uptake of Cl^- when the hydrolysis of ATP is associated with the generation of sufficient OH^- to supply the TET-mediated anion/OH^- exchanger.[339]

The TET syndrome can be mimicked in rats by 2'-chloro-2,4-dinitro-5',6-di(trifluoromethyl)diphenylamine (CDTD), a potent uncoupler of oxidative phosphorylation in brain mitochondria.[224] CDTD, an acaricide for the control of red spider mites, increases the water content of brain and spinal cord, causes hindlimb weakness, and intramyelinic vacuolation with splitting of myelin at the intraperiod line. [Na^+], but not [K^+], was increased in the edematous CNS. ATP synthesis was depressed in isolated mitochondria. The CNS edema was reversible. It is curious that CDTD, which is structurally unrelated to trialkyltin, would induce such a strikingly similar neuropathy.[224] Lock[223] reported an intramyelinic edema in rats associated with 3:5-diiodo-4'-chlorosalicylanilide (DCSA), another uncoupler of oxidative phosphorylation. Like TET, DCSA causes an inhibition of ATP synthesis and stimulation of pyruvate oxidation in isolated rat brain and liver mitochondria. The similarities of TET, hexachlorophene and DCSA in producing a unique form of intramyelinic edema suggest a selective mitochondrial effect. However, since pentachlorophenol and triethyllead do not produce CNS edema, but have similar action on mitochondrial function, other factors must be considered regarding the exact mechanism for TMT-induced myelinogenic edema production.

Aldridge and co-workers have presented a series of papers over the last 30 years which explore the actions of triorganotin compounds on energy metabolism by the mitochondrion.[4,8,11,13,15,17] In fact, the organotin compounds were employed as biochemical tools to distinguish the coupling of oxidative phosphorylation from the metabolism of substrates. In 1955, Aldridge and Cremer[9] demonstrated that TET sulfate inhibited oxidative phosphorylation in rat brain and rat liver mitochondria. Other studies revealed that trialkyltins inhibited oxidative phosphorylation induced by 2,4-dinitrophenyl stimulation of pyruvate oxidation,[4] stimulated the hydrolysis of ATP,[5,8] inhibited ATP synthesis,[4,8,10] caused mitochondrial swelling,[5,8] and enjoyed specific mitochondrial binding sites.[12,13] The high affinity binding sites were shown to be linked to ATP synthesis.[12,13] In 1976, Aldridge reviewed the effects of various organotin species on oxidative phosphorylation and included a discussion of possible mechanisms.[5] By this time, it was known that the trialkyltin compounds not only uncouple oxidative phosphorylation by interaction with ATP synthase, but discharge the OH^-/Cl^- gradient across the mitochondrial membrane, leading to swelling.[8,11,16] Mitochondrial exposure to TET was known to increase ATP hydrolysis, O_2 uptake, swelling, and uptake of Cl^- into mitochondria and to inhibit ATP production.[5] Thus, many other investigations have used trialkyltin for the purpose of studying the coupling of oxidative phosphorylation and ATP production.[89]

Low concentrations of TET inhibit glucose oxidation by rat brain cortical slices in the presence of Cl^-.[222] Inhibition of pyruvate oxidation requires high doses. It was proposed that TET, by promoting Cl^- entry into mitochondria, would interfere with the malate/aspartate shuttle and would decrease mitochondrial uptake of pyruvate.[222] This would lead to anaerobic conversion of pyruvate to lactate. Indeed, TET increased lactate formation coincident with a fall in pyruvate concentrations and increased glucose uptake into rat brain slices under aerobic conditions. On the other hand, TET depressed both glucose uptake and lactate formation under anaerobic conditions.[35] TET also caused increased lactate production in *Escherichia coli*,[215] indicating an impairment in oxidative phosphorylation.

Cremer[103] showed that TET *in vivo* decreases the rate at which pyruvate formed from glucose is oxidized in rat brain; glycolysis is not inhibited. The brain prefers glucose as a substrate rather than pyruvate.[101] However, TET inhibited the oxidation of glucose by more than 70% in rat brain slices, but did not alter pyruvate oxidation. Lactate levels were greater when glucose rather than pyruvate was used as a substrate; TET increased lactate levels to twice normal.[102]

Mitochondrial pyruvate dehydrogenase (PDH) from rabbit brain is partially inactivated by TET-induced phosphorylation *in vitro*.[266] This would block the movement of acetyl-CoA into the TCA cycle and inhibit aerobic metabolism, creating a deficiency in cellular energy. The synthesis of ACh would also be inhibited since choline acetyltransferase requires acetyl-CoA. Interference with PDH *in vivo* would therefore have a profound effect on the function of excitable tissue in general and cholinergic neurotransmission in particular.

Neumann and Taketa[266] have demonstrated that TET (1 to 50 μM) *in vitro* affects the phosphorylation of specific proteins in subcellular fractions of rat brain. Increased phosphorylation was detected in proteins identified as the alpha subunit of pyruvate dehydrogenase (PDH) and the neural protein synapsin. Decreased phosphorylation was found in a 52,000 mol wt protein thought to be the regulatory subunit of cyclic AMP-dependent protein kinase. Phosphorylation of mitochondrial PDH partially inactivates the PDH complex in rabbit brain. An inhibition of PDH by TET could have deleterious effects on a number of neuronal systems. The brain does not have excess PDH like other organs. Since the brain requires primarily the aerobic oxidation of glucose for its energy needs, a failure of PDH to introduce pyruvate into the tricarboxylic acid cycle in mitochondria would result in an energy deficit. In addition, partial inactivation of PDH would reduce the supply of acetyl-CoA required for ACh synthesis. A TET-induced deficit in this neurotransmitter during motoneuron stimulation has been demonstrated.[43] Furthermore, calcium management in the nerve terminal relies, in part, on a pyruvate oxidation-dependent calcium sequestration mechanism in mitochondria. Interference with this mechanism may alter neurotransmitter release.[41,43] Finally, the ability of TET to affect phosphorylation of synapsin may also lead to alterations in neurotransmission.

The ATP synthases, known as ATPases, of energy-transducing membranes are made up of an integral membrane protein complex, F_0 which mediates the translocation of protons, and a peripheral protein complex, F_1, which in its soluble form catalyzes the net hydrolysis of ATP.[60] The organotin compounds appear to selectively inhibit the ATP synthase enzyme complex. TET is able to inhibit ATPase at a concentration of 1 μM when added to an *in vitro* assay.[250] TMT appears to be 300 to 500 times less potent in inhibiting this enzyme.[56] This capacity of trialkyltins to inhibit ATPases *in vitro* has led many authors to attribute the toxic effects of the trialkyltin compounds to a shortage of ATP for energy dependent reactions via inhibition of the mitochondrial ATP synthase complex. Indeed, a number of trialkyltins are highly specific inhibitors of this enzyme at concentrations close to the higher levels found during severe intoxication *in vivo*. The F_0,F_1-ATPase complex provides a high affinity binding site for trialkyltins.[65,107,194,207,276,353] Dibutyltin inhibits a lipoic acid cofactor in the mitochondrial inner membrane associated with the F_0 component of the ATPase which results in inhibition of oxidative phosphorylation.[66] Purification of the proton conductor, F_0, has been accomplished[87] permitting a more detailed study of the F_0,F_1-ATPase complex. In addition to inhibiting the lipoic co-factor, TET also affects a high affinity site within the ATP synthase complex (i.e., the F_0 ATPase) membrane component.[65] Not all organotins may inhibit the mitochondrial ATPase and oxidative phosphorylation by inhibition of the same binding site.[67] Nonetheless, the end results seem to be the same, that is, a reduction in the production of ATP.

Venturicidin, a toxic antibiotic, and TET both bind to the F_0 subunit of the ATPase complex. Strains of *Saccharomyces cerevisiae* which are resistant to venturicidin also show resistance to TET. It appears that these strains have modified binding sites rather than a major alteration in the membrane enzyme component, permeability, or detoxification mechanisms.[160]

Other ATP utilizing enzymes are also affected by organotins.[74,394] Jacobs et al.[187] reported inhibition of ATPase by TET *in vitro* in brain and liver homogenates. Both mitochondrial and Na$^+$/K$^+$ ATPase were inhibited. However, TET-exposed rats failed to exhibit brain levels of TET which were high enough to inhibit ATPases in brain; although marginal inhibition was observed in liver. Chang et al.[84] demonstrated that TMT caused intracellular edema with severe distention of the cytoplasmic membrane in hippocampal granule and pyramidal neurons. The intraneuronal edema with subsequent vacuolation was postulated to be the result of suppression of ATP-driven pumps in the plasma membrane which would lead to rapid fluid and electrolyte imbalances in the neuron.[84]

Stoner and Threlfall[359] measured ATP, ADP, AMP, NADH, and creatine phosphate in brain tissue from rats exposed to fatal doses of TET (12 mg/kg, i.p.) and found no alteration in these compounds using [^{32}P]-labeling method 20 h post-treatment. Using a very sensitive HPLC technique, Bierkamper and Buxton (unpublished observations) have recently found reduced ATP levels in skeletal muscle samples taken from LE rats exposed to TET (10 mg/l of drinking water) for 3 weeks or more. It seems clear for whatever reason that the energy currency of excitable tissue, ATP, is diminished during chronic TET exposure and that this may contribute to the neuromuscular weakness ascribed to this toxic agent.

VI. CONCLUSIONS

In reviewing the toxicology of the organotin compounds we have attempted to demonstrate that a wide range of mammalian organ systems are affected. The principal mechanisms of toxicity appear to be a set of most intriguing effects of the organotins on the synthesis and utilization of ATP. This fundamental theme is both a useful mechanism of action when attempting to explain the effects of these compounds on the brain, as well as a primer on the usefulness of the organotins in the study of ATP utilizing enzymes. Although some studies have not fit results from *in vitro* work with likely concentrations found *in vivo*, it is not unlikely that these studies suffer from problems with sample preparation and the inevitable difficulty of relating the test tube to the animal.

The range of effects of organotins is particularly wide, encompassing alteration in membrane permeability of cells and organelles, changes in the electrical potential of excitable cell membranes, and alterations in the phosphorylation of unique proteins by effectors coupled to receptor-regulated pathways. If the theme of alteration in ATP synthesis and utilization is at the root of an understanding of the discrete mechanism of action of organotins, then it is the subsystem in which these changes take place that is most useful in describing the neurotoxicology of these compounds. Indeed, both at the level of neuronal signal conduction and biochemical signal-transduction, the organotins exert unique effects.

ACKNOWLEDGMENTS

We wish to thank our colleagues for their support, interest, and patience during the preparation of this manuscript. In particular, we are extremely grateful to Dr. David Westfall for his scientific leadership and his unselfish guidance of our careers. We are grateful to Eileen Randolph and Laura Chappel for their secretarial assistance. This chapter was finished by ILOB following the untimely death of George Bierkamper in December of 1988.

REFERENCES

1. **Adinarayana, M., Singh, U. S., and Dwivedi, T. S.,** A biochromatographic technique for the quantitative estimation of triphenyltin fungicides, *J. Chromatogr.*, 435, 210, 1988.
2. **Aizenman, E., Stanley, E. F., and Bierkamper, G. G.,** The effect of triethyltin intoxication on proximal and distal peripheral motor and sensory nerve conduction velocities, *Soc. Neurosci. Abstr.*, 8, 954, 1982.
3. **Alajouanine, T., Derobert, L., and Thieffry, S.,** Etude clinique d'ensemble de 210 cas d'intoxication par les sels organiques d'etain, *Rev. Neurol.*, 98, 85, 96, 1958.
4. **Aldridge, W. N.,** The biochemistry of organotin compounds. Trialkyltins and oxidative phosphorylation. *Biochem. J.*, 69, 367, 1958.
5. **Aldridge, W. N.,** The influence of organotin compounds on mitochondrial functions, *Adv. Chem. Ser.*, 157, 186, 1976.
6. **Aldridge, W. N.,** The toxicology and biological properties of organotin compounds, in: *Tin as a Vital Nutrient*, Cardarelli, N. F., Ed., CRC Press, Boca Raton, 1986, p. 245.
7. **Aldridge, W. N., Brown, A. W., Brierly, J. B., Verschoyle, R. D., and Street, B. W.,** Brain damage due to trimethyltin compounds, *Lancet*, 8248, 692, 1981.
8. **Aldridge, W. N., Casida, J. E., Fish, R. H., Kimmel, E. C., and Street, B. W.,** Action on mitochondria and toxicity of metabolites of tri-n-butyltin derivatives, *Biochem. Pharmacol.*, 26, 1997, 1977.
9. **Aldridge, W. N. and Cremer, J. E.,** The biochemistry of organo-tin compounds, *Biochem. J.*, 61, 406, 1955.
10. **Aldridge, W. N. and Rose, M. S.,** Mechanism of oxidative phosphorylation: A hypothesis derived from studies of trimethyltin and triethyltin compounds, *FEBS Lett.*, 4, 61, 1969.
11. **Aldridge, W. N. and Street, B. W.,** Oxidative phosphorylation: Biochemical effects and properties of trialkyltins, *Biochem. J.*, 91, 287, 1964.
12. **Aldridge, W. N. and Street, B. W.,** Oxidative phosphorylation: The specific binding of trimethyltin and triethyltin to rat liver mitochondria, *Biochem. J.*, 118, 171, 1970.
13. **Aldridge, W. N. and Street, B. W.,** Oxidative phosphorylation: The relation between specific binding of trimethyltin and triethyltin to mitochondria and their effect on various mitochondrial functions, *Biochem. J*, 124, 221, 1971.
14. **Aldridge, W. N. and Street, B. W.,** Spectrophotometric and fluorimetric determination of tri- and di-organotin and -organolead compounds using dithizone and 3-hydroxyflavone, *Analyst*, 106, 60, 1981.
15. **Aldridge, W. N., Street, B. W., and Noltes, J. G.,** The action of 5-coordinate triorganotin compounds on rat liver mitochondria, *Chem. Biol. Interact*, 34, 223, 1981.
16. **Aldridge, W. N., Street, B. W., and Skilleter, D. N.,** Oxidative phosphorylation: Halide-dependent and halide-independent effects of triorganotin and triorganolead compounds on mitochondrial function, *Biochem. J.*, 168, 353, 1977.

17. **Aldridge, W. N.,Verschoyle, R. D., Thompson, C. A., and Brown, A. W.,** The toxicity and neuropathology of dimethylethyltin and methyldiethyltin in rats, *Neuropathol. Appl. Neurobiol.,* 13, 55, 1987.
18. **Aleu, F. P., Katzman, R., and Terry, R. D.,** Fine structure and electrolyte analyses of cerebral edema induced by alkyl tin intoxication, *J. Neuropathol. Exp. Neurol.,* 22, 403, 1963.
19. **Ali, A. A., Upreti, R. K, and Kidwai, A. M.,** Interaction of di- and tributyltin chloride with human erythrocyte membrane, *Toxicol. Lett.,* 38, 13, 1987.
20. **Ali, S. F., Cranmer, J. M., Goad, P. T., Slikker, W., Jr., Harbison, R. D., and Cranmer, M. F.,** Trimethyltin induced changes of neurotransmitter levels and brain receptor binding in the mouse, *Neurotoxicology,* 4, 29, 1983.
21. **Ali, S. F., Newport, G. D., Slikker, W., Jr., and Bondy, S. C.,** Effect of trimethyltin on ornithine decarboxylase in various regions of the mouse brain, *Toxicol. Lett.,* 36, 67, 1987.
22. **Ali, S. F., Slikker, W., Jr., Newport, G. D., and Goad, P. T.,** Cholinergic and dopaminergic alterations in the mouse central nervous system following acute trimethyltin exposure, *Acta Pharmacol. Toxicol.,* 59, 179, 1986.
23. **Allen, C. N. and Fonnum, F.,** Trimethyltin inhibits the activity of hippocampal neurons recorded in vitro, *Neurotoxicology,* 5, 23, 1984.
24. **Allen, J. E., Gage, P. W., Leaver, D. D., and Leow, A. C. T.,** Triethyltin depresses evoked transmitter releases at the mouse neuromuscular junction, *Chem. Biol. Interact.,* 31, 227, 1980.
25. **Ally, A. I., Vierira, L., and Reuhl, K. R.,** Trimethyltin as a selective adrenal chemosympatholytic agent in vivo: Effect precedes both clinical and histopathological evidence of toxicity, *Toxicology,* 40, 215, 1986.
26. **Amochaev, A., Johnson, R. C., Salamy, A., and Shah, S. N.,** Brain stem auditory evoked potentials and myelin changes in triethyltin-induced edema in young adult rats, *Exp. Neurol.,* 66, 629, 1979.
27. **Antler, M.,** Organotins as additives in lubricants, *Ind. Eng. Chem.,* 6, 753, 1959.
28. **Arakawa, Y., Wada, O., and Yu, T. H.,** Dealkylation and distribution of tin compounds, *Toxicol. Appl. Pharmacol.,* 60, 1, 1981.
29. **Armstrong, D. L., Read, H. L., Cork, A. E., Montemayor, F., and Wayner, M. J.,** Effects of iontophoretic application of trimethyltin on spontaneous neuronal activity in mouse hippocampus slices., *Neurobehav. Toxicol. Teratol.,* 8, 637, 1986.
30. **Armstrong, D. L., Read, H. L., Cork, A. E., Montemayor, F., and Wayner, M. J.,** Effects of trimethyltin on evoked potentials in mouse hippocampal slices, *Neurotoxicol. Teratol.,* 9, 359, 1987.
31. **Babich, H. and Borenfreund, E.,** Structure-activity relationships for diorganotins, chlorinated benzenes, and chlorinated anilines established with bluegill sunfish BF-2 cells, *Fund. Appl. Toxicol.,* 10, 295, 1988.
32. **Bakay, L.,** Morphological and chemical studies in cerebral edema: Triethyltin induced edema, *J. Neurol. Sci.,* 2, 52, 1965.
33. **Barbieri, R. and Musmeci, M. T.,** A ^{119}Sn mossbauer spectroscopic study on the interaction of dimethyltin (IV) derivatives with rat hemoglobin, and of related model systems in aqueous solution, *J. Inorg. Chem.,* 32, 89, 1988.
34. **Barnes, J. M. and Stoner, H. B.,** Toxic properties of some dialkyl and trialkyltin salts, *Br. J. Ind. Med.,* 15, 15, 1958.
35. **Barnes, J. M. and Stoner, H. B.,** The toxicology of tin compounds, *Pharmacol. Rev.,* 11, 211, 1959.
36. **Benedek, G., Szikszay, M., Zoltan, T. O., and Obal, F.,** Metabolic and EEG alterations during the early phase of triethyltin sulfate intoxication, *Acta Physiol. Acad. Sci. Hung.,* 54, 381, 1979.
37. **Benedek, G., Turani, K., Ozswar, A., Jo, F., Ors, T. Z., and Obal, F.,** Body temperature in experimental cerebral oedema induced by triethyltin sulphate, *Acta Physiol. Acad. Sci. Hung.,* 44, 299, 1973.
38. **Benoy, C. J., Hooper, P. A., and Schneider, R.,** The toxicity of tin in canned fruit juices and solid foods, *Food. Cosmet. Toxicol.,* 9, 645, 1971.
39. **Bentue-Ferrer, D., Reymann, J. M., Van den Driessche, J., Allain, H., and Bagot, H.,** Effect of triethyltin chloride on the central aminergic neurotransmitters and their metabolites: relationship with pathophysiology of aging, *Exp. Aging Res.,* 11, 137, 1985.
40. **Besser, R., Kramer, G., Thumler, R., Bohl, J., Gutmann, L., and Hopf, H. C.,** Acute trimethyltin limbic-cerebellar syndrome, *Neurology,* 37, 945, 1987.
41. **Bierkamper, G. G.,** In vitro assessment of neuromuscular toxicity, *Neurobehav. Toxicol. Teratol.,* 4, 597, 1982.
42. **Bierkamper, G. G. and Cenedella, R. J.,** Cerebral cortical cholesterol changes in cobalt-induced epilepsy, *Epilepsia,* 19, 155, 1978.
43. **Bierkamper, G. G. and Valdes, J. J.,** Triethyltin intoxication alters acetylcholine release from rat phrenic nerve-hemidiaphragm, *Neurobehav. Toxicol. Teratol.,* 4, 251, 1982.
44. **Blair, E. H.,** Biodegradation of tricyclohexyltin hydroxide, *Environ. Qual. Saf.,* Suppl.3, 406, 1975.
45. **Blair, W. R., Parks, E. J., Olson, G. J., Brinckman, F. E., M. C. Valeiras-Price, M. C., and Bellama, J. M.,** Characterization of organotin species using microprobe and capillary liquid chromatographic techniques with an epifluorescence microscope as a novel imaging detector, *J. Chromatogr.,* 410, 383, 1987.
46. **Blaker, W. D., Krigman, M. R., Thomas, D. J., Mushak, P., and Morell, P.,** Effect of triethyltin on myelination in the developing rat, *J. Neurochem.,* 36, 44, 1981.

47. **Bland, B. H.,** The physiology and pharmacology of hippocampal theta rhythms, *Prog. Neurobiol.,* 26, 1, 1986.
48. **Borenfreund, E. and Babich, H.,** In vitro cytotoxicity of heavy metals acrylamide and organotin salts to neural cells and fibroblasts, *Cell Biol. Toxicol.,* 3, 63, 1987.
49. **Bouldin, T. W., Gaines, N. D., Bagnell, C. R., and Krigman, M. R.,** Pathogenesis of trimethyltin neuronal toxicity: ultrastructural and cytochemical observations, *Am. J. Pathol.,* 104, 237, 1981.
50. **Boulton, A. P., Huggins, A. K., and Munday, K. A.,** Effects of organometallic antifouling agents on the metabolism of the barnacle *elminius modestus, Toxicol. Appl. Pharmacol.,* 20, 487, 1971.
51. **Boyd, A. L. and Jones, J. M.,** Strain and sex differences in susceptibility of rats to dioctyltin dichloride, *Toxicol. Lett.,* 30, 253, 1986.
52. **Boyes, W. K. and Dyer, R. S.,** Pattern reversal and flash evoked potentials following acute triethyltin exposure, *Neurobehav. Toxicol. Teratol.,* 5, 571, 1983.
53. **Bridges, J. W., Davies, D. S., and Williams, R. T.,** The fate of ethyltin and diethyltin derivatives in the rat, *Biochem. J.,* 105, 1261, 1967.
54. **Brinckman, F. E., Jackson, J. A., Blair, W. R., Olson, G. J., and Iverson, W. P.,** Ultratrace speciation and biogenesis of methyltin transport species in estuarine waters, in *Trace Metal in Sea Water,* Wong, C. S., Boyle, E., Bruland, K. W., Burton, J. D., and Goldberg, E. D, Eds, New York, Plenum Press, 1983, 39.
55. **Brown, A. W., Aldridge, W. N., Street, B. W., and Verschoyle, R. D.,** The behavioral and neuropathologic sequelae of intoxication by trimethyltin compounds in the rat, *Am. J. Pathol.,* 97, 59, 1979.
56. **Brown, A. W., Cavanagh, J. B., Verschoyle, R. D., Gysbers, M. F., Jones, H. B., and Aldridge, W. N.,** Evolution of the intracellular changes in neurons caused by trimethyltin, *Neuropath. Appl. Neurobiol.,* 10, 267, 1984.
57. **Brown, A. W., Verchoyle, B. W., Street, B. W., Aldridge, W. N., and Grindley, H.,** The neurotoxicity of trimethyltin chloride in hamsters, gerbils and marmosets, *J. Appl. Toxicol.,* 4, 12, 1984.
58. **Buckton, G. B.,** Further remarks on the organometallic radicals and observations more particularly directed to the isolation of mercuric, plumbic and stannic ethyl, *Proc. Royal Soc.,* 9, 309, 1858.
59. **Bueding, E. and Swartzwelder, C.,** Anthelmintics. *Pharmacol. Rev.,* 9, 329, 1957.
60. **Bullough, D. A. and Allison, W. S.,** Three copies of the B subunit must be modified to achieve complete inactivation of the bovine mitochondrial F1-ATPase by 5'-p-fluorosulfonylbenzoyladenosine, *J. Biol. Chem.,* 261, 5722, 1986.
61. **Burns, D. T., Glockling, F., and Harriott, M.,** Investigation of the determination of tin tetraalkyls and alkyltin chlorides by atomic-absorption spectrometry after separation by gas-liquid or high-performance liquid-liquid chromatography, *Analyst,* 106, 921, 1981.
62. **Byington, K. H. and Hansbrough, E.,** Inhibition of the enzymatic activity of ligandin by organogermanium, organolead or organotin compounds and the biliary excretion of sulfobromophthalein by the rat., *J. Pharmacol. Exp. Ther.,* 208, 248, 1979.
63. **Byington, K. H., Yeh, R. Y., and Forte, L. R.,** The hemolytic activity of some trialkyltin and triphenyltin compounds, *Toxicol. Appl. Pharmacol.,* 27, 230, 1974.
64. **Byrd, J. T. and Andreae, M. O.,** Tin and methyltin species in seawater: Concentrations and fluxes, *Science,* 218, 565, 1982.
65. **Cain, K. and Griffiths, D. E.,** Studies of energy-linked reactions—localization of the site of action of trialkyltin in yeast mitochondria, *Biochem. J.,* 162, 575, 1977.
66. **Cain, K., L. Hyams, L., and Griffiths, D. E.,** Studies on energy-linked reactions: Inhibition of oxidative phosphorylation and energy-linked reactions by dibutyltin dichloride, *FEBS Lett.,* 82, 23, 1977.
67. **Cain, K., Partid, M. D., and Griffiths, D. E.,** Dibutylchloromethyltin chloride, a covalent inhibitor of the adenosine triphosphate synthetase complex, *Biochem. J.,* 166, 593, 1977.
68. **Calley, D., Guess, W. L., and Autian, J.,** Ultrastructural hepatotoxicity induced by an organotin ester, *J. Pharm. Sci.,* 56, 1267, 1967.
69. **Cammer, W.,** Uncoupling of oxidative phosphorylation in vitro by the neurotoxic fragrance compound acetylethyltetramethyltetralin and its putative metabolite, *Biochem. Pharmacol.,* 29, 1531, 1980.
70. **Cammer, W., Rose, A. L., and Norton, W. T.,** Biochemical and pathological studies of myelin in hexachlorophene intoxication, *Brain Res.,* 98, 547, 1975.
71. **Cardarelli, N. F.,** *Tin as a Vital Nutrient: Implications in Cancer Prophylaxis and Other Physiological Processes,* CRC Press, Boca Raton, 1986.
72. **Casida, J. E., Kimmel, E. C., Holm, B., and Widmark, G.,** Oxidative dealkylation of tetra-, tri-, and dialkyltins and tetra- and trialkylleads by liver microsomes, *Acta Chem. Scand.,* 25, 1497, 1971.
73. **Cenedella, R. J., Galli, C., and Paoletti, R.,** Brain free fatty acid levels in rats sacrificed by decapitation versus focused microwave irradiation, *Lipids,* 10, 290, 1975.
74. **Chambers, J. P., Rizopoulos, E., Armstrong, D. L., Wayner, M. J., and Valdes, J. J.,** The effects of trimethyltin on the Ca^{++}, Mg^{++} and Ca^{++} and Mg^{++}-dependent ATPases of human neuroblastoma GM 3320, *Brain Res. Bull.,* 18, 569, 1987.
75. **Chang, L. W.,** Hippocampal lesions induced by trimethyltin in the neonatal rat brain, *Neurotoxicology,* 5, 205, 1984.

76. **Chang, L. W.,** Neuropathology of Trimethyltin: A proposed pathogenesis mechanism, *Fund. Appl. Toxicol.,* 6, 217, 1986.
77. **Chang, L. W. and Dyer, R. S.,** Trimethyltin induced pathology in sensory neurons, *Neurobehav. Toxicol. Teratol.,* 5, 673, 1983.
78. **Chang, L. W. and Dyer, R. S.,** A time-course study of trimethyltin induced neuropathology in rats, *Neurobehav. Toxicol. Teratol.,* 5, 443, 1983.
79. **Chang, L. W. and Dyer, R. S.,** Trimethyltin induced zinc depletion in rat hippocampus, in *The Neurobiology of Zinc, Part B. Deficiency, Toxicity, and Pathology,* Frederickson, C. J., Howell, G. A., and Kasarskis, E. J., Eds, Alan R. Liss, New York, 1984, 275.
80. **Chang, L. W. and Dyer, R. S.,** Septotemporal gradients of trimethyltin-induced hippocampal lesions, *Neurobehav. Toxicol. Teratol.,* 7, 43, 1985.
81. **Chang, L. W. and Dyer, R. S.,** Early effects of trimethyltin on the dentate gyrus basket cells: a morphological study, *J. Toxicol. Environ. Hlth.,* 16, 641, 1985.
82. **Chang, L. W., Tiemeyer, T. M., Wenger, G. R., and McMillan, D. E.,** Neuropathology of trimethyltin intoxication. III. Changes in the brain stem neurons, *Environ. Res.,* 30, 399, 1983.
83. **Chang, L. W., Tiemeyer, T. M., Wenger, G. R., and McMillan, D. E.,** Neuropathology of mouse hippocampus in acute trimethyltin intoxication, *Neurobehav. Toxicol. Teratol.,* 4, 149, 1982.
84. **Chang, L. W., Tiemeyer, T. M., Wenger, G. R., McMillan, D. E., and Reuhl, K. R.,** Neuropathology of trimethyltin intoxication. II. Electron microscopic study of the hippocampus, *Environ. Res.,* 29, 445, 1982.
85. **Chang, L. W., Tiemeyer, T. M., Wenger, G. R., McMillan, D. E., and Reuhl, K. R.,** Neuropathology of trimethyltin intoxication. I. Light microscopic study, *Environ. Res.,* 29, 435, 1982.
86. **Chang, L. W., Wenger, G. R., and McMillan, D. E.,** Neuropathology of trimethyltin intoxication. IV. Changes in the spinal cord, *Environ. Res.,* 34, 123, 1984.
87. **Chang, L. W., Wenger, G. R., McMillan, D. E., and Dyer, R. S.,** Species and strain comparison of acute neurotoxic effects of trimethyltin in mice and rats, *Neurobehav. Toxicol. Teratol.,* 5, 337, 1983.
88. **Chester, A. E. and Meyers, F. H.,** Central Sympathoplegic and Norepinephrine-Depleting Effects of Antioxidants, *Proc. Soc. Exp. Biol. Med.,* 187, 62, 1988.
89. **Coleman, J. O. D. and Palmer, J. M.,** Influence of pH on the inhibition of oxidative phosphorylation and electron transport by triethyltin, *Biochim. Biophys. Acta,* 245, 313, 1971.
90. **Cook, L., Jacobs, K. S., and Reiter, L. W.,** Tin distribution in adult and neonate rat brain following exposure to triethyltin, *Toxicol. Appl. Pharmacol.,* 72, 75, 1984.
91. **Cook, L. L., Heath, S. M., and O'Callaghan, J. P.,** Distribution of tin brain subcellular fractions following the administration of trimethyltin and triethyltin to the rat, *Toxicol. Appl. Pharmacol.,* 73, 564, 1984.
92. **Cook, L. L., Stine, K. E., and Reiter, L. W.,** Tin distribution in adult rat tissues after exposure of trimethyltin and triethyltin, *Toxicol. Appl. Pharmacol.,* 76, 344, 1984.
93. **Cossa, P., Duplay, Fischgold, Arfel-Capdevielle, Lafon, Passouant, Minvielle, and Radermecker, J.,** Encephalopathies toxiques au stalinon, *Rev. Neurologique,* 98, 97, 1958.
94. **Costa, L. G., Doctor, S. V., and Murphy, S. D.,** Antinociceptive and hypothermic effects of trimethyltin, *Life. Sci.,* 31, 1093, 1982.
95. **Costa, L. G. and Sulaiman, R.,** Inhibition of protein synthesis by trimethyltin, *Toxicol. Appl. Pharmacol.,* 86, 189, 1986.
96. **Cotta-Ramusino, M. and Doci, A.,** Acute toxicity of brestan and fentin acetate on some freshwater organisms, *Bull. Environ. Contam. Toxicol.,* 38, 647, 1987.
97. **Coulon, J. F., Lacroix, P., Linee, P. H., and David, J. C.,** Effects of triethyltin on brain octopamines and their metabolism in the rat, *Biochim. Biophys. Acta,*135, 53, 1987.
98. **Cremer, J. E.,** The metabolism in vitro of tissue slices from rats given triethyltin compounds, *Biochem. J.,* 67, 87, 1957.
99. **Cremer, J. E.,** The biochemistry of organotin compounds. The conversion of tetraethyltin into triethyltin in mammals, *Biochem. J.,* 68, 685, 1958.
100. **Cremer, J. E.,** Amino acid metabolism in rat brain studied with 14C-labeled glucose, *J. Neurochem.,* 11, 165, 1964.
101. **Cremer, J. E.,** Studies on brain cortex slices. The influence of various inhibitors on the retention of potassium ions and amino acids with glucose or pyruvate as substrate, *Biochem. J.,* 104, 223, 1967.
102. **Cremer, J. E.,** Brain cortex slices. Differences in the oxidation of 14C-labeled glucose and pyruvate revealed by the action of triethyltin and other toxic agents, *Biochem. J.,* 104, 212, 1967.
103. **Cremer, J. E.,** Selective inhibition of glucose oxidation by triethyltin in rat brain in vivo, *Biochem. J.,* 119, 95, 1970.
104. **Cremer, J. E.,** Specific Toxic effects on the nervous system, in *Organ Directed Toxicity: Chemical Indices and Mechanisms,* Brown, S. S. and Davis, D. S., Eds., Pergamon Press, Oxford, 1981, 213.
105. **Davidoff, F., and Carr, S.,** Interaction of triethyltin with pyruvate kinase, *Biochemistry,* 12, 1415, 1973.
106. **Davis, M., Gendelman, D., Tischler, M., and Gendelman, P.,** A primary acoustic startle circuit: lesions and stimulation studies, *J. Neurosci.,* 2, 791, 1982.

107. **Dawson, A. P., Farrow, B. G., and Selwyn, M. J.,** Studies on the nature of the high affinity trialkyltin binding site of rat liver mitochondria, *Biochem. J.,* 202, 163, 1982.

108. **Dawson, A. P. and Selwyn, M. J.,** The action of tributyltin on energy coupling-factor-deficient submitochondrial particles, *Biochem. J.,* 152, 333, 1975.

109. **Dehaven, D. L., Krigman, M. R., and Mailman, R. B.,** Temporal changes in dopaminergic and serotonergic function caused by administration of trimethyltin to adult rats, *Neurobehav. Toxicol. Teratol.,* 8, 475, 1986.

110. **Dehaven, D. L., Walsh, T. J., and Mailman, R. B.,** Effects of trimethyltin on dopaminergic and serotonergic function in the central nervous system, *Toxicol. Appl. Pharmacol.,* 75,182, 1984.

111. **Dehaven, D. L., Wayner, M. J., Barone, F. C., and Evans, S. M.,** Effects of triethyltin on schedule dependent and schedule induced behaviors under different schedules of reinforcement, *Neurobehav. Toxicol. Teratol.,* 4, 231, 1982.

112. **Dehaven, D. L., Wayner, M. J., Barons, F. C., and Evans, S. M.,** Effects of triethyltin on ingestive behavior at ad lib, reduced, and recovered body weight, *Neurobehav. Toxicol. Teratol.,* 4, 217, 1982.

113. **Di Nucci, A., Gregotti, C., and Manzo, L.,** Triphenyl tin hepatotoxicity in rats, *Arch. Toxicol. Suppl.,* 9, 402, 1986.

114. **Doctor, S. V., Costa, L. G., Kendall, D. A., and Murphy, S. D.,** Trimethyltin inhibits uptake of neurotransmitters into mouse forebrain synaptosomes, *Toxicology,* 25, 213, 1982.

115. **Doctor, S. V., Costa, L. G., and Murphy, S. D.,** Effect of trimethyltin on chemically-induced seizures, *Toxicol. Lett.,* 13, 217, 1982.

116. **Doctor, S. V. and Fox, D. A.,** Effects of organotin compounds on maximal electroshock seizures (MES) responsiveness in mice. II. Tricyclohexyltin and triphenyltin, *J. Toxicol. Environ. Health,* 10, 53, 1982.

117. **Doctor, S. V. and Fox, D. A.,** Effects of organotin compounds on maximal electroshock seizure (MES) responsiveness in mice. I. Tri-(n-alkyl)tin compounds, *J. Toxicol. Environ. Health,* 10, 43, 1982.

118. **Doctor, S. V. and Fox, D. A.,** On the role of carbonic anhydrase in the anticonvulsant effects of triethyltin (TET), *Experientia,* 38, 824, 1982.

119. **Doctor, S. V. and Fox, D. A.,** Immediate and long-term alterations in maximal electroshock seizure responsiveness in rats neonatally exposed to triethyltin bromide, *Toxicol. Appl. Pharmacol.,* 68, 268, 1983.

120. **Doctor, S. V., Sultatos, L. G., and Murphy, S. D.,** Effect of trimethyltin on hepatic and extra-hepatic nonprotein sulfhydryl levels in the mouse, *Toxicol. Appl. Pharmacol.,* 70, 165, 1983.

121. **Doctor, S. V., Sultatos, L. G., and Murphy, S. D.,** Distribution of trimethyltin in various tissues of the male mouse, *Toxicol. Lett.,* 17, 43, 1983.

122. **Duncan, J.,** The toxicology of molluscicides. The organotins, *Pharmac. Ther.,* 10, 407, 1980.

123. **Dyer, R. S. and Boyes, W. K.,** Trimethyltin reduces recurrent inhibition in rats, *Neurobehav. Toxicol. Teratol.,* 6, 369, 1984.

124. **Dyer, R. S., Deshields, T. L., and Wonderlin, W. F.,** Trimethyltin-induced changes in gross morphology of the hippocampus, *Neurobehav. Toxicol. Teratol.,* 4, 141, 1982.

125. **Dyer, R. S., and Howell, W. E.,** Triethyltin: ambient temperature alters visual system toxicity, *Neurobehav. Toxicol. Teratol.,* 4, 267, 1982.

126. **Dyer, R. S., and Howell, W. E.,** Acute triethyltin exposure: Effects on the visual evoked potential and hippocampal afterdischarge, *Neurobehav. Toxicol. Teratol.,* 4, 259, 1982.

127. **Dyer, R. S., Howell, W. E., and Reiter, L. W.,** Neonatal triethyltin exposure alters adult electrophysiology in rats, *Neurotoxicology,* 2, 609, 1981.

128. **Dyer, R. S., Howell, W. E., and Wonderlin, W. F.,** Visual system dysfunction following acute trimethyltin exposure in rats, *Neurobehav. Toxicol. Teratol.,* 4, 191, 1982.

129. **Dyer, R. S., Walsh, T. J., Swartzwelder, H. S., and Wayner, M. J.,** Neurotoxicology of the alkyltins, special issue, *Neurobehav. Toxicol. Teratol.,* 4, 1, 1982.

130. **Dyer, R. S., Walsh, T. J., Wonderlin, W. F., and Bercegeay, M.,** The trimethyltin syndrome in rats, *Neurobehav. Toxicol. Teratol.,* 4, 127, 1982.

131. **Dyer, R. S., Wonderlin, W. F., and Walsh, T. J.,** Increased seizure susceptibility following trimethyltin administration in rats, *Neurobehav. Toxicol. Teratol.,* 4, 203, 1982.

132. **Eastman, C. L., Young, J. S., and Fechter, L. D.,** Trimethyltin ototoxicity in albino rats, *Neurotoxicol. Teratol.,* 9, 329, 1987.

133. **Elferink, J. G., Deierkauf, M., and Van Steveninck, J.,** Toxicity of organotin compounds for polymorphonuclear leukocytes: The effect on phagocytosis and exocytosis, *Biochem. Pharmacol.,* 35, 3727, 1986.

134. **Elliott, B. M. and Aldridge, W. N.,** Binding of triethyltin to cat hemoglobin and modification of the binding by diethylpyrocarbonate, *Biochem. J.,* 163, 583, 1977.

135. **Elliott, B. M., Aldridge, W. N., and Bridges, J. W.,** Triethyltin binding to cat hemoglobin. Evidence for two chemically distinct sites and a role for histidine and cysteine residues, *Biochem. J.,* 177, 461, 1979.

136. **Erickson, K. L.,** Effect of triethyltin bromide on glucose catabolism in the rat, *Diss. Abstr. Int. B,* 34, 2747, 1973.

137. **Eto, Y. and Suzuki, K.,** Lipid composition of rat brain myelin in triethyltin-induced edema, *J. Lipid Res.,* 12, 570, 1971.

138. **Evans, H. L., Bushnell, P. J., Taylor, J. D., Monico, A., Teal, J. J., and Pontecorvo, M. J.,** A system for assessing toxicity of chemicals by continuous monitoring of homecage behaviors, *Fund. Appl. Toxicol.,* 6, 721, 1986.

139. **Evans, J. G., Scott, M. P., and Miller, K.,** The effect of pregnancy on dioctyltin dichloride-induced thymic injury, *Thymus,* 8, 319, 1986.

140. **Evans, W. H., Cardarelli, N. F., and Smith, D. J.,** Accumulation and excretion of [1-^{14}C]bis(tri-n-butyltin)oxide in mice, *J. Toxicol. Environ. Health,* 5, 871, 1979.

141. **Fechter, L. D. and Young, J. S.,** Discrimination of auditory from nonauditory toxicity by reflex modulation audiometry: effects of triethyltin, *Toxicol. Appl. Pharmacol.,* 70, 216, 1983.

142. **Fechter, L. D., Young, J. S., and Nuttal, A. L.,** Trimethyltin ototoxicity: evidence for a cochlear site of injury, *Hear. Res.,* 23, 275, 1986.

143. **Feldman, R. G.,** Central and peripheral nervous system effects of metals: a survey, *Occup. Neurol.,* 66, 143, 1982.

144. **Fish, R. H.,** Bioorganotin chemistry: a commentary on the reactions of organotin compounds with a cytochrome P-450 dependent monooxygenase enzyme system, *Neurotoxicology,* 5, 159, 1984.

145. **Fish, R. H., Casida, J. E., and Kimmel, E. C.,** Bioorganotin chemistry: sites and stereoselectivity in the reaction of cyclohexyltinphenyltin with a cytochrome P-450 dependent monooxygenase enzyme system, *Tetrahedron Lett.,* 40, 3515, 1977.

146. **Fish, R. H., Kimmel, E. C., and Casida, J. E.,** Bioorganotin chemistry: biological oxidation of tri-butyltin derivatives, *J. Organomet. Chem.,* 93, C1-C5, 1975.

147. **Fish, R. H., Kimmel, E. C., and Casida, J. E.,** Bioorganotin chemistry: biological oxidation of organotin compounds, in *Adv. Chem. Ser,* 157th ed., Zuckerman, J. J., Ed., Washington, D.C., American Chemical Society, Washington, D.C., 1976, 197.

148. **Fish, R. H., Kimmel, E. C., and Casida, J. E.,** Bioorganotin chemistry: reactions of tributyltin derivatives with a cytochrome P-450 dependent monooxygenase enzyme system, *J. Organomet. Chem.,* 118, 41, 1976.

149. **Foncin, J. F. and Gruner, J. E.,** Tin neurotoxicity, in *Unknown,* 1980, 279.

150. **Fortemps, E., Amand, G., Bomboir, A., Lauwerys, R., and Laterre, E. C.,** Trimethyltin poisoning: report of two cases, *Int. Arch. Occup. Environ. Health,* 41, 1, 1978.

151. **Fox, D. A.,** Pharmacological and biochemical evaluation of triethyltin's anticonvulsant effects, *Neurobehav. Toxicol. Teratol.,* 4, 273, 1982.

152. **Garber, A. R., Armstrong, D. L., and Wayner, M. J.,** Intracellular recording of CA1 pyramidal cell responses following exposure to trimethyltin, *The Toxicologist,* 8(Abstr.), 228, 1988.

153. **Gerren, R. A., Groswald, D. E., and Luttges, M. W.,** Triethyltin toxicity as a model for degenerative disorders, *Pharmac. Biochem. Behav.,* 5, 299, 1976.

154. **Gordon, C. J., Long, M. D., and Dyer, R. S.,** Effect of triethyltin on autonomic and behavioral thermoregulation of mice, *Toxicol. Appl. Pharmacol.,* 73, 543, 1984.

155. **Graham, D. I., Bonilla, E., Gonatas, N. K., and Schotland, D. L.,** Core formation in the muscle of rats intoxicated with triethyltin sulfate, *J. Neuropath. Exp. Neurol.,* 35, 1, 1976.

156. **Graham, D. I., De Jesus, P. V., Pleasure, D. E., and Gonatas, N. K.,** Triethyltin sulfate-induced neuropathy in rats, *Arch. Neurol.,* 33, 40, 1976.

157. **Graham, D. I. and Gonatas, N. K.,** Triethyltin sulfate-induced splitting of peripheral myelin in rats, *Lab. Invest.,* 29, 628, 1973.

158. **Graham, D. I., Gonatas, N. K., and Gasser, D. L.,** Experimental allergic encephalomyelitis in triethyltin sulfate-treated brown Norwegian and Lewis rats, *Lab. Invest.,* 31, 24, 1974.

159. **Graham, D. I., Kim, S. U., Gonatas, N. K., and Guyotte, L.,** The neurotoxic effects of triethyltin (TET) sulfate on myelinating cultures of mouse spinal cord, *J. Neuropathol. Exp. Neurol.,* 34, 401, 1975.

160. **Griffiths, D. E., Houghton, R. L., Lancashire, W. E., and Meadows, P. A.,** Studies on energy-linked reactions: isolation and properties of mitochondrial venturicidin-resistant mutants of *Saccraharomyces cerevisiae, Eur. J. Biochem.,* 51, 393, 1975.

161. **Guess, W. L. and Stetson, J. B.,** Tissue reactions to organotin-stabilized polyvinyl chloride (PVC) catheters, *JAMA,* 204, 580, 1968.

162. **Guta-Socaciu, C., Giurgea, R., and Rosioru, C.,** Modifications in thymus and bursa fabricii, induced by n-butyltin pesticides treatment in chickens, *Agressologie,* 27, 123, 1986.

163. **Guta-Socaciu, C., Giurgea, R., and Rosioru, C.,** Thymo-bursal and adrenal modifications induced by triphenyltin compounds in chickens, *Arch. Exp. Vet. Med.,* 40, 307, 1986.

164. **Hallas, L. E., Means, J. C., and Cooney, J. J.,** Methylation of tin by estuarine microorganisms, *Science,* 215, 1505, 1982.

165. **Han, J. S. and Weber, J. H.,** Speciation of methyl- and butyltin compounds and inorganic tin in oysters by hydride generation atomic absorption spectrometry, *Anal. Chem.,* 60, 316, 1988.

166. **Handelman, G. E. and Olton, D. S.,** Spatial memory following damage to hippocampal CA3 pyramidal cells with kainic acid: impairment and recovery with preoperative training, *Brain Res.,* 217, 41, 1981.

167. **Hanin, I., Krigman, M. R., and Mailman, R. B.,** Central neurotransmitter effects of organotin compounds: trials, tribulations and observations, *Neurotoxicology,* 5, 267, 1984.

168. **Hardy, J. A., Boakes, R. J., Thomas, D. J. E., Kidd, A. M., Edwardson, J. A., Virmani, M., Turner, J., and Dodd, P. R.,** Release of aspartate and glutamate by chloride reduction in synaptosomal incubation media, *J. Neurochem.,* 42, 875, 1984.

169. **Harry, G. J. and Tilson, H. A.,** Postpartum exposure to triethyltin produces long-term alterations in responsiveness to apomorphine, *Neurotoxicology,* 3, 64, 1982.

170. **Harry, G. J. and Tilson, H. A.,** The effects of postpartum exposure to triethyltin on the neurobehavioral functioning of rats, *Neurotoxicology,* 2, 283, 1981.

171. **Hasan, Z., Woolley, D. E., and Zimmer, L.,** Effects of trimethyltin on potentials evoked in the limbic system of the rat, *Proc. West. Pharm. Soc.,* 25, 449, 1982.

172. **Hasan, Z., Zimmer, L., and Woolley, D. E.,** Time course of the effects of trimethyltin on limbic evoked potentials and distribution of tin in blood and brain in the rat, *Neurotoxicology,* 5, 217, 1984.

173. **Henry, R. A. and Byington, K. H.,** Inhibition of glutathione-S-aryltransferase from rat liver by organogermanium, lead, and tin compounds, *Biochem. Pharmacol.,* 25, 2291, 1976.

174. **Hirano, A., Dembitzer, H. M., Becker, N. H., and Zimmerman, H. M.,** The distribution of peroxidase in the triethyltin-intoxicated rat brain, *J. Neuropath. Exp. Neurol.,* 28, 507, 1969.

175. **Hirano, A., Zimmerman, H. M., and Levine, S.,** Intramyelinic and extracellular spaces in triethyltin intoxication, *J. Neuropathol. Exp. Neurol.,* 27, 571, 1968.

176. **Holwerda, D. A. and Herwig, H. J.,** Accumulation and metabolic effects of di-n-butyltin dichloride in the freshwater clam, anodonta anatina, *Bull. Environ. Contam. Toxicol.,* 36, 756, 1986.

177. **Howell, W. E., Walsh, T. J., and Dyer, R. S.,** Somatosensory dysfunction following acute trimethyltin exposure, *Neurobehav. Toxicol. Teratol.,* 4, 197, 1982.

178. **Hoy, M. S., Conley, J., and Robinson, W.,** Cyhexatin and fenbutatin-oxide resistance in pacific spider mite (Acari: tetranychidae): stability and mode of inheritance, *J. Econ. Entomol.,* 81, 57, 1988.

179. **Hudson, P. B., Sanger, G., and Sproul, E. E.,** Effective system of bactericidal conditioning for hospitals, *JAMA,* 169, 1549, 1959.

180. **Hultstrom, D., Forssen, M., Pettersson, A., Tengvar, C., Jarild, M., and Olsson, Y.,** Vascular permeability in acute triethyltin-induced brain edema studied with FITC-dextrans, sodium fluorescein and horseradish peroxidase as tracers, *Acta Neurol. Scand.,* 69, 255, 1984.

181. **Ingham, R. K., Rosenberg, S. D., and Gilman, H.,** Organotin compounds, *Chem. Rev.,* 60, 459, 1960.

182. **Ingelesfield, C.,** Fenbutatin oxide and chlorfenvinphos effects on the entomophagous arthropod fauna of citrus, *Bull. Environ. Contam. Toxicol.,* 38, 813, 1987.

183. **Ishaava, I., Engel, J. L., and Casida, J. E.,** Dietary triorganotins affect lymphatic tissues and blood composition of mice, *Pestic. Biochem. Physiol.,* 6, 270, 1976.

184. **Iwai, H. and Wada, O.,** Dealkylation of tetraalkyltin compounds in the intestinal mucosa of rabbits, *Ind. Health,* 19, 247, 1981.

185. **Jackson, J. A., Blair, W. R., Brinckman, F. E., and Iverson, W. P.,** Gas-chromatographic speciation of methylstannanes in the Chesapeake Bay using purge and trap sampling with a tin-selective detector, *Environ. Sci. Technol.,* 16, 110, 1982.

186. **Jacobs, J. M., Cremer, J. E., and Cavanagh, J. B.,** Acute effects of triethyltin on the rat myelin sheath, *Neuropathol. Appl. Neurobiol.,* 3, 169, 1977.

187. **Jacobs, K. S., Lemasters, J. J., and Reiter, L. W.,** Inhibition of ATPase activities of brain and liver homogenates by triethyltin (TET), in *Developments in the Science and Practice of Toxicology,* Hayes, A., Schnell, R. C., and Miya, T. S., Eds., Elsevier Science Publishers, New York, 1983.

188. **Jang, J. J., Takahashi, M., Furukawa, F., Toyoda, K., Hasegawa, R., Sato, H., and Hayashi, Y.,** Inhibitory effect of dibutyltin dichloride on pancreatic adenocarcinoma development by N-nitrosobis(2-oxopropyl)-amine in the syrian hamster, *Jpn., J. Cancer Res.,* 77, 1091, 1986.

189. **Jewett, K. L. and Brinckman, F. E.,** Speciation of trace di- and triorganotins in water by ion-exchange HPLC-GFAA, *J. Chromat. Sci.,* 19, 583, 1981.

190. **Jo'o, F., Varga, L., Domjan, G., Borcsok, E., Rott, S., Rosta, M., Zoltan, O. T., Csillik, B., and Foldi, M.,** Collapse of the blood-brain barrier in lymphostatic cerebral hemangiopathy and in triethyltin poisoning, *Med. Exp.,* 19, 342, 1969.

191. **Jo'o, F., Zoltan, O. T., Csillik, B., and Foldi, M.,** Studies on triethyltin sulfate induced brain edema in rats. 5. Effect of triethyltin sulfate on the permeability of blood-brain barrier, *Arzneim. Forsch.,* 19, 296, 1969.

192. **Johnson, C. T., Dunn, A. R., and Swartzwelder, H. S.,** Disruption of learned and spontaneous alteration in the rat by trimethyltin: chronic effects, *Neurobehav. Toxicol. Teratol.,* 6, 337, 1984.

193. **Jolyet, F. and Cahours, A.,** Recherches sur l'action physiologique des stanethyles et des stanmethyles. *C. R. Acad. Sci. Paris,* 68, 1276, 1869.

194. **Kagawa, K.,** Proton motive ATP synthesis, in *Bioenergetics,* Ernster, L., Ed., Elsevier, New York, 1984, 149.

195. **Kanner, B. I.,** Active transport of gamma-aminobutyric acid by membrane vesicles isolated from rat brain, *Biochemistry,* 17, 1207, 1978.

196. **Kansouh, A. S. H.,** Toxicity synergism of Du-Ter to spodoptera littoralis (Boisd.) in relation to the interaction of organotin compounds with carbohydrate metabolism, *Bull. Entomol. Soc. Egypt Econ. Ser.,* 9, 349, 1975.

197. **Katzman, R., Aleu, F. P., and Wilson, C.,** Further observations on triethyltin edema, *Arch. Neurol.,* 9, 178, 1963.

198. **Kerr, K. B.,** Butynorate, an effective and safe substance for the removal of *Raillietins cesticillus* from chickens, *Poult. Sci.,* 31, 328, 1952.

199. **Kimbrough, R. D.,** Toxicity and health effects of selected organotin compounds: a review, *Environ. Health. Perspect.,* 14, 51, 1976.

200. **Kimbrough, R. D. and Gains, T. B.,** Hexachlorophene effects on rat brain, study of high doses by light and electronmicroscopy, *Arch. Environ. Health,* 23, 114, 1971.

201. **Kimmel, E. C., Casida, J. E., and Fish, R. H.,** Bioorganotin chemistry. Microsomal monooxygenase and mammalian metabolism of cyclohexyltin compounds including miticide cyhexatin, *J. Agric. Food Chem.,* 28, 117, 1980.

202. **Kimmel, E. C., Fish, R. H., and Casida, J. E.,** Bioorganotin chemistry: metabolism of organotin compounds in microsomal monooxygenase systems and in mammals, *J. Agric. Food Chem.,* 25, 1, 1977.

203. **Kirschner, D. A. and Sapirstein, V. S.,** Triethyl tin-induced myelin oedema: An intermediate swelling state detected by X-ray diffraction, *J. Neurocytol.,* 11, 559, 1982.

204. **Klimmer, O. R.,** Use of organotin compounds from an experimental-toxicological point of view, *Arzneim. Forsch.,* 19, 934, 1969.

205. **Knowles, C. O. and Johnson, T. L.,** Influence of Organotins on Rat Platelet Aggregation Mechanisms, *Environ. Res.,* 39, 172, 1986.

206. **Krigman, M. R. and Silverman, A. P.,** General toxicology of tin and its organic compounds, *Neurotoxicology,* 5, 129, 1984.

207. **Lancashire, W. E. and Griffiths, D. E.,** Studies on energy-linked reactions: isolation, characterization and genetic analysis of trialkyl-tin-resistant mutants of *Saccharomyces cerevisiae, Eur. J. Biochem.,* 51, 377, 1975.

208. **Lapresle, J. and Fardcan, M.,** The central nervous system and carbon monoxide poisoning. II. Anatomical study of brain lesions following intoxication with carbon monoxide (22 cases), in *Carbon Monoxide Poisoning. Progress in Brain Research,* 24th ed., Bour, H. and Ledingham, I., Ed., Elsevier, New York, 1964, 31.

209. **Laughlin, R. and Linden, O.,** Fate and effects of organotin compounds, *Ambio,* 14, 88, 1985.

210. **Leander, J. D. and Gau, B. A.,** Flavor aversions rapidly produced by inorganic lead and triethyltin, *Neurotoxicology,* 1, 6350, 1980.

211. **Lee, J. C. and Bakay, L.,** Ultrastructural changes in the edematous central nervous system. I. Triethyltin edema, *Arch. Neurol.,* 13, 48, 1965.

212. **Legrain, Y. and Mackenzie, E. T.,** Various indices of brain metabolism and activity in a model of chronic neurological dysfunction: Triethyltin intoxication in the rat, *Eur. Neurol.,* 20, 183, 1981.

213. **Lehotzky, K., Szeberenyi, J. M., Gonda, Z., Horkay, F., and Kiss, A.,** Effects of prenatal triphenyltin exposure on the development of behavior and conditioned learning in rat pups, *Neurobehav. Toxicol. Teratol.,* 4, 247, 1982.

214. **Leow, A. C. T., Anderson, R., Little, R. A., and Leaver, D. D.,** A sequential study of changes in the brain and cerebrospinal fluid of the rat following triethyltin poisoning, *Acta Neuropathol.,* 47, 117, 1979.

215. **Leow, A. C. T. and Leaver, D. D.,** Effect of triethyltin on escherichia coli k-12, *Chem. Biol. Interact,* 19, 339, 1977.

216. **Leow, A. C. T., Towns, K. M., and Leaver, D. D.,** Effect of organotin compounds and hexachlorophene on brain adenosine cyclic 3′, 5′-monophosphate metabolism, *Chem. Biol. Interact,* 27, 125, 1979.

217. **Leow, A. C. T., Towns, K. M., and Leaver, D. D.,** The effects of triehtyltin in the rat following systemic and intracerebroventricular injection, *Chem. Biol. Interact,* 31, 233, 1980.

218. **Lijinsky, W. and Aldridge, W. N.,** Increase in cerebral fluid in rats after treatment with triethyltin, *Biochem. Pharmacol.,* 24, 481, 1975.

219. **Lipscomb, J. C., Paule, M. G., and Slikker, W.,** Fetomaternal kinetics of 14C-trimethyltin, *Neurotoxicol,* 7, 581, 1986.

220. **Liu, D. and Thompson, K.,** Biochemical responses of bacteria after short exposure to alkyltins, *Bull. Environ. Contam. Toxicol.,* 36, 60, 1986.

221. **Lock, E. A.,** Increase in cerebral fluids in rats after treatment with hexachlorophene or triethyltin, *Biochem. Pharmacol.,* 25, 1455, 1976.

222. **Lock, E. A.,** The action of triethyltin on the respiration of rat brain cortex slices, *J. Neurochem.,* 26, 887, 1976.

223. **Lock, E. A.,** The production of an oedematous change in the white matter of the central nervous system of the rat by 3:5-diiodo-4′-chlorosalicylanilide, *Proc. Int. Soc. Neurochem.,* 6, 357, 1977.

224. **Lock, E. A.,** Toxic action of 2′-chloro-2,4-dinitro-5′,6-di(trifluoromethyl)-diphenylamine in the rat, *Chem. Biol. Interact,* 28, 35, 1979.

225. **Lock, E. A. and Aldridge, W. N.,** The binding of triethyltin to rat brain myelin, *J. Neurochem.,* 25, 871, 1975.

226. **Loullis, C. C., Dean, R. L., Lippa, A. S., D.E. Clody, D. E., and Coupet, J.,** Hippocampal muscarinic receptor loss following trimethyltin administration, *Pharmac. Biochem. Behav.,* 22, 147, 1985.

227. **Lowig, C.,** Production of triethyltin bromide, 84, 308, 1852.
228. **Luijten, J. G. A. and Klimmer, O. R.,** Toxicological data on organotin compounds, in *Annalen der Chemie,* I.T.R.I. Publ. No. 538, Appendix D, Greenford, Middlesex: International Tin Research Institute, 1978, 11.
229. **Luttges, M. W., Kelly, P. T., and Gerren, R. A.,** Degenerative changes in mouse sciatic nerves: electro-phoretic and electrophysiological characteristics, *Exp. Neurol.,* 50, 706, 1976.
230. **Machado-Saks, J. P. and Scheibel, A. B.,** Limbic system of the aged mouse, *Exp. Neurol.,* 63, 347, 1979.
231. **Macovschi, O., Prigent, A. -F., Nemoz, G., and Pacheco, H.,** Effects of an extract of ginkgo biboba on the 3′, 5′-cyclic AMP phosphodiesterase activity of the brain of normal and triethyltin intoxicated rats, *J. Neurochem.,* 49, 107, 1987.
232. **Macphail, R. C.,** Studies on the flavor aversions induced by trialkyltin compounds, *Neurobehav. Toxicol. Teratol.,* 4, 225, 1982.
233. **Magee, P. N., Stoner, H. B., and Barnes, J. M.,** The experimental production of edema in the central nervous system of the rat by triethyltin compounds, *J. Pathol. Bacteriol.,* 73, 107, 1957.
234. **Maguire, R. J., Tkaez, R. J., Chau, Y. K., Bengert, G. A., and Wong, P. T. S.,** Occurrence of organotin compounds in water and sediment in Canada, *Chemosphere,* 15, 253, 1986.
235. **Mailman, R. B., Krigman, M. R., Frye, G. D., and Hanin, I.,** Effects of postnatal trimethyltin or triethyltin treatment on CNS catecholamine, GABA and actetylcholine systems in the rat, *J. Neurochem.,* 40, 1423, 1983.
236. **Mailman, R. B. and Lewis, M. H.,** Neurotoxicants and central catecholamine systems, *Neurotoxicology,* 8, 123, 1987.
237. **Mair, R. G., Langlais, P. J., Mazurek, M. F., Beal, M. F., Martin, J. B., and McEntee, W. J.,** Reduced concentrations of arginine vasopressin and MHPG in lumbar CSF of patients with Korsakoff's psychosis, *Life Sci.,* 38, 2301 1986.
238. **Majlathova, L.,** Chronic feeding experiments with some organostannic compounds on mice, *Cs. Hyg.,* 21, 60, 1976.
239. **Mastropaolo, J. P., Dacanay, R. J., and Riley, A. L.,** Effects of trimethyltin chloride on the LiC1 dose-response function for conditioned taste aversions in rats, *Neurobehav. Toxicol. Teratol.,* 8, 297, 1986.
240. **Mastropaolo, J. P., Dacanay, R. J., Luna, B. H., Tuck, D. L., and Riley, A. L.,** Effects of trimethyltin chloride on differential-reinforcement-of-low-rate responding, *Neurobehav. Toxicol. Teratol.,* 6, 193, 1984.
241. **McGuinness, J. D.,** Migration from packaging materials into foodstuffs: a need for more fundamental information, *Food Additives and Contaminant,* 3, 95, 1986.
242. **McMillan, D. E., Chang, L. W., Idemudia, S. O., and Wenger, G. R.,** Effects of trimethyltin and triethyltin on lever pressing, water drinking and running in an activity wheel: associated neuropathology, *Neurobehav. Toxicol. Teratol.,* 8, 499, 1986.
243. **McMillan, D. E. and Wenger, G. R.,** Neurobehavioral toxicology of trialkyltins, *Pharmacol. Rev.,* 37, 365, 1985.
244. **Means, J. C. and Hulebak, K. L.,** A methodology for speciation of methyltins in mammalian tissues, *Neurotoxicology,* 4, 37, 1983.
245. **Messing, R. B. and Sparber, S. B.,** Increased forebrain beta-adrenergic ligand binding induced by trimethyltin, *Toxicol. Lett.,* 32, 107, 1986.
246. **Miller, D. B., Eckerman, D. A., Krigman, M. R., and Grant, L. D.,** Chronic neonatal organotin exposure alters radial-arm maze performance in adult rats, *Neurobehav. Toxicol. Teratol.,* 4, 185, 1982.
247. **Miller, K., Maisey, J., and Nicklin, S.,** Effect of orally administered dioctyltin dichloride on murine immunocompetence, *Environ. Res.,* 39, 434, 1986.
248. **Miller, K., Scott, M. P., Hutchinson, A. P., and Nicklin, S.,** Suppression of thymocyte proliferation in vitro by a dioctyltin dichloride-induced serum factor, *Int. J. Immunopharmac.,* 8, 237, 1986.
249. **Montgomery, R. L. and Christian, E. L.,** Pathologic effects of antimetabolites on the hippocampus of diet controlled mice, *Brain Res. Bull.,* 1, 255, 1976.
250. **Moore, K. E. and Brody, T. M.,** The effect of triethyltin on mitochondrial swelling, *Biochem. Pharmacol.,* 6, 134, 1961.
251. **Moore, K. E. and Brody, T. M.,** The effect of triethyltin on tissue amines, *J. Pharmacol. Exp. Ther.,* 132, 6, 1961.
252. **Morot-Gaudry, Y.,** Inhibition by trimethyltin, probenecid and phenylglyoxal on depolarization-induced acetylcholine release in torpedo synaptosomes, *Neurochem. Int.,* 6, 553, 1984.
253. **Murthy, L., Menden, E. E., Eller, P. M., and Petering, H. G.,** Atomic absorption determination of zinc, copper, cadmium and lead in tissues solubilized by aqueous tetramethylammonium hydroxide, *Anal. Biochem.,* 53, 365, 1973.
254. **Mushak, P.,** The analysis of total and form-variable tin in various media, *Neurotoxicology,* 5(2), 163, 1984.
255. **Mushak, P., Krigman, M. R., and Mailman, R. B.,** Comparative organotin toxicity in the developing rat: somatic and morphological changes and relationship to accumulation of total tin, *Neurobehav. Toxicol. Teratol.,* 4, 209, 1982.

256. **Naalsund, L. U., Allan, C. N., and Fonnum, F.,** Changes in neurobiological parameters in the hippocampus after exposure to trimethyltin, *Neurotoxicology,* 6(3), 145, 1985.

257. **Naalsund, L. U. and Fonnum, F.,** The effect of trimethyltin on three glutamergic and gabaergic transmitter parameters in vitro: high affinity uptake, release and receptor building, *Neurotoxicology* 7(3), 53, 1986.

258. **Naalsund, L. U. and Fonnum, F.,** Differences in anionic dependence of the synaptic efflux of d-aspartic acid and gamma-aminobutyric acid, *J. Neurochem.,* 47, 687, 1986.

259. **Nadler, J. V. and Cuthbertson, G. J.,** Kainic acid neurotoxicity toward hippocampal formation: dependence on specific excitatory pathways, *Brain Res.,* 195, 47, 1980.

260. **Nadler, J. V., Perry, B. W., Gentry, C., and Cotman, C. W.,** Degeneration of hippocampal CA3 pyramidal cells induced by intraventricular kainic acid, *J. Comp. Neurol.,* 192, 333, 1980.

261. **Nagara, H., Suzuki, K., and Tiffany, C. W.,** Resistance of quaking mouse CNS to triethyltin edema, *Brain Res.,* 221, 441, 1981.

262. **Nagara, H., Suzuki, K., and Tiffany, C. W.,** Triethyltin does not induce intramyelinic vacuoles in the CNS of the quaking mouse, *Brain Res.,* 225, 413, 1981.

263. **Nation, J. R., Bourgeois, A. E., Clark, D. E., and Eliasalde, M.,** The effects of acute trimethyltin exposure on appetitive acquistion and extinction performance in the adult rat, *Behav. Neurosci.,* 98, 919, 1984.

264. **Nelson-Krause, D. C. and Howard, B. D.,** Energy utilization in the induced release of gamma-aminobutyric acid from synaptosomes, *Brain Res.,* 147, 91, 1978.

265. **Neuman, W. P.,** *The Organic Chemistry of Tin,* John Wiley & Sons, New York, 1970, 38.

266. Neumann, P. E. and Taketa, F., Effects of triethyltin bromide on protein phosphorlation in subcellular fractions from rat and rabbit brain, *Mol. Brain. Res.,* 2, 83, 1987.

267. **Nimni, M.,** Sensitive biological tests for medical grade plastics. I. Toxicity of organotin stabilizers, *J. Pharm. Sci.,* 53, 1262, 1964.

268. **Noland, E. A., Taylor, D. H., and Bull, R. J.,** Monomethyl- and trimethyltin compounds induce learning deficiencies in young rats, *Neurobehav. Toxicol. Teratol.,* 4, 539, 1982.

269. **O'Callaghan, J. P. and Miller, D. B.,** Acute exposure of the neonatal rat to triethyltin results in persistent changes in neurotypic and gliotypic proteins, *J. Pharmacol. Exp. Ther.,* 244, 368, 1988.

270. **O'Callaghan, J. P., Miller, D. B., and Reiter, L. W.,** Acute postnatal exposure to triethyltin in the rat: effects on specific protein composition of subcellular fractions from developing and adult brain, *J. Pharmacol. Exp. Ther.,* 224, 466, 1983.

271. **O'Shaughnessy, D. J. and Losos, G. J.,** Peripheral and central nervous system lesions caused by triethl- and trimethyltin salts in rats, *Toxicol. Pathol.,* 14(2), 141, 1986.

272. **Opacka, J. and Sparrow, S.,** Nephrotoxic effect of trimethyltin in rats, *Toxicol. Lett.,* 27, 97, 1985.

273. **Orren, D. K., Braswell, W. M., and Mushak, P.,** Quantitative analysis of ethyltin compounds in mammalian tissue using HPLC/FAAS, *J. Anal. Toxicol.,* 10, 93, 1986.

274. **Owsianowski, M.,** Cholesteral biosynthesis by rat brain in triethyltin intoxication, *Neuropathol. Pol.,* 13, 433, 1975.

275. **Papazian, D. M., Rahamimoff, H., and Goldin, S. M.,** Partial purification and functional identification of a calmodulin-activated adenosine 5′-triphosphate-dependent calcium pump from synaptic plasma membranes, *J. Neurosci.,* 4, 1933, 1984.

276. **Partis, M. D., Griffiths, D. G., and Beechey, R. B.,** Discrimination between the binding sites of modulators of the H⁺-translocating ATPase activity in rat liver mitochondrial membranes, *Arch. Biochem. Biophys.,* 232, 610, 1984.

277. **Patritti-Laborde, N., Lachant, N. A., Tomoda, A., Chan, M. H., and Tanaka, K. R.,** Effects of triethyltin on enzyme activity in human adult and cord red cells, *Enzyme,* 35, 87, 1986.

278. **Paule, M. G., Reuhl, K., Chen, J. J., Ali, S. F., and Slikker, W., Jr.,** Developmental toxicology of trimethyltin in the rat, *Toxicol. Appl. Pharmacol.,* 84, 412, 1986.

279. **Penninks, A. H., Hilgers, L., and Seinen, W.,** The absorption, tissue distribution and excretion of di-n-octyltin dichloride in rats, *Toxicology,* 44, 107, 1987.

280. **Penninks, A. H. and Seinen, W.,** Toxicity of organotin compounds. IV. Impairment of energy metabolism of rat thymocytes by various dialkyltin compounds, *Toxicol. Appl. Pharmacol.,* 56, 221, 1980.

281. **Penninks, A. H. and Seinen, W.,** Comparative toxicity of alkyltin and estertin stabilizers, *Fed. Chem. Toxic.,* 20, 909, 1982.

282. **Pettibone, G. W. and Cooney, J. J.,** Effects of organotins on fecal pollution indicator organisms, *Appl. Environ. Microbiol.,* 52, 562, 1986.

283. **Pieper, G. R. and Casida, J. E.,** House fly adenosine triphosphatases and their inhibitions by insecticidal organotin compounds, *J. Econ. Entomol.,* 58, 392, 1965.

284. **Pinel, R. and Madiec, H.,** Determination of "heavy" organotin pollution of water and shellfish by a modified hydride atomic absorption procedure, *Int. J. Environ. Anal. Chem.,* 27, 265, 1986.

285. **Piver, W. T.,** Organotin compounds: Industrial applications and biological investigation, *Environ. Heath. Perspect.,* 4, 61, 1973.

286. **Pluta, R. and Ostrowska, B.,** Acute poisoning with triethyltin in the rat. Changes in cerebral blood flow, cerebral oxygen consumption, arterial and cerebral venous blood gases, *Exp. Neurol.,* 98, 67, 1987.

287. **Prough, R. A., Stalmach, M. A., Wiebkin, P., and Bridges, J. W.,** The microsomal metabolism of the organometallic derivatives of the group-IV elements, germanium, tin and lead, *Biochem. J.,* 196, 763, 1981.

288. **Prull, G. and Rompel, K.,** EEG changes in acute poisoning with organic tin compounds, *Electroenceph. Clin. Neurophysiol.,* 29, 215, 1979.

289. **Ray, D. E.,** Electroencephalographic and evoked response correlates of trimethyltin induced neuronal damage in the rat hippocampus, *J. Appl. Toxicol.,* 1, 145, 1981.

290. **Reed, D. J., Woodbury, D. M., and Holtzer, R. L.,** Brain edema, electrolytes and extracellular space: effect of triethyltin on brain and skeletal muscle, *Arch. Neurol.,* 10, 604, 1964.

291. **Reiss, D. S., M.B. Lees, M. B., and Sapirstein, V. S.,** Is Na$^+$,K$^+$ ATPase a myeline-associated enzyme?, *J. Neurochem.,* 36, 1418, 1981.

292. **Reiter, L., Kidd, K., Heavner, G., and Ruppert, P.,** Behavioral toxicity of acute and subacute exposure to triethyltin in the rat. *Neurotoxicology,* 2, 97, 1980.

293. **Reiter, L. W., Heavner, G., Dean, K. F., and Ruppert, P.,** Developmental and behavioral effects of early postnatal exposure to triethyltin in rats, *Neurobehav. Toxicol. Teratol.,* 3, 285, 1981.

294. **Reiter, L. W. and Ruppert, P. H.,** Behavioral toxicity of trialkyltin compounds: a review, *Neurotoxicology,* 5, 177, 1984.

295. **Renhof, M., Kretzer, U., Schuermeyer, T., Skopuik, H., and Kemper, F. H.,** Toxicity of organotin compounds in chicken and rats, *Arch. Toxicol. Suppl.,* 4, 148, 1980.

296. **Reuhl, K. R. and Cranmer, J. M.,** Developmental neuropathology of organotin compounds, *Neurotoxicology,* 5, 187, 1984.

297. **Reuhl, K. R., Gilbert, S. G., Mackenzie, B. A., Mallett, J. E., and Rice, D. C.,** Acute trimethyltin intoxication in the monkey (*Macaca fascicularis*), *Toxicol. Appl. Pharmacol.,* 79, 436, 1985.

298. **Reuhl, K. R., Smallridge, E. A., Chang, L. W., and Mackenzie, B. A.,** Developmental effects of trimethyltin intoxication in the neonatal mouse. I. Light microscopic studies, *Neurotoxicology,* 4, 19, 1983.

299. **Rey, C. H., Reinecke, H. J., and Besser, R.,** Methyltin intoxication in six men: toxicological and clinical aspects, *Vet. Hum. Toxicol.,* 26, 121, 1984.

300. **Richman, E. A. and Bierkamper, G. G.,** Histopathology of spinal cord, peripheral nerve, and soleus muscle of rats treated with triethyltin bromide, *Exp. Neurol.,* 86, 122, 1984.

301. **Riley, A. L., Dacancy, R. J., and Mastropaolo, J. P.,** The effects of trimethyltin chloride on the acquistion of long delay conditioned taste aversion learning in the rat, *Neurotoxicology,* 5, 291, 1984.

302. **Robertson, D. G., Gray, R. H., and De La Iglesia, F. A.,** Quantitative assessment of trimethyltin induced of the hippocampus, *Toxicol. Pathol.,* 15, 7, 1987.

303. **Robertson, D. G., Gray, R. H., Kim, S. N., and De La Iglesia, F. A.,** Nephrotoxicity of trimethyltin chloride (TMTC) in rats, *Toxicologist,* 4, 33, 1984.

304. **Robertson, D. G., Gray, R. H., Richard, R. J., and De La Iglesia, F. A.,** The effects of trimethyltin chloride (TMTC) on acetylcholinesterase (AChE) in the hippocampus of rat brain, *Soc. Neurosci. Abstr.,* 8(Abstr.), 118, 1982.

305. **Robertson, D. G., Sang-Nam Kim, Gray, R. H., and De La Iglesia, F. A.,** The pathogenesis of trimethyltin chloride-induced nephrotoxicity, *Fund. Appl. Toxicol.,* 8, 147, 1987.

306. **Robinson, I. M.,** Effects of some organtin compounds on tissue amine levels in rats, *Food. Cosmet. Toxicol.,* 7, 47, 1969.

307. **Rocco, R. M., Blumberg, J. B., and Sapirstein, V. S.,** Metabolic abnormalities in triethyltin treated rats, *Soc. Neurosci. Abstr.,* 9(Abstr.), 1245, 1983.

308. **Rocco, R. M. and Sapirstein, V. S.,** Biochemical effects of triethyltin in developing rats, *Soc. Neurosci. Abstr.,* 8(Abstr.), 82, 1982.

309. **Rose, G.P., Taylor, J. M., Waters, J., and Kemp, J.,** The use of the trimethyltin model of CNS neurotoxicity to study biochemical indices of neuropathological change in the rat, *Neurotoxicology,* 5, 279, 1984.

310. **Rose, M. S.,** Evidence for histidine in the triethyltin binding site of rat hemoglobin, *Biochem. J.,* 111, 129, 1969.

311. **Rose, M. S. and Aldridge, W. N.,** Triethyltin and the incorporation of (^{32}P) phosphate into rat brain phospholipids, *J. Neurochem.,* 13, 103, 1966.

312. **Rose, M. S. and Aldridge, W. N.,** The interaction of triethyltin with components of animal tissue, *Biochem. J.,* 106, 821, 1968.

313. **Rose, M. S. and Lock, E. A.,** The interaction of triethyltin with a component of guinea pig supernatant. Evidence for histidine in the binding sites, *Biochem. J.,* 120, 151, 1970.

314. **Rosenberg, D. W., Drummond, G. S., Cornish, H. C., and Kappas, A.,** Prolonged induction of hepatic haemoxygenase and decreases in cytochrome P-450 content by organotin compounds, *Biochem. J.,* 190, 465, 1980.

315. **Ross, A.,** Industrial applications of organotin compounds, *N.Y. Acad. Sci. Ann.,* 125, 107, 1965.

316. **Ross, W. D., Emmett, E. A., Steiner, J., and Tureen, R.,** Neurotoxic effects of occupational exposure to organotins, *Am. J. Psychiatry,* 138, 1092, 1981.

317. **Rossner, W. and Nusstein, R.,** Course of action of bis(triethyltin) sulfate on the permeability of the blood-brain barrier, *Arzneim. Forsch.,* 22, 1372, 1972.
318. **Rothlein, J. E. and Parson, S. M.,** Specificity of association of a Ca^{++}/Mg^{++} ATPase with cholinergic synaptic vesicles from torpedo electric organ, *Biochem. Biophys. Res. Comm.,* 88, 1069, 1979.
319. **Rothman, S. M. and Olney, J. W.,** Glutamate and the pathophysiology of hypoxic-ischemic brain damage, *Ann. Neurol.,* 19, 105, 1986.
320. **Ruppert, P. H., Dean, K. F., and Reiter, L. W.,** Trimethyltin disrupts acoustic startle responding in adult rats, *Toxicol. Lett.,* 22, 33, 1984.
321. **Ruppert, P. H., Walsh, T. J., Reiter, L. W., and Dyer, R. S.,** Trimethyltin-induced hyperactivity: time course and pattern, *Neurobehav. Toxicol. Teratol.,* 4, 135, 1982.
322. **Sasaki, K., Ishizaka, T., Suzuki, T., and Saito, Y.,** Determination of tri-n-butyltin and di-n-butyltin compounds in fish by gas chromatography with flame photometric detection, *J. Assoc. Off. Anal. Chem.,* 71, 360, 1988.
323. **Sato, S. M., Frazier, J. M., and Goldberg, A. M.,** Perturbation of a hippocampal zinc-binding pool after postnatal lead exposure in rats, *Exp. Neurol.,* 85, 620, 1984.
324. **Sato, S. M., Frazier, J. M., and Goldberg, A. M.,** A kinetic study of the in vivo incorporation of 65Zn into rat hippocampus, *J. Neurosci.,* 4, 1671, 1984.
325. **Sax, N. I. and Lewis, R. J., Sr.,** *Hazardous Chemicals Desk Reference,* Van Nostrand Reinhold, New York, 1987, 116.
326. **Scheinberg, L. C., Taylor, J. M., Herzog, I., and Mandell, S.,** Optic and peripheral nerve response to triethyltin intoxication in the rabbit: biochemical and ultrastructural studies, *J. Neuropath. Exp. Neurol.,* 25, 202, 1066.
327. **Seifter, J.,** Pharmacology of metal alkyls. III. Tetramethyltin, *J. Pharmacol. Exp. Ther.,* 66, 32, 1939.
328. **Seinen, W., Vos, J. G., Van Krieken, R., Penninks, A., Brands, R., and Hooykaas, H.,** Toxicity of organotin compounds. III. Suppression of thymus-dependent immunity in rats by di-n-butyltindichloride and di-n-octyltindichloride, *Toxicol. Appl. Pharmacol.,* 42, 213, 1977.
329. **Seinen, W., Vos, J. G., Van Spanje, I., Snoek, M., Brands, R., and Hooykaas, H.,** Toxicity of organotin compounds. II. Comparative in vivo and in vitro studies with various organotin and organolead compounds in different animal species with special emphasis on lymphocyte cytotoxicity., *Toxicol. Appl. Pharmacol.,* 42, 197, 1977.
330. **Seinen, W. and Willems, M. I.,** Toxicity of organotin compounds. I. Atrophy of thymus and thymus-dependent lymphoid tissue in rats fed di-n-octyltindichloride, *Toxicol. Appl. Pharmacol.,* 35, 63, 1976.
331. **Selwyn, M. J., Dawson, A. P., Stockdale, M., and Gains, N.,** Chloride-hydroxide exchange across mitochondrial, erythrocyte and artificial lipid membranes mediated by trialkyl- and triphenyltin compounds, *Eur. J. Biochem.,* 14, 120, 1970.
332. **Sherman, L. R. and Carlson, T. L.,** A modified phenylfluorone method for determining organotin compounds in the parts per billion and sub-parts per billion range, *J. Anal. Toxicol.,* 4, 31, 1980.
333. **Siebenlist, K. R. and Taketa, F.,** Inhibition of red cell and yeast hexokinase by triethyltin bromide, *Biochem. Biophys. Res. Comm.,* 95, 758, 1980.
334. **Siebenlist, K. R. and Taketa, F.,** The effect of temperature on the inhibition of trout, carp and human red cell hexokinase by triethyltin bromide, *Comp. Biochem. Physiol.,* 70C, 261, 1981.
335. **Siebenlist, K. R. and Taketa, F.,** Inactivation of yeast hexokinase by triethyltin bromide and reactivation by dithiothreitol and glucose, *Biochemistry,* 22, 4642, 1983.
336. **Siebenlist, K. R. and Taketa, F.,** The effects of triethyltin bromide on red cell and brain cyclic AMP-dependent protein kinases, *J. Biol. Chem.,* 258, 11384, 1983.
337. **Siebenlist, K. R. and Taketa, F.,** Organotin-protein interactions: binding of triethyltinbromide to cat haemoglobin, *Biochem. J.,* 233, 471, 1986.
338. **Skilleter, D. N.,** The decrease of mitochondrial substrate uptake caused by trialkyltin and trialkyllead compounds in chloride media and its relevance to inhibition of oxidative phosphorylation, *Biochem. J.,* 146, 465, 1975.
339. **Skilleter, D. N.,** The influence of adenine nucleotides and oxidizable substrates on triethyltin-mediated chloride uptake by rat liver mitochondria in potassium chloride media, *Biochem. J.,* 154, 271, 1976.
340. **Slikker, W., Jr., Ali, S. F., Lipscomb, J., and Denton, R.,** Time course alterations in tremor and muscarinic receptor binding produced by trimethyltin, *Proc. West. Pharm. Soc.,* 28, 139, 1985.
341. **Sloviter, R. S.,** 'Epileptic' brain damage in rats induced by sustained electrical stimulation of the perforant path. I. acute electrophysiological and light microscopic studies, *Brain Res.,* 10, 675, 1983.
342. **Sloviter, R. S.,** A selective loss of hippocampal mossy fiber Timm stain accompanies granule cell seizure activity induced by perforant path stimulation, *Brain Res.,* 330, 150, 1985.
343. **Sloviter, R. S. and Damiano, B. P.,** On the relationship between kainic acid-induced epileptiform activity and hippocampal neuronal damage, *Neuropharmacology,* 20, 1003, 1981.

344. **Sloviter, R. S., Von Knebel Doeberitz, C., Walsh, T. J., and Dempster, D. W.,** On the role of seizure activity in the hippocampal damage produced by trimethyltin, *Brain Res.,* 367, 169, 1986.

345. **Smith, J. F., McLaurin, R. L., Nichols, J. B., and Asbury, A.,** Studies in cerebral edema and cerebral swelling. I. The changes in lead encephalopathy in children compared with those of alkyltin poisoning in animals, *Brain,* 83, 411, 1960.

346. **Smith, M. E.,** Studies on the mechanism of demyelination: triethyl tin-induced demyelination, *J. Neurochem.,* 21, 357, 1973.

347. **Smith, P. J.,** Toxicological Data on Organotin Compounds, in *Annalen der Chemie,* Greenford, Middlesex: International Tin Research Institute, 1978.

348. **Snoeij, N. J., Penninks, A. H., and Seinen, W.,** Biological activity of organotin compounds: an overview, *Environ. Res.,* 44, 335, 1987.

349. **Snoeij, N. J., Van Iersel, A. A. J., Penninks, A. H., and Seinen, W.,** Triorganotin-induced cytotoxicity to rat thymus, bone marrow, and red blood cells as determined by several in vitro assays, *Toxicology,* 39, 71, 1986.

350. **Snoeij, N. J., Van Iersel, A. J., Penninks, A. H., and Seinen, W.,** Toxicity of triorganotin compounds: comparative in vivo studies with a series of trialkyltin compounds and triphenyltin chloride in male rats, *Toxicol. Appl. Pharmacol.,* 81, 274, 1985.

351. **Soderquist, C. J. and Crosby, D. G.,** Degradation of triphenyltin hydroxide in water, *J. Agric. Food Chem.,* 28, 111, 1980.

352. **Sone, N. and Hagihara, B.,** Inhibitory action of trialkyltin compounds on oxidative phosphorylation in mitochondria, *J. Biochem. (Tokyo),* 56, 151, 1964.

353. **Sone, N., Yoshida, M., Hirata, H., and Kagawa, Y.,** Purification and properties of a dicyclohexylcarbodiimide-sensitive adenosine triphosphatase from a thermophilic bacterium, *J. Biol. Chem.,* 250, 7917, 1975.

354. **Sparber, S. B., Cohen, C. A., and Messing, R. B.,** Reversal of a trimethyltin-induced learning deficit by desglycinamide-8-arginine vasopressin, *Life. Sci.,* 42, 171, 1987.

355. **Squibb, R. E., Carmichael, N. G., and Tilson, H. A.,** Behavioral and neuromorphological effects of triethyl tin bromide in adult rats, *Toxicol. Appl. Pharmacol.,* 55, 188, 1980.

356. **Stockdale, M., Dawson, A. P., and Selwyn, M. J.,** Effects of trialkyltin and triphenyltin compounds on mitochondrial respiration, *Eur. J. Biochem.,* 15, 342, 1970.

357. **Stoner, H. B.,** Toxicity of triphenyltin, *Br. J. Ind. Med.,* 23, 222, 1966.

358. **Stoner, H. B., Barnes, J. M., and Duff, J. I.,** Studies on the toxicity of alkyltin compounds, *Brit. J. Pharmacol.,* 10, 16, 1955.

359. **Stoner, H. B. and Threlfall, C. J.,** The biochemistry of organotin compounds: effect of triethyltin sulphate on tissue phosphates in the rat, *Biochem. J.,* 69, 376, 1958.

360. **Studer, R. K., Siegel, B. A., Morgan, J., and Potchen, E. J.,** Dexamethasone therapy of triethyltin induced cerebral edema, *Exp. Neurol.,* 38, 429, 1973.

361. **Sullivan, J. J., Torkelson, J. D., Wekell, M. M., Hollingsworth, T. A., Saxton, W. L., Miller, G. A., Panaro, K. W., and Uhler, A. D.,** Determination of tri-n-butyltin and di-n-butyltin in fish as hydride derivatives by reaction gas chromatography, *Anal. Chem.,* 60, 626, 1988.

362. **Sumner, P. R. and Hirsch, J. D.,** Trimethyltin induced changes in ·H-QNB binding in various rodent brain areas, *Soc. Neurosci. Abstr.,* 8(Abstr.), 310, 1982.

363. **Suzuki, K.,** Some new observations on triethyl-tin intoxication in rats, *Exp. Neurol.,* 31, 207, 1971.

364. **Swartzwelder, H. S., Dyer, R. S., Holahan, W., and Myers, R. D.,** Activity changes in rats following acute trimethyltin exposure, *Neurotoxicology,* 2, 589, 1981.

365. **Swartzwelder, H. S., Hepler, J. S., Holahan, W., King, S., Leverenz, H., P. Miller, P., and Myers, R. D.,** Severely impaired maze performance in the rat caused by trimethyltin treatment: Problem-solving deficits and perseveration, *Neurobehav. Toxicol. Teratol.,* 4, 169, 1982.

366. **Szikszay, M., Benedek, G. Y., Zoltan, T. O., and Oobal, F.,** Oedema induced by triethylstannous sulfate, *Acta. Physiol. Acad. Sci. Hung.,* 49, 354, 1977.

367. **Taketa, F., Siebenlist, K., Kasten-Jolly, J., and Palosaari, N.,** Interaction of triethyltin with cat hemoglobin: identification of binding sites and effects on hemoglobin function, *Arch. Biochem. Biophys.,* 203(Abstr.), 466, 1980.

368. **Taketa, F. and Siebenlist, K. R.,** Activation of protein kinases by triethyltin, *Neurotoxicology,* 3, 132, 1982.

369. **Tan, L. P. and Ng, M. L.,** The toxic effects of trialkyltin compounds on nerve and muscle, *J. Neurochem.,* 29(Abstr), 689, 1977.

370. **Tan, L. P., Ng, M. L., and Kumar Das, V. G.,** The effect of trialkyltin compounds on tubulin polymerization, *J. Neurochem.,* 31, 1035, 1978.

371. **Tilson, H. A. and Burne, T. A.,** Effects of triethyl tin on pain reactivity and neuromotor function of rats, *J. Toxicol. Environ. Health.,* 8, 317, 1981.

372. **Tipping, E., Ketterer, B., Christodoulides, L., Elliott, B. M., Aldridge, W. N., and Bridges, J. W.,** The interactions of triethyltin with rat glutathione-S-transerases A, B, and C. Enzyme inhibition and equilibrium dialysis studies, *Chem. Biol. Interact.,* 24(Abstr.), 317, 1979.

373. **Toews, A. D., Ray, R., Goines, N. D., and Boulding, T. W.,** Increased synthesis of membrane macromolecules is an early response of retinal neurons to trimethyltin intoxication, *Brain. Res.,* 398(Abstr.), 298, 1986.

374. **Toll, L., Gundersen, C. B., Jr., and Howard, B. D.,** Energy utilization in the uptake of catecholamines by synaptic vesicles and adrenal chromaffin granules, *Brain Res.,* 136, 59, 1977.

375. **Torack, R., Gordon, J., and Prokop, J.,** Pathobiology of acute triethyltin intoxication, *Int. Rev. Neurobiol.,* 12, 45, 1970.

376. **Torack, R. M.,** The relationship between adenosine-triphosphatase activity and triethyltin toxicity in the production of cerebral edema of the rat, *Am. J. Pathol.,* 46(Abstr.), 245, 1965.

377. **Torack, R. M., Terry, R. D., and Zimmerman, H. M.,** The fine structure of cerebral fluid accumulation. II. Swelling produced by triethyltin poisoning and its comparison with that in the human brain, *Am. J. Pathol.,* 36, 273, 1960.

378. **Towfighi, J.,** Hexachlorophene, in *Experimental and Clinical Neurotoxicology,* Spencer, P. S. and Schaumburg, H. H., Eds., Williams & Wilkins, Baltimore, 1980, 440.

379. **Tsuda, T., Nakanishi, H., Aoki, S., and Takebayashi, J.,** Determination of butyltin and phenyltin compounds in biological and sediment samples by electro-capture gas chromatography, *J. Chromatogr.,* 387, 361, 1987.

380. **Ungar, E. and Bodlander, G.,** Ueber die toxischen Wirkungen des Zinns, *Z. Hyg.,* 2, 241, 1887.

381. **Valdes, J. J., Cory, R., Bierkamper, G. G., Goldberg, A. M., and Annau, Z.,** Triethyltin intoxication induces selective deficits in serotonergic and cholinergic neurotransmission in rat brain synaptosomes, *Soc. Neurosci. Abstr.,* 6(Abstr.), 799, 1980.

382. **Valdes, J. J., Mactutus, C. F., Santos-Anderson, R. M., Dawson, R., and Annau, Z.,** Selective neurochemical and histological lesions in rat hippocampus following chronic trimethyltin exposure, *Neurobehav. Toxicol. Teratol.,* 5, 357, 1983.

383. **Van Der Kerk, G. J. M. and Luijten, J. G. A.,** Investigations on organo-tin compounds. III. The biocidal properties of organo-tim compounds, *J. Appl. Chem.,* 4, 314, 1954.

384. **Varkony, T., Zoltan, O. T., and Foldi, M.,** Brain edema in the rat induced by triethyltin sulfate. VIII. Effect of theophylline on the electro-microscopic picture of experimental triethyltin sulfate poisoning in rats, *Arzneim. Forsch.,* 20, 1594, 1970.

385. **Varkonyi, T., Csillik, B., Zoltan, O. T., and Foldi, M.,** Studies on triethyltin sulfate induced brain edema in rats. IV. Electronoptic studies, *Arzneim. Forsch.,* 19, 293, 1969.

386. **Veronesi, B. and Bondy, S.,** Triethyltin-induced neuronal damage in neonatally exposed rats, *Neurotoxicology,* 7, 69, 1986.

387. **Walsh, G. E., McLaughlin, L. L., Louie, M. K., Deans, C. H., and Lores, E. M.,** Inhibition of arm regeneration by ophioderma brevispine by tributyltin oxide and triphenyltin oxide, *Ecotoxicol. Environ. Safety,* 12, 95, 1986.

388. **Walsh, J. M., Curley, M. D., Burch, L. S., and Kurlansik, L.,** The behavioral toxicity of a tributyltin ester in the rat, *Neurobehav. Toxicol. Teratol.,* 4, 241, 1982.

389. **Walsh, T. J. and Dehaven, D. L.,** Neurotoxicity of the alkyltins, in *Metal Neurotoxicity,* Bondy, S. C. and Prasad, K. N., Eds., CRC Press, Boca Raton, 1988, 87.

390. **Walsh, T. J. and Emerich, D. F.,** The hippocampus as a common target of neurotoxic agents, *Toxicology,* 49, 137, 1988.

391. **Walsh, T. J., Gallagher, M., Bostock, E., and Dyer, R. S.,** Trimethyltin impairs retention of a passive avoidance task, *Neurobehav. Toxicol. Teratol.,* 4, 163, 1982.

392. **Walsh, T. J., McLamb, R. L., and Tilson, H. A.,** Organometal-induced antinociception: a time- and dose-response comparison of triethyl and trimethyllead and tin, *Toxicol. Appl. Pharmacol.,* 73, 295, 1984.

393. **Walsh, T. J., Miller, D. B., and Dyer, R. S.,** Trimethyltin, a selective limbic system neurotoxicant impairs radial-arm maze performance, *Neurobehav. Toxicol. Teratol.,* 4, 177, 1982.

394. **Wassenaar, J. S. and Kroon, A. M.,** Effects of triethyltin on different ATPases, 5′-nucleotidase and phosphodiesterase in grey and white matter of rabbit brain and their relation with brain edema, *Europ. Neurol.,* 10, 349, 1973.

395. **Watanabe, I.,** Organotins (Triethyltin), in *Experimental and Clinical Neurotoxicology,* Spencer, P. S. and Schaumburg, H. H., Eds., Williams & Wilkins, Baltimore, 1980, 545.

396. **Webber, R. J., Dollins, S. C., Harris, M., and Hough, A. J., Jr.,** Effect of alkyltins on rabbit articular and growth plate chondrocytes in monolayer culture, *J. Toxicol. Environ. Health,* 16, 229, 1985.

397. **Wender, M., Filipek-Wender, H., and Stanislawska, J.,** Cholesteryl esters and nonesterified fatty acids of rat brain in triethyltin-induced edema, *Neuropathol. Pol.,* 14(Abstr.), 217, 1976.

398. **Wender, M., Kozik, M., and Piechowski, A.,** Effect of chronic triethyltin intoxication on the morphological and histoenzymatic pattern of rat brain, *Neuropathol. Pol.,* 11(Abstr.), 323, 1973.

399. **Wender, M., Mularek, O., and Piechowski, A.,** The effects of triethyltin intoxication at the early stage of extrauterine life on cerebral myelination, *Neuropathol. Pol.,* 12, 13, 1974.

400. **Wender, M., Mularek, O., and Piechowski, A.,** Protective influence of cerebrolysine against the effects of TET intoxication at an early stage of extrauterine life, *Neuropathol. Pol.,* 13(Abstr.), 441, 1975.

401. **Wender, M., Suiatala-Kamasa, M., and Piechowski, A.,** Myelin of the optic nerve in rats chronically intoxicated with triethyltin sulfate, *Neuropathol. Pol.,* 20(Abstr.), 223, 1982.

402. **Wender, M., Zgorzalewicz, J., Piechowski, A., Spieszalski, W., and Buchols, M.,** The pattern of myelin proteins in triethyltin (TET) intoxication, *Exp. Pathol.,* 23, 193, 1983.

403. **Wenger, G. R., McMillan, D. E., and Chang, L. W.,** Behavioral toxicology of acute trimethyltin exposure in the mouse, *Neurobehav. Toxicol. Teratol.,* 4, 157, 1982.

404. **Wenger, G. R., McMillan, D. E., and Chang, L. W.,** Behavioral effect of trimethyltin in two strains of mice. I. Spontaneous motor activity, *Toxicol. Appl. Pharmacol.,* 73, 78, 1984.

405. **Wenger, G. R., McMillan, D. E., Chang, L. W., Zitaglio, T., and Hardwick, W. C.,** The effects of triethyltin and trimethyltin in rats responding under a DRL schedule of reinforcement, *Toxicol. Appl. Pharmacol.,* 78(Abstr.), 248, 1985.

406. **Westendorf, J., Marquardt, H., and Marquardt, H.,** DNA interaction and mutagenicity of the plastic stabilizer di-n-octyltin dichloride, *Arzneim. Forsch.,* 36, 1263, 1986.

407. **White, T. P.,** Ueber die wirkungen des zinns auf den tierisches organismus, *Naunyn-Schmiedeberg's Arch. Exp. Pathol. Pharmakol.,* 13, 53, 1880.

408. WHO Tin and Organotin compounds. A preliminary review, *Environ. Health Crit.,* 15, 1980.

409. **Wiebkin, P., Prough, R. A., and Bridges, J. W.,** The metabolism and toxicity of some organotin compounds in isolated rat hepatocytes, *Toxicol. Appl. Pharmacol.,* 62, 409, 1982.

410. **Wilkinson, R. R.,** Technoeconomic and environmental assessment of industrial organotin compounds, *Neurotoxicology,* 5, 141, 1984.

411. **Wilson, W. E., Hudson, P. M., Kanamatsu, T., Walsh, T. J., Tilson, H. A., Hong, J. S., Marenpot, R. R., and Thompson, M.,** Trimethyltin-induced alterations in brain amino acids, amines and amine metabolites: relationship to hyperammonemia, *Neurotoxicology,* 7(3), 63, 1986.

412. **Yamada, J., Oishi, K., Tatsuguchi, K., and Watanabe, T.,** Studies on the antimicrobial action of trialkyltin compounds. IV. Interaction of trialkyltin chloride with inorganic phosphate and phospholipids, *Agric. Biol. Chem.,* 43(Abstr.), 1015, 1979.

413. **Young, J. S. and Fechter, L. D.,** Triemthyltin exposure produces an unusual form of toxic auditory damage in rats, *Toxicol. Appl. Pharmacol.,* 82(Abstr.), 87, 1986.

414. **Zaczek, R., Nelson, M., and Coyle, J. T.,** Kainic acid neurotoxicity and seizures, *Neuropharmacology,* 20, 183, 1981.

415. **Zimmer, J.,** Changes in the Timm sulfide silver staining pattern of the rat hippocampus and fascia dentata following early postnatal deafferentation, *Brain Res.,* 64, 313, 1973.

416. **Zimmer, L., Hasan, Z., Woolley, D., and Chang, L. W.,** Evoked potentials in the limbic system of the rat reveal sites of trimethyltin toxicity, *Neurotoxicology,* 3, 135, 1982.

417. **Zimmer, L., Woolley, D., and Chang, L. W.,** Does phenobarbital protect against trimethyltin-induced neuropathology of limbic structures?, *Life. Sci.,* 36(Abstr.), 851, 1985.

418. **Zoltan, O. T., Fekete, E., Domonkos, H., and Foldi, M.,** Studies on triethyltin sulfate induced brain edema in rats. I. Changes of the water content of the brain substances and their therapeutic modification by aescin, *Arzneim. Forsch.,* 19(Abstr.), 287, 1969.

419. **Zoltan, O. T. and Foldi, M.,** Studies on triethyltin sulfate induced brain edema in rats. II. Effect of aescin on the mortality, *Arzneim. Forsch.,* 19(Abstr.), 288, 1969.

420. **Zoltan, O. T. and Foldi, M.,** Studies on triethyltin sulfate induced brain edema in rats. III. Damage of the function of the central nervous system and therapeutic effect of aescin, *Arzneim. Forsch.,* 19(Abstr.), 290, 1969.

421. **Zoltan, O. T. and Gorini, S.,** Relation between triethyltin sulfate induced brain edema and cerebrospinal fluid pressure in dogs, *Arzneim. Forsch.,* 20(Abstr.), 1939, 1970.

422. **Zoltan, O. T. and Gorini, S.,** Cerebrospinal fluid pressure in brain edema experimentally induced by triethyltin sulfate. Effect of aescin therapy, *Arzneim. Forsch.,* 20(Abstr.), 1812, 1970.

423. **Zoltan, O. T. and Gyori, I.,** Studies of triethyltin-sulfate-induced brain edema in rats. VII. Therapeutic effect of silymarin, theophyllin and mannitol on triethyltin sulfate poisoning in rats in a conditioned reflex test, *Arzneim. Forsch.,* 20(Abstr.), 1248, 1970.

Chapter 6

RETINAL DAMAGE FOLLOWING DEVELOPMENTAL LEAD EXPOSURE: A ROD SELECTIVE DEFICIT

Donald A. Fox

TABLE OF CONTENTS

I. INTRODUCTION

The eye, and especially the retina, of man is extremely sensitive to chemical and/or environmental insult.[1-5] Nearly all categories of systemically administered drugs, environmental toxicants, and occupational chemicals are able to cause adverse ocular effects, and all ocular structures appear to be vulnerable. As with any toxic effect, the determinants of the degree of adverse retinal reactions are both chemical-dependent (e.g., nature of the chemical, dose, route, and duration exposure) and organism-dependent (e.g., nutritional status, genetic and hormonal factors, pathophysiological variables, age). In addition, in retinal photoxicity there are optical radiation-dependent factors both in the presence[5] and absence[6] of chemicals.

Most of the reported visual deficits in man following chemical exposure involve alterations in visual acuity or color vision.[1-6] This is not surprising since changes in these capacities, which are initially mediated by the cone photoreceptor system, are the easiest to recognize since they function optimally under daylight or photopic luminance conditions. Occasionally, however, there are reports of more subtle alterations, such as decreased visual sensitivity or altered dark adaptation, following chemical exposure.[7-10] These changes occur under dim light or scotopic luminance conditions which are initially mediated by the rod photoreceptor system. In spite of the functional visual impairments produced by chemical agents, precious little information exists on the sites and/or mechanisms of action accounting for these deficits. Furthermore, the effects of low-level chemical exposure, especially during development, remain a virtually unexplored area of neurotoxicological research. Therefore, the purpose of this chapter is to describe experiments aimed at elucidating the retinal sites and mechanisms of action responsible for the scotopic deficits following low-level developmental lead exposure.

Long-term scotopic deficits following low-level lead exposure during early postnatal development were first reported in monkeys and hooded rats.[9,11-13] Although some of these deficits are possibly due to CNS alterations,[11,14,15] early retinal involvement is suggested by reports describing a decrease in the amplitude of the ERG b-wave and its oscillatory potentials in workers exposed to low levels of lead.[16,17] This idea is further supported by the findings of Fox and Sillman[18] who demonstrated that micromolar concentrations of lead selectively depress the amplitude and sensitivity of the rod, but not cone, photoreceptor potential in isolated, perfused bullfrog retinas. Intracellular recording from isolated axolotl rods exposed to lead validated these original findings.[19] More recently, electroretinographic (ERG) and quantitative histological findings by Fox and co-workers have confirmed that developmental lead exposure in hooded rats produces a long-term selective rod photoreceptor deficit.[20,21]

Following the chronic administration of lead to neonatal rats, degenerative changes in the photoreceptor layer are observed.[22] These morphological alterations are similar to those reported in studies of inherited and drug-induced retinal degenerations which have suggested that elevated levels of cyclic GMP (cGMP) are responsible for the observed selective rod photoreceptor degenerations.[23,24] The pathological similarities with lead-exposed rats and the demonstration that the concentration of cAMP and cGMP are different in rod-dominated and mixed rod-cone retinas[24] led us to conduct cyclic nucleotide metabolism studies in conjunction with our ERG analysis. Selective changes in cGMP metabolism, however, may not totally account for our rod selective deficits[20,21] since it has been suggested that cGMP is also the internal transmitter in cones.[25] Our observed rod selective alterations in the ERG may also result from the loss and/or alteration of rhodopsin. Therefore we determined the rhodopsin content and its relation to the ERG threshold in the developmentally lead-exposed rats.[26]

In normal man and animals there is a log-linear relation between ERG threshold and rhodopsin content,[27-29] and a correlation between rhodopsin content and rod outer segment (ROS) length.[30,31] Similar relationships between retinal sensitivity, rhodopsin content and ROS length exist in humans with retinitis pigmentosa[32] and in experimental animals with vitamin A deficiency,[33] inherited retinal dystrophy,[34,35] and continuous light exposure.[36] For example, degeneration of the outer nuclear layer (ONL) following vitamin A deficiency or continuous light exposure leads to a decrease in the retinal rhodopsin content and an increase in the log relative threshold of the rat ERG b-wave.[33,36] Quantitative histological studies have revealed selective rod degenerations following both vitamin A deficiency[37,38] and continuous light exposure.[39]

Quantitative histological and ultrastructural experiments were conducted.[21,26] to determine whether there is a selective rod degeneration following low-level lead exposure during early postnatal developmental, as suggested by ERG and biochemical studies.[20,26] The quantitative studies were also used to determine if there are regional differences in susceptibility, changes in ROS length, or changes in total retinal thickness. In addition, attempts were made to correlate the histological changes with the biochemical findings (i.e., rhodopsin content) and ultrastructural observations. Such studies are important for three reasons. First, if low-level lead exposure produces scotopic deficits, there are over one quarter million children potentially at risk. Second, if the scotopic ERG is demonstrated to be a sensitive and reliable indicator of early lead exposure, it may be employed as an early noninvasive screening tool. Finally, if lead is demonstrated to be a selective rod toxicant, it can be utilized as a neurobiological tool, like pharmacological agents and inherited retinal degenerations, to study biochemical mechanisms and structure-function relations in rod photoreceptos and rod-cone interactions.

II. TECHNIQUES FOR ASSESSING RETINAL RESPONSES TO NEUROTOXICANTS

To determine the retinal sites and mechanisms of action responsible for suspected scotopic deficits requires highly specialized electrophysiological, biochemical, and morphological techniques and equipment. For example, to assess scotopic visual function electrophysiologically and biochemically requires a light-tight room containing all the necessary equipment, light sources calibrated for intensity and wavelength, safe (long wavelength) lights, etc. Furthermore, as with all experiments assessing visual system function, special care must be taken to control and monitor the intensity and diurnal nature of the ambient lighting environment. This is done to assure the investigator that light-induced damage does not contribute to the observed results.[3,5,6] We describe here in detail the experimental procedures used in our laboratory.

A. EXPERIMENTAL ANIMALS AND TREATMENT

Timed-pregnant Long-Evans hooded rats (Charles River, Boston, MA) are singly housed in stainless steel hanging cages and fed Purina® chow (Code No. 5001) and water *ad libitum* until parturition. The animals are housed in a room with ambient temperature of 22 ± 1 °C and maintained on a 12:12 light:dark cycle with illumination of 5 to 10 lx. Pregnant dams are randomly divided into two groups: control and lead-exposed. On the day of birth (day 0), (1) the tap water used for the lead-exposed group is replaced by 0.2% lead acetate (Fischer Scientific L-63) solution (1090 ppm lead), (2) the number, sex, and weight of the offspring are recorded, and (3) the litter is culled to eight pups. Only female pups are used in these experiments in order to directly compare our present results with our previous results.[12,13,15,21] The lead drinking solution, made from boiling distilled water, is changed every other day and provided *ad libitum* to the dams throughout lactation. No alterations in body weight or fluid consumption are observed in the lead-exposed dams compared to controls. Thereby the pups are exposed to lead only via the mother's milk from parturition (day 0) to weaning (day 21). The neonates consume approximately 300 µg of lead during the suckling period, resulting in an exposure of 0.5 µg/g of body weight per day.[40] Further details of this low-level lead model during early postnatal development are described in earlier papers.[40,41]

At weaning the rats are transferred to hanging stainless steel cages for the duration of the experiment and fed Purina® chow and water *ad libitum*. At 90 d of age animals are utilized for acute ERG recordings, cyclic nucleotide metabolism studies, ultrastructural studies, and quantitative histology.[20,21,26] The rhodopsin experiments and all quantitative histological studies are conducted in control and lead-exposed rats reared in cyclic lighting conditions and sacrificed between 1.5 and 2 h after light onset, the peak of ROS disc shedding.[39] Blood and retinal lead concentrations are determined at 21 d (end of exposure) and 90 d (time of experiments) of age.

B. ELECTRORETINOGRAPHIC PROCEDURES AND ANALYSIS

Typical procedures involve opthalmoscopic examination, prior to dark adaptation, to assure that all eyes are free from cataracts and other gross anomalies, and examination and/or control of the pupil diameter following dark-adaptation to assure that there are no differences between control and experimental subjects. Under all lighting conditions the pupil diameter in the lead-exposed rats is the same as in controls. Animals are dark-adapted overnight and all surgical procedures are carried out under dim red illumination (wavelength > 650 nm). The rats are anesthetized with urethane (1.7 g/kg, i.p.), positioned in a Kopf stereotaxic apparatus and placed on a heating pad that maintains core body temperature at $37.0 \pm 0.5°C$. The eyelid of the experimental eye is drawn back with sutures, the cornea is anesthetized with 0.5% proparacaine hydrochloride, the pupil is dilated with 1% atropine sulfate, and the eye is regularly irrigated with saline to prevent corneal edema.

A circular platinum-iridium recording electrode is positioned on the cornea close to the limbus. The platinum-iridium reference and ground electrodes are placed in the tongue and ear, respectively. The ERG signals are amplified by a Grass differential a.c. amplifier, filtered (bandwidth: 0.1 Hz to 1 kHz), and displayed on a Tektronix 502 oscilloscope. The amplified signals are fed into a Nicolet 1074 signal averaging computer and stored on a Nicolet 2090 digital oscilloscope with diskette, and then plotted on a X-Y plotter (H-P Model 7044B). Measures of amplitude are obtained from autoscaled displays of stored responses by manual cursor adjustment.

The light stimulation apparatus consists of two channels. The source for the background light adaptation is a 100-W solid tungsten lamp (GE) in a light-tight housing. Its duration is controlled by an electronic Uniblitz shutter. The test source is a xenon strobe (Grass PS22) which produces a short duration (10 μs) flash of white light. The two light sources are positioned on the pupillary axis and can be varied independently with respect to intensity by Oriel neutral density filters. To produce scotopically balanced flashes, neutral density filters are used in combination with Oriel 10 nm bandpass interference filters. The beams from both sources uniformly illuminate the same area (4 cm in diameter) of a quartz opal glass translucent diffuser placed 10 mm in front of the rat's eye. This full-field (ganzfeld) stimulation results in a relatively homogeneous stimulation to almost the entire retina. The luminance of the unattenuated test flash, as measured by an EG and G radiometer, is 2.4×10^4 lx. The background illuminance for the light-adapted conditions is 220 lx.

For each animal, rod and cone functions should be assessed independently. Single-flash ERGs, recorded over an 8-log unit range in fully dark-adapted rats, are used to generate voltage-log intensity (V-log I) and latency-log intensity (L-log I) curves for the a- and b-wave. The interval between single flashes is chosen, from preliminary experiments, so that it does not affect the absolute sensitivity of the b- and/or a-wave. The amplitude of the b-wave response is measured from the baseline to the peak of its response in the absence of the a-wave, or from the trough of the a-wave to the peak of the b-wave. Similarly, the amplitude of the a-wave is measured from the baseline to the peak of its response. The time interval from the onset of the test stimulus flash and the peak of the a- or b-wave is used for the measurement of latency or implicit time. To examine the rod photoreceptor system, scotopically balanced blue (470 nm) and yellow-green (570 nm) stimuli are utilized in dark-adapted rats. Equal amplitude (b-wave) ERGs are produced, averaged, and examined at several non-saturated stimulus intensities. To further assess rod function and to examine cone function, (b-wave) critical flicker fusion frequency (CFF) at different stimulus intensities is measured. The rat b-wave CFF curve consists of two plateaus which represent the scotopic and photopic receptor mechanisms present in the rat retina.[42-44] The CFF is defined as the highest stimulus frequency at which an individual flash is followed by a b-wave. Care is taken to insure that flash intensity is equivalent at all stimulus frequencies. To aid the CFF analysis, a stimulus marker pulse is generated on a separate channel for each light flash and plotted directly under the trace. The b-wave rod and cone responses are averaged to obtain each data point in the function. Finally, to examine the rat cone photoreceptor system, cone b-wave latencies (implicit times) of averaged responses are examined under conditions of light adaptation and rod saturation, 2 log units above the point where no a-wave is elicited,[44] by unattenuated white flashes presented at 30 Hz.

For retinal sensitivity experiments, single-flash rod ERGs also are recorded in fully dark-adapted rats. The threshold intensity of white light required to reliably elicit a b-wave of 30 μv (i.e., constant amplitude criterion method), which is about 3% of the maximum amplitude of the b-wave, is determined. The normal threshold of control rats at 90 d of age is measured and is set arbitrarily as unity so that the normal adult log threshold equals zero.[27] The interval between single flashes is chosen, from preliminary experiments, so that it does not affect the absolute sensitivity of the b-wave. The amplitude of the b-wave response is measured from the baseline to peak of its response.

C. CYCLIC NUCLEOTIDE METABOLISM STUDIES
1.Cyclic AMP and Cyclic GMP Analysis

Cyclic nucleotides are assayed following the procedure described by Farber and Lolley.[45] Experimental and control animals are dark-adapted and light-adapted for 3 h and are sacrificed in the appropriate lighting condition. Each dissected retina is homogenized in 0.1 N HCl, an aliquot of the homogenate is removed for the determination of protein by the method of Lowry et al.,[46] and the rest of the homogenate is boiled and centrifuged to separate the precipitated protein. The supernatant of each sample is then appropriately diluted with 50 mM sodium acetate buffer, pH 6.2 and three serial dilutions are assayed in triplicate. The cyclic nucleotides in each sample are acetylated before radioimmunoassay. Once the assay is completed and the ^{125}I in each tube is counted, nonspecific binding is subtracted from the samples and the total counts per minute (cpm) per cpm bound in each standard are plotted against increasing fmol of standard assayed. The cyclic nucleotide concentration in the tissue samples is read from the linear plot. These values are corrected with the appropriate dilution factors, divided by the protein content of each sample and expressed as picomoles per milligram of protein.

2. Guanylate Cyclase Activity

Retinal tissue from dark-adapted or light-adapted control and lead-exposed animals is homogenized in 50 mM Tris-HCl buffer, pH 7.6. Guanylate cyclase activity is measured as described by Farber and Lolley,[47] using 7×10^{-4} M GTP as substrate. Briefly, the assay mixture contains 50 mM Tris-HCl (pH 7.6), 3.3 mM MnSO$_4$, 0.4% Triton X-100, 8×10^{-5} M isobutylmethylxanthine (IBMX), an ATP regenerating system, 1 mM ^{3}H cGMP (to account for recovery) and the alpha labeled [^{32}P] GTP. For every sample studied, duplicate tubes are incubated with the retinal protein for 1, 3, and 5 min at 30°C. The reaction is stopped with 30 mM EDTA and the reaction mixture is diluted with 1 ml of 50 mM Tris-HCl buffer and applied to a column containing 1 g of neutral alumina. Cyclic GMP is eluted with the same buffer. The total cpm of ^{32}P in the 1, 3, and 5 min samples, after subtraction of the blank, are adjusted for cGMP degradation and recovery from the column, and plotted as a function of time. Guanylate cyclase activity is calculated from the slope of the plot.

3. Cyclic GMP-Phosphodiesterase (PDE) Activity

The reaction is carried out as described previously,[47,48] using a two step procedure. Briefly, retinal tissue from light-adapted control and lead-exposed animals is incubated for 5, 10, and 15 min at 37°C with 5×10^{-4} M [^{3}H] cGMP. This is approximately the K_m value for cGMP obtained with cGMP-PDE from rat retinas.[48] The reaction is terminated by boiling and after cooling, the samples are incubated for 10 min at 37°C with 0.3 units of alkaline phosphatase. This second reaction is stopped by addition of 1.0 ml of a 1:3 slurry of resin (Bio-Rad AG1-X2, C1- form; 50-100 mesh). The tubes are left standing for 30 min and then centrifuged at $3500 \times g$: 250 µl of each supernatant fraction are counted after the addition of scintillation cocktail. In each experiment, the percentage of retention of guanosine by the resin slurry is determined spectrophotometrically. The corresponding factor to correct for nucleoside retention is introduced into the calculation of enzyme activity, which is expressed as nanomoles of cGMP hydrolyzed per minute per milligran of protein.

D. RHODOPSIN MEASUREMENTS

The entire dark-adapted procedure must be carried out under dim red (wavelength > 650 nm) light. Rats, dark-adapted for 16 to 20 h, are decapitated, the retinas are rapidly removed[49] and then assayed individually using a modification of the Fulton et al. procedure.[50] Briefly, all the visual pigment is extracted from excised neuroretina in 2% Emulphogene BC-720 (Gaf Corp.), and pre-bleach and post-bleach spectra are taken from 700 to 350 nm scans with a Beckman MVII recording spectrophotometer. The absorbances of rhodopsin at its λ_{max} (497 to 500 nm) are obtained from the difference spectra. The rhodopsin values are expressed as nanomoles per eye (rhodopsin content per eye).

E. HISTOLOGICAL PROCEDURES
1. General Procedures

Details of the fixation, embedding, and sectioning are provided in Fox and Chu.[21] Briefly, for both light and electron microscopic studies, the eyes are fixed by cardiac perfusion with a mixture of 2% paraformaldehyde, 3% glutaraldehyde, and 0.1% $CaCl_2$ in 0.1 M phosphate buffer (pH 7.4). All perfusions are conducted in deeply anesthetized (sodium pentobarbital, 60 mg/kg) rats 1.5 to 2.0 h after light onset, the peak of ROS disc shedding.[39] After decapitation, a suture is secured to the superior portion of Tenon's capsule for identification of retinal quadrants. The corneas are slit and the deskinned heads are kept overnight in the same fixative (4°C). On the following day, the eyes are dissected from the head and bisected along the vertical meridian from the optic nerve head to the ora serata. Subsequently, all eyes are processed using a modification of the techniques described by LaVail and Battelle.[35] The eye hemispheres are rinsed in a phosphate-sucrose buffer, post-fixed in 1% osmium tetroxide on 0.1 M phosphate buffer (pH 7.4), dehydrated in an ethanol series, and embedded in pure Spurr's epoxy medium. For light microscopy, semi-thin sections (1 to 1.2 µm) are stained with 1% toluidine blue in 1% sodium borate. This procedure facilitates photoreceptor identification by enhancing nuclear heterochromatin staining.[51] For electron microscopy, the tissues are treated as described above except that ultra-thin sections stained with 3.5% uranyl acetate and Reynold's lead citrate are examined in a JEOL 100-C transmission electron microscope. Monoparticulate glycogen beta-particles are distinguished electron microscopically by their size (15 to 35 nm in diameter), shape, affinity for lead stains, and coherence.[52,53]

2. Rod and Cone Photoreceptor Nuclei Counting

Counts of rod and cone photoreceptor cells (nuceli) are made both at the posterior pole of the eye and in the far periphery of both the superior and inferior temporal quadrants using a modification of the procedure described by LaVail.[51] Rod and cone nuclei are distinguished by their nuclear morphology, size, shape, and location in the ONL. Only those photoreceptor nuclei with an apparent diameter of approximately 4 µm or greater are included in the counts in order to avoid errors based on tangential sections through the nuclei. Counts are made directly from slides using a 100×oil immersion objective lens (total magnification of 1000×) and a calibrated Filar micrometer eyepiece (Reichert Scientific Instruments, Buffalo, NY). Twenty fields, each 100 μm in length, are examined in each of three sections: ten beginning 300 μm from the optic nerve head and moving peripherally and another ten beginning 300 μm from the ora serata and moving posteriorly. The number of rod and cone photoreceptor cells is recorded in each region and the mean is calculated for each animal in the control and lead-exposed groups. Then the overall group mean, standard error of the mean (SEM) and percentage of each cell type are calculated for each treatment group.

3. Rod Outer Segment Length Measurements

Measurements of ROS length in a 2000 μm length of posterior or central retina are made directly from slides using a 100 × oil immersion objective lens (total magnification of 1000 ×) and a calibrated Filar micrometer eyepiece (Reichert Scientific Instruments, Buffalo, NY). Ten consecutive fields, each 100 μm in length, are examined on each side of the optic nerve head in each of three sections from control and experimentals. The ROS length, in superior and inferior posterior (central) retina, is recorded and the mean for each animal in the control and lead-exposed groups is calculated. The overall group mean and standard error of the mean (SEM) is then calculated for each treatment group.

F. BLOOD AND RETINAL LEAD CONCENTRATIONS

Blood and retinal lead concentrations are determined from individual control and lead-exposed rats at 21 (end of lead exposure) and 90 (time of experiment) d of age employing 7 to 10 rats per treatment group. All samples are analyzed by Environmental Sciences Associates, Inc. (Bedford, MA), a clinical health laboratory licensed by the Department of Health and Human Services, Center for Disease Control, utilizing anodic stripping voltammetry.

G. STATISTICAL ANALYSIS

All group data are studied using the appropriate analysis of variance (ANOVA). Statistical significance levels for this data are based on post-hoc multiple comparisons using Tukey's Honestly Significant Difference test. In comparisons involving only two means, the Student's t-test is employed. Differences between groups, for all data, are regarded as significant if P values are equal to or less than 0.05. The statistical theory and method of application are described by Winer.[54] No more than two animals per litter are used for any measurement throughout this series of experiments.

III. RECENT FINDINGS

The techniques described in the previous section have permitted the collection of significant new information in the past few years. The relevance and applicability of these date, however, to subclinical pediatric lead poisoning has yet to be established.

A. BLOOD AND RETINAL LEAD CONCENTRATIONS[20]

At 21 d of age (end of lead exposure), blood and retinal lead concentrations were 0.01 and 0.02 ppm in controls and 0.59 and 0.61 ppm (3.0×10^{-6} M) in lead-exposed rats, respectively. At 90 d of age, blood and retinal lead concentrations were 0.05 and 0.04 ppm in controls and 0.07 and 0.12 ppm (5.8×10^{-7} M) in lead-exposed rats, respectively.

B. ELECTRORETINOGRAPHIC OBSERVATIONS[20,26]

Representative single-flash ERGs from fully dark-adapted adult control and lead-exposed Long-Evans hooded rats are shown in Figure 1. In controls, the b-wave reaches its maximum

CONTROL   LEAD

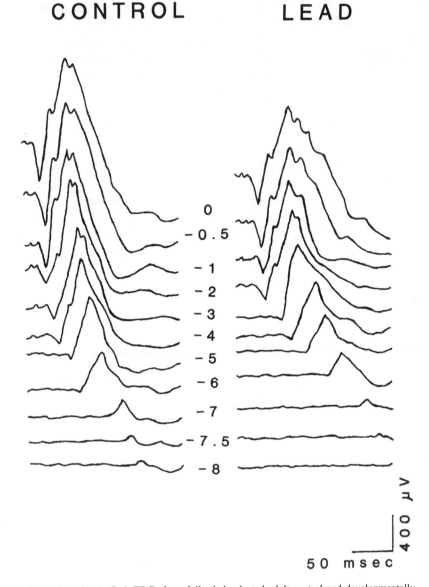

FIGURE 1. Single-flash ERGs from fully dark-adapted adult control and developmentally lead-exposed rats. ERGs were produced by a 10-μs white flash. The relative intensitites of the stimuli, in log units, are indicated between the columns. In response to weak flashes, only a corneal-negative b-wave is elicited. Stronger stimuli evoke an earlier, positive a-wave. At the lowest stimulus intensity (-8 ND), no measurable ERG was produced in the lead-exposed rats. Calibrations are indicated. (From Fox, D. A. and Farber D. B., *Exp. Eye Res.,* 46, 597, 1988. With permission.)

amplitude (Rmax) and a plateau at -0.5 log relative light intensity (ND) whereas the a-wave has not yet reached a plateau. The Rmax (mean ± SEM) of the a-wave and b-wave in 12 control rats at O ND is 416.4 ± 14.0 μV and 1352.8 ± 30.8 μV, respectively, resulting in an a-wave/b-wave Rmax ratio of 0.31 (Figure 2). At this light intensity a-wave and b-wave latencies (mean ± SD) are 21.5 ± 2.0 and 51.8 ± 4.3 ms and decrease to 55.7 ± 4.4 and 143.8 ± 9.1 ms at threshold intensities, respectively (Figure 3). The a-wave threshold is 3 log units above the b-wave threshold (Figure 2).

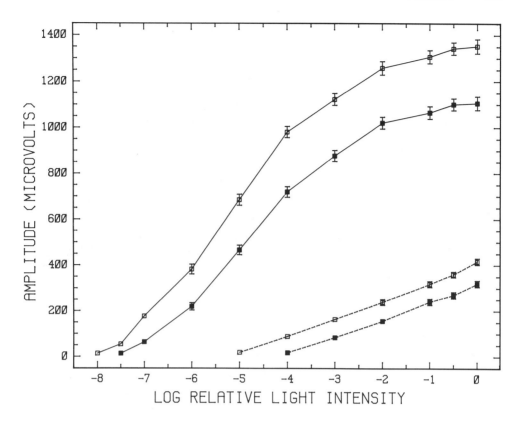

FIGURE 2. Single-flash voltage-log intensity a-wave (dashed lines) and b-wave (solid lines) functions in contol (open squares) and developmentally lead-exposed (filled squares) rats on a semi-log plot. Flash duration was 10 µs. Values represent the mean ± SEM for 13 control and 15 lead-exposed rats. All values in the lead-exposed group were significantly different from controls at $p < 0.05$. (From Fox, D. A. and Farber, D. B., *Exp. Eye Res.,* 46, 597, 1988. With permission.)

Three main effects of lead exposure on the ERG a-wave and b-wave are readily observable in Figure 1. First, the amplitudes of both waves are significantly decreased at all luminance intensities. Second, the latencies of both waves are significantly increased at all luminance intensities. Third, the absolute sensitivity is signficantly decreased for both waveforms. No gross differences in the ERG waveforms occurred in the lead-exposed rats, except for a slight increase in the b-wave duration at half-amplitude at the highest intensities. Likewise, the general shape of the V-log I (Figure 2) and L-log I (Figure 3) curves for the controls and lead-exposed rats were similar. In 13 lead-exposed rats, the mean amplitude of the a- and b-wave was decreased 74 to 79% at the lowest measurable intensity and 18 to 23% at maximum stimulus intensity (Figure 2). Overall, the mean amplitude decrease was 40% in the a-wave and 27% in the b-wave. However, the a-wave/b-wave maximum amplitude ratio (0.29) was not significantly different from controls. Mean latency of the a- and b-wave was increased 61 and 32% at the lowest measurable intensity and 39 and 24% at maximum stimulus intensity, respectively (Figure 3). Overall, the mean increase in a- and b-wave latency was 47 and 29%, respectively. Absolute sensitivity was decreased 1.0 log unit for the a-wave and 0.5 log unit for the b-wave (Figure 2). In addition, the mean log relative b-wave threshold (i.e., retinal sensitivity) in lead-exposed rats was elevated 1.1 to 1.2 log units or 13 to 16 times at 90 d of age. Thus, following lead exposure during early postnatal development single-flash ERGs reveal significant amplitude, latency and sensitivity alterations in both the a- and b-wave with larger deficits occurring in the a-wave. These data suggest that low-level lead exposure during early postnatal develop-

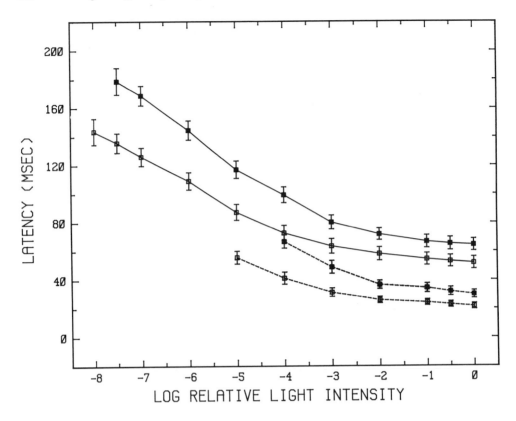

FIGURE 3. Single-flash latency-log intensity a-wave (dashed lines) and b-wave (solid lines) functions in control (open squares) and developmentally lead-exposed (filled squares) rats on a semi-log plot. Flash duration was 10 μs. Values represent the mean ± SD for 13 control and 15 lead-exposed rats. All values in the lead-exposed group were significantly different from controls at $p < 0.05$. (From Fox, D. A. and Farber, D. B., *Exp. Eye Res.*, 46, 597, 1988. With permission.)

ment produces a long-term selective rod deficit. To further delineate and establish the selective rod deficit, scotopically balanced ERGs, cone ERGs, and scotopic and photopic flicker fusion functions were examined in control and lead-exposed rats.

Scotopically balanced b-wave ERGs were examined over a 4 log unit range in fully dark-adapted control and lead-exposed rats. These waveforms, elicited with blue (470 nm) and yellow-green (570 nm) stimuli, are composed of a relatively large amplitude b-wave and almost no a-wave (see inset in Figure 4) and are similar to those obtained with a white light stimulus and a 4 to 5 ND filter (see Figure 1). In lead-exposed rats, the mean amplitude of the scotopic b-wave was significantly decreased using either the blue (-43%: range -63% at lowest intensity to -29% at highest intensity) or yellow-green (-41%: range -59% at lowest intensity to -28% at highest intensity) stimulus (Figure 4). These results, which further support the presence of a rod deficit, are similar to those observed in the single-flash ERG experiments (Figure 2), although of a slightly larger magnitude (-42% vs. -27%).

Full-field cone ERGs were elicited by using 30 Hz white flashes in the presence of a white background adapting light. The cone b-wave implicit time, used as an indicator of normal cone function, was not significantly different in control and lead-exposed rats. The mean (± SD) latency in controls and lead-exposed rats was 29.4 ± 1.7 and 30.1 ± 1.5 ms, respectively. Although cone b-wave amplitudes were not quantified, they do not appear to be different in the two groups and range from 4 to 10 μV (see Figure 5 in Fox and Farber[20]).

The b-wave CFF curves in our control and lead-exposed rats possess two distinct (rod and

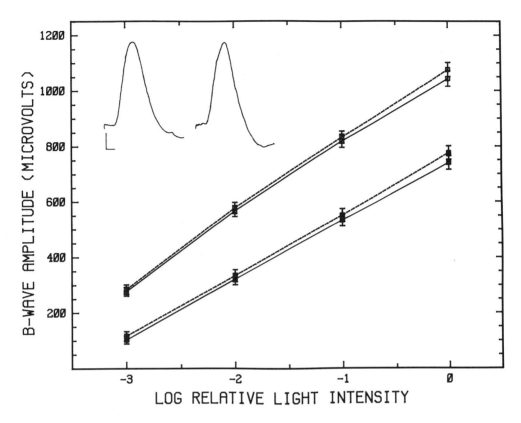

FIGURE 4. Voltage-log intensity b-wave functions for scotopically balanced blue (solid lines) and yellow-green (dashed lines) stimuli in control (open squares) and developmentally lead-exposed (filled squares) rats on a semi-log plot. Flash duration was 10 μs. Values represent the mean ± SEM for 10 control and 11 lead-exposed rats. All values in the lead-exposed group were significantly different from controls at $p < 0.05$. The inset illustrates the scotopically balanced ERG waveform, at maximal stimulus intensity, in a control rat for the blue (left) and yellow-green (right) stimuli. The calibration marker in the lower left represents 200 μV vertically and 25 ms horizontally. (From Fox, D. A. and Farber, D. B., *Exp. Eye Res.,* 46, 597, 1988. With permission.)

cone) plateaus (Fiugre 5), as previously reported in controls.[42-44] At low (scotopic) stimulus intensities the CFF was between 10 and 20 flashes per second, while at high (photopic) stimulus intensities the CFF increased between 30 and 40 flashes per second. At the four lowest stimulus intensities the mean CFF was significantly decreased (-18%) after lead exposure. In contrast, at the highest three intensities the mean CFF was increased (+10%) in the lead-exposed rats. These results, in combination with those from the cone ERG studies, further support the selectivity of the rod deficit and interestingly suggest a possible enhancement of cone functioning.

C. STUDIES ON CYCLIC NUCLEOTIDE METABOLISM[20]

The concentrations of cAMP and cGMP in the dark-adapted and light-adapted state were determined in 12 control and 14 lead-exposed retinas. As shown in Figure 6, in controls the cAMP concentration (mean ± SEM) was 16.2 ± 1.1 and 12.0 ± 2.0 pmol/mg protein in the dark-adapted and light-adapted states, respectively, while the cGMP concentration (mean ± SEM) was 37.3 ± 2.6 and 23.6 ± 2.4 pmol/mg protein in the dark-adapted and light-adapted state, respectively. No significant differences in cAMP concentrations were found in either state in the lead-exposed rats. In marked contrast, the cGMP concentration was significantly increased in lead-exposed retinas in both the dark-adapted (+42%) and light-adapted state (+23%). The

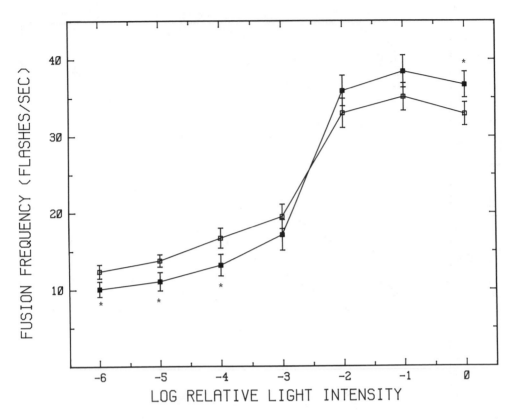

FIGURE 5. Critical flicker fusion frequency as a function of stimulus intensity in control (open squares) and developmentally lead-exposed (filled squares) rats. The CFF was defined as the highest stimulus frequency at which an individual 10 μs equivalent intensity flash was followed by a b-wave. Values represent the mean ± SD for 10 control and 11 lead-exposed rats. All values in the lead-exposed group marked by an asterisk (*) were significantly different from mean control values at $p < 0.05$. (From Fox, D. A. and Farber, D. B., *Exp. Eye Res.,* 46, 597, 1988. With permission.)

increase in the retinal cGMP concentration in the lead-exposed rat retinas could be the result of an activation of guanylate cyclase and/or an inhibition of cGMP-PDE. Therefore we investigated the activity of these enzymes in control and lead-exposed retinas. The guanylate cyclase activity (mean ± SEM), measured in six control (1.05 ± 0.16 nmol cGMP formed per minute per milligram of protein) and six lead-exposed (1.08 ± 0.03 nmol cGMP formed per minute per milligram of protein) retinas which are either dark-adapted or light-adapted, did not significantly differ between treatment groups. However, the cGMP-PDE activity (mean ± SEM), measured in light-adapted retinas, was significantly decreased (-40%) in the lead-exposed retinas compared to controls (6.69 ± 1.30 vs. 11.15 ± 1.15 nmol of cGMP hydrolyzed per minute per milligram of protein).

In order to determine if the effects of lead on the enzymes of cyclic nucleotide metabolism were direct, *in vitro* retinal studies were carried out. The activity of guanylate cyclase and cGMP-PDE was measured in adult control retinas incubated with several concentrations of lead for 5 min. Note that the highest *in vivo* retinal lead concentration was $3 \times 10^{-6} M$, while at the time of ERG testing and biochemical analysis it was about $6 \times 10^{-7} M$. Incubating with 10^{-7} to $10^{-5} M$ lead did not effect guanylate cyclase activity. At 1×10^{-4} and $5 \times 10^{-4} M$ lead, however, guanylate cyclase activity was significantly increased 20 and 158%, respectively (Table 1). The activity of cGMP-PDE was more sensitive to the effects of added lead, which is similar to what is observed with *in vivo* lead exposure. Below $10^{-7} M$, lead does not affect cGMP-PDE activity.

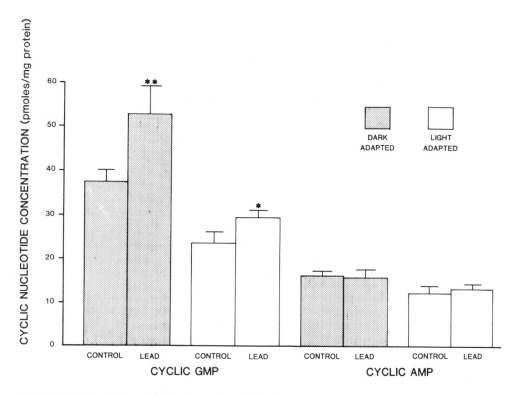

FIGURE 6. Cyclic AMP and GMP concentrations of dark-adapted (filled bars) and light-adapted (open bars) retinas from control and developmentally lead-exposed rats. Animals were dark-adapted or light-adapted for 3 h prior to sacrifice in the appropriate lighting condition. Values represent the mean ± SEM of 14 determinations for the control animals and 12 for the lead-exposed rats. Values in the lead-exposed group were significantly different from controls at the following levels of significance: *$p < 0.05$; ** $p < 0.01$. (From Fox, D. A. and Farber, D. B., *Exp. Eye Res.*, 46, 597, 1988. With permission.)

However, at 10^{-6} (-10%), 10^{-5} (-25%) and 10^{-4} M (-40%) lead there were significant decreases in activity (Figure 7.)

D. RHODOPSIN CONTENT AND RETINAL SENSITIVITY[26]

The rhodopsin content in adult control rats is 1.99 nmol per eye. Compared to controls, the eyes from 90 d old rats exposed to lead only during development contained 34% less rhodopsin. No change in the λ_{max} of rhodopsin was seen in the retinas from lead-exposed rats.

There is a log-linear relationship between the ERG b-wave threshold and rhodopsin content for both experimental conditions: control and lead exposure. That is, in both control and lead-exposed rats a 1, 3, and 12 months of age, the log threshold rises linearly as the rhodopsin content decreases (See Figure 4 in Fox and Rubinstein[26]). The slopes of the regression lines, determined by the method of least squares, are not significantly different for the controls and lead-exposed rats. Therefore the data were combined to yield a single regression line. The equation for the regression line is $y = -29.6x + 99.6$ with a standard estimate of error of 1.1 and a correlation coefficient (r) of -0.99.

E. LIGHT MICROSCOPY AND QUANTITATIVE HISTOLOGY[21,26]

The retinas from the adult control rats were normal as shown in Figure 8A. The rod outer and inner segments are compactly organized, the ONL is 10 to 12 nuclei thick and generally arranged in vertical columns, the INL is 3 to 5 nuclei thick, and the inner plexiform layer is compact and

TABLE 1
In Vitro Effects of Lead on Guanylate Cyclase Activity

Lead Concentration (M)	Guanylate Cyclase Activity[a] (nmol cGMP formed per min/mg protein)
Control	0.88 ± 0.03[b]
1×10^{-7}	0.78 ± 0.04
1×10^{-6}	0.81 ± 0.06
1×10^{-5}	0.91 ± 0.05
1×10^{-4}	1.06 ± 0.03[c]
5×10^{-4}	2.27 ± 0.09[c]

[a]   Aliquots of homogenates of adult control retinas were incubated at 30°C with lead acetate concentrations ranging from 5×10^{-4} to 1×10^{-7} M. After 5 min all the components of the guanylate cyclase reaction were added to the tubes and the assay was carried as described in the methods.

[b]   Values represent the mean $\pm$ SEM for triplicate samples of three homogenates.

[c]   Significantly different from controls at $p < 0.01$.

From Fox, D. A. and Farber D. B., *Exp. Eye Res.,* 46, 597, 1988. With permisssion.

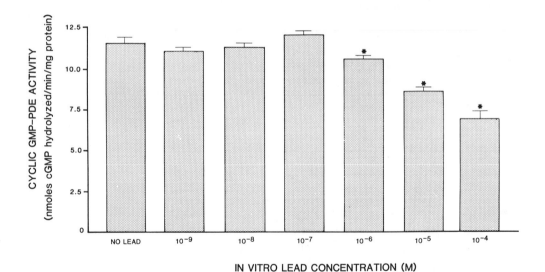

FIGURE 7.   Cyclic GMP phosphodiesterase activity in adult control retinas following incubation in 10^{-9} to $10^{-4} M$ lead for five minutes. Values represent the mean $\pm$ SEM of at least six determinations for each data point. Mean values in the lead-exposed group marked by an asterisk (*) were significantly different from controls at $p < 0.05$. (From Fox, D. A. and Farber, D. B., *Exp. Eye Res.,* 46, 597, 1988. With permission.)

uniformally stained. Total retinal thickness ranges from 180 to 200 m. The rod and cone nuclei are distinguished by their nuclear morphology, size, shape, and location in the ONL. Although the absolute number of rod cells declines from the posterior pole of the retina to the periphery, on a percentage basis the rod cells represents approximately 98% of the total photoreceptor population (Table 2). No difference in the number of rod or cone photoreceptor cells is observed in superior compared to the inferior hemisphere of the control temporal retina; the values presented in Table 2 therefore represent the mean of both hemispheres.

Light microscopic examination of adult retinas from developmentally lead-exposed rats reveal several distinct alterations (Figure 8B). The two most notable are the distended and disorganized ROSs and the thinned ONL (only 6 to 8 nuclei thick). In 11 of 12 lead-exposed rats we observed random patches of swollen and disrupted ROSs primarily occurring in the proximal one half to one third of the ROS. This alteration is not observed in any of the controls from this or any other study we have conducted. Interestingly, the RPE and the intimate contact between the RPE cells and outer segments appear relatively normal in all animals. Additional changes observed are a thinned INL (2 to 3 cells thick), a thinned and vacuolized inner plexiform layer, gliosis, and glycogen accumulation in retinal neurons. Total retinal thickness is decreased 14.5% in the superior retina (201.50 ± 8.07 µm in controls vs. 172.23 ± 6.28 µm in lead-exposed rats) and 20.8% in the inferior retina (187.24 ± 4.46 µm in controls vs. 148.35 ± 5.03 µm in lead-exposed rats).

In agreement with our ERG and cyclic nucleotide metabolism studies demonstrating a selective rod deficit in rats exposed to lead during early postnatal development,[20] our quantitative histological analysis reveals that lead exposure during development produces a long-term selective loss of rods (Table 2). As documented in Table 2, the loss of rod nuclei is more extensive in the inferior retinal hemisphere than in the superior retinal hemisphere (-24.9% vs. -14.7%) and in the posterior than in the peripheral retina (-22.3% vs. -17.3%). Thus, there is also a central-peripheral gradient of rod degeneration. The classification of cell types based on nuclear morphology is more difficult in lead-exposed rats due to degenerative and pyknotic changes in the rod nuclei (e.g., see Figure 11). Based on the findings of a similar absolute number of cone nuclei in control and lead-exposed rat retinas, however, the present results appear valid. Thus, the selective loss of rods (overall approximately 20%) results in an increase in the percentage of cones from 1.8 to 2.2%.

The ROS lengths in the superior and inferior posterior (central) retina of 90 d old control and lead-exposed rats are shown in Table 3. The ROS lengths (mean $\pm$ SEM) in the superior and inferior retinas of adult controls are 33.55 ± 1.21 and 27.49 ± 0.82 µm, respectively, and in adult lead-exposed rats are 30.44 ± 1.04 and 23.50 ± 1.05 µm, respectively. In both treatment groups, the ROSs are longer in the superior retina (22 to 30%) than in the inferior retina. In the lead-exposed rats, however, there is a consistently greater decrease in ROS length in the inferior (15%) than in the superior (9%) retina which is consistent with the above rod nuclei data showing a greater loss in the inferior than superior retina. This differential effect on ROS length in the inferior retina of lead-exposed rats results in a greater mean superior/inferior ROS length ratio in lead-exposed rats (1.30) compared to controls (1.22). These results, in combination with the selective rod loss, most likely account for the 30 to 34% loss of rhodopsin.[26]

F. ELECTRON MICROSCOPY[21]

The retinas of control and developmentally lead-exposed rats were also examined by electron microscopy. Figures 8C and D illustrate, at relatively low magnification, the ROS in representative control and lead-exposed rats, respectively. The most obvious lead-induced alteration (Figure 8D) is the consistently observed ROS swelling and disorganization which primarily occurs in the proximal one half of the outer segment. At higher magnification (Figure 9), examination of swollen ROSs suggests that this alteration is due to a recent shearing force (e.g., osmotic swelling during perfusion) rather than a long-term degeneration. This conclusion is enhanced by the lack of occurrence of ROS vesiculation in these lead-exposed rats as compared to vitamin A deficit rats[37] or light-damaged rats.[55]

As shown in Figures 9, 10, and 11, the reintas from lead-exposed rats contain large accumulations of monoparticulate (beta-type) glycogen particles, predominantly localized in the distal retina (i.e., ellipsoid region of the RIS, ONL, and outer plexiform layer: OPL) (Figures 9, 10, and 11). Surprisingly, in the distal retina these large glycogen accumulations are localized

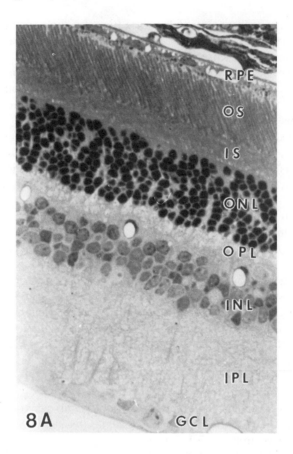

FIGURE 8. Light micrographs (A and B: X 1000) and a montage of low power electron micrographs (C and D: X 5300) of inferior posterior retina from control (A and C) and lead-exposed (B and D) rats. (A) Control retina showing principal retinal features including the compact organization of the outer nuclear layer (10 to 12 nuclei thick) and inner nuclear layer (4 to 5 nuclei thick). RPE, retinal pigment epithelium; OS, photoreceptor outer segments; IS, photoreceptor inner segments; ONL, outer nuclear layer; OPL, outer plexiform layer; INL, inner nuclear layer; IPL, inner plexiform layer; GCL, ganglion cell layer. (B) Lead-treated retina illustrating retinal thinning in the ONL, INL, and IPL and swollen rod OSs (ROS). (C) Control RPE and photoreceptor cell OS and IS layer. b, basal infoldings; a, apical folds; and c, cell body of retinal pigment epithelium. (D) RPE and photoreceptor OS layer in lead-exposed retina. Random patches of swollen and disrupted ROSs (arrowheads) which were primarily located in the proximal one half of the ROS. A monocyte was present in the choroidal vasculature (arrow). (From Fox, D. A. and Chu, L. W. -F., *Exp. Eye Res.*, 46, 613, 1988. With permission.)

exclusively in rod mitochondria. For example, Figures 9 and 10B illustrate the accumulation of large glycogen deposits in RIS. Glycogen particles are also observed in the mitochondria of rod axons (Figure 11B and insert) and rod synaptic terminals (Figures 10C and D). Two specific ultrastructural features of the glycogen containing mitochondria are worthy of note. The first is that the enclosing double membrane of mitochondria is always clearly present, although it appears to be broken down in one or more points. Second, in most cases the volume of glycogen is so large that the cristae either are pushed to the periphery or are no longer present. However, occasionally a mitochondrion is observed (e.g., rod synaptic) containing only a few glycogen particles in the intracristal space (Figure 10D).

A morphological comparison of the external limiting membrane and ONL from control and lead-exposed rats was made (Figure 11). As previously reported, the intercellular junctional

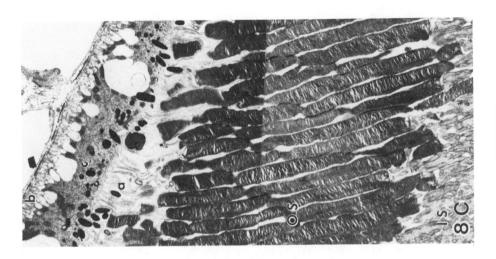

FIGURE 8C.

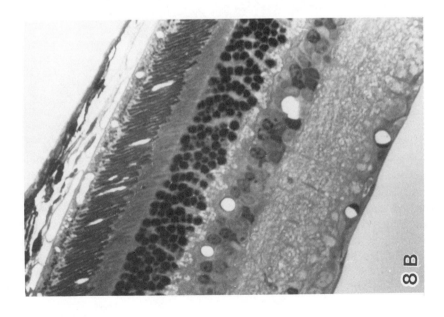

FIGURE 8B.

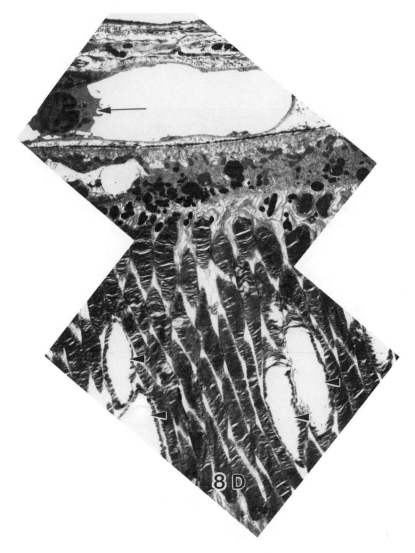

FIGURE 8D.

complexes, or zonula adherentes, between Muller cells are characterized by an increased electron density of the membranes and neighboring cytoplasm.[56] Although the Muller cells are slightly swollen in the lead-exposed rats, the structure of the external limiting membrane remains relatively normal (Figure 11B). In contrast, the ONL exhibits a selective loss of rod photoreceptor nuceli (vide supra), disrupted cytoarchitecture (i.e., increased inter-nuclei spacing) and gliosis (Figure 11B). In addition, most of the remaining rod photoreceptor nuclei are shrunken (Figure 11B). Furthermore, the outer plexiform layer in the lead-exposed rats contains swollen mitochondria and appears more electron lucent (Figure 11B).

IV. SIGNIFICANCE OF RECENT FINDINGS

Previous behavioral and electrophysiological studies have suggested that lead exposure during development produces long-term selective scotopic visual deficits. However, no direct analyses of rod and cone functions had been performed. Our recent studies on rod and cone ERGs, cyclic nucleotide metabolism, and rhodopsin, in combination with the morphological

TABLE 2
Rod and Cone Nuclei in Sections of Superior and Inferior Temporal Retina from Control and Developmentally Lead-Exposed Rats[a]

	Control		Lead	
	Superior	**Inferior**	**Superior**	**Inferior**
Posterior retina:				
Rod nuclei	140.2		116.5	101.5
	$\pm 3.2^b$		$\pm 1.8^c$	$\pm 1.8^c$
Cone nuclei	2.6			2.5
	± 0.3			0.4
% rod nuclei remaining			83.1	72.4
% cone nuclei remaining				96.2
Peripheral retina:				
Rod nuclei	81.0		70.9	63.1
	± 1.3		$\pm 1.7^c$	$\pm 1.5^c$
Cone nuclei	1.5			1.5
	± 0.2			± 0.4
% rod nuclei remaining			87.5	77.9
% cone nuclei remaining				100.0

[a] The rod and cone nuclei counts are based on 30-100 μm lengths; 10 consecutive lengths in each of three sections from 11 control and 12 lead-exposed rats as described in the Methods.
[b] Values represent the mean ± SEM per 100 μm of retina.
[c] Significantly different from controls at $p < 0.01$.

From Fox, D. A. and Chu, L. W. -F., *Exp. Eye Res.,* 46, 613, 1988. With permission.

TABLE 3
Rod Outer Segment Lengths (μm) in the Posterior Superior and Inferior Retina of Long-Evans Hooded Control and Lead-Exposed Rats[a]

Retinal Quadrant	Control	Lead-Exposed[b]
Superior	33.55 ± 1.21^c	30.44 ± 1.04
Inferior	27.49 ± 0.82	23.50 ± 1.05

[a] The rod outer segment length was determined in ten consecutive fields, each 100 μm in length, on each side of the optic nerve head in each of three sections from controls and lead-exposed rats.
[b] Values represent the mean ± SEM for 5 to 12 rats per treatment.
[c] Values in the superior and inferior retina of lead-exposed rats are significantly different form similar matched areas in controls at $p < 0.05$.

From Fox, D. A. and Rubinstein, S. D., *Exp. Eye Res.,* 48, 237, 1989. With permission.

and quantitative histological studies[20,21,26] demonstrate, in a well-validated animal model of subclinical pediatric lead poisoning, that low-level lead exposure during early postnatal development produces long-term selective deficits in rod photoreceptor function and biochemistry and selective rod degeneration. In addition, lead exposure during development causes some general retinal damage as evidenced by the mitochondrial glycogen accumulation.

The V-log I data for the a-wave and b-wave and the retinal sensitivity data reveal a marked reduction of the light response and Rmax, as well as a loss of relative and absolute sensitivity.

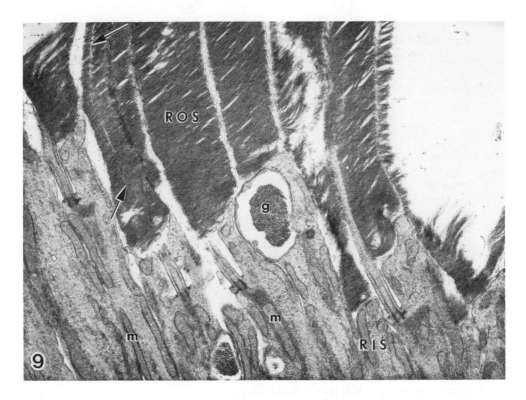

FIGURE 9. A higher magnification of rod outer (ROS) and inner segments (RIS) showing both the swollen and disrupted ROSs and the large intramitochondrial glycogen accumulations in the RIS. In the ROS to the left, the incisures in most of the discs are in register (between arrows). × 21,000. ROS, rod outer segment; RIS, rod inner segment; m, mitochondria; g, beta-glycogen particles. (Fox, D. A. and Chu, L. W.-F., *Exp. Eye Res.*, 46, 613, 1988. With permission.)

Several possible, although not mutually exclusive, mechanisms could account for these ERG alterations. One possible explanation involves our findings of a lead-induced elevation of cGMP levels in both the dark-adapted and light-adapted states, and of inhibition of light-activated cGMP-PDE. Initially, elevated cGMP levels in the dark would open more light-sensitive sodium ion channels in the ROS,[57] leading to an increased depolarization of the ROS plasma membrane.[58] Then the lead-induced inhibition of light-activated cGMP-PDE would prevent closure of all the open sodium ion channels, thereby decreasing the absolute and relative sensitivity and reducing the amplitude of the light response. Similar rod ERG and photoreceptor changes occur in toad, cat, and man following exposure to PDE inhibitors.[59-61] Second, these ERG alterations could result from loss of the visual pigment rhodopsin and the consequent reduction in quantum catch. The sensitivity loss and reduced Rmax found in our lead-exposed rats requires that the rhodopsin concentration be reduced to 70 to 75% of the adult control value.[27,30] Our findings of a 30 to 35% reduction in rhodopsin concentration in similarly lead-treated rats are consistent with this hypothesis.[26] Third, lead-induced changes in ROS calcium homeostasis may account for the observed ERG alterations. Large decreases in absolute sensitivity and Rmax occur following prolonged superfusion of isolated, dark-adapted rat and toad retinas with low calcium Ringer's solution.[62,63] Observations from many diverse studies suggest that the primary toxic action of lead is the disruption of calcium transport, storage, and/or metabolism.[64] Finally, these ERG deficits may be due to a decrease in the conductance of ROS cation channels, a change reported in other neuronal preparations following lead exposure.[65]

The L-log I and CFF functions both reveal a slowing of the time course of the rod flash response. The increased latencies and decreased CFF values are most likely due to the lead-induced inhibition of light-activated cGMP-PDE and/or increase of cGMP concentration. Experiments by several investigators have demonstrated that the latency (and time course) of the rod response is increased by inhibition of PDE[19,60,61,66] or by elevation of the intracellular levels of cGMP.[58] Alternatively, the increased latencies and slowed rod responses may be due to the alterations in ROS calcium homeostasis (vide supra). Similar temporal changes occur in isolated, dark-adapted toad retinas superfused with low calcium Ringer's solution.[67] Interestingly, the cone plateau phase of the CFF function in the lead-exposed rats shows a slight enhancement. This effect probably results from a release (decrease) of tonic rod inhibition on the cones, as suggested by flicker detection experiments in man.[68,69] Occasionally, a small cone enhancement is observed in the isolated, perfused bullfrog retina preparation, but only when using high concentrations of lead that maximally depress the rod response.[70]

Selective degeneration of normal rod photoreceptors has been produced in amphibian and human retinas in culture following the application of IBMX, a cGMP-PDE inhibitor.[71-73] This effect is demonstrated to depend on the accumulation of excess cGMP. Similarly, a deficiency in cGMP-PDE activity that results in the accumulation of cGMP is involved in the process of retinal degeneration (inherited) that preferentially affects rod photoreceptors in several species of animals.[24] Finding from our studies, demonstrating selective functional rod deficits and selective rod degeneration,[20,21,26] also appear to result from the inhibition of cGMP-PDE and the accumulation of excess cGMP. The exact nature of the deficiency in cGMP-PDE activity has yet to be determined. Interestingly, the functional and morphological deficits observed in the recent studies occur at lower concentrations of excess cGMP and with less inhibition of cGMP-PDE than in those studies cited above. This suggests that lead may have additional sites and mechanisms of action. The thinning of the INL may be due to transneuronal degeneration.

As suggested by our *in vitro* lead data, the increased *in vivo* levels of cGMP in the dark-adapted and light-adapted states also may result from a direct stimulation of guanylate cyclase activity by lead. In addition, the selective stimulation of guanylate cyclase activity *in vitro* and possibly *in vivo* may partially result from an indirect action of lead on calcium metabolism. As previously noted, lead can compete with the divalent calcium ion and alter its effect concentration.[64] For example, it has been demonstrated that dark-adapted mice retinas incubated in low calcium solution have significantly increased levels of cGMP with no change in cAMP.[74] More recently, this selective rod effect has been shown in dark-adapted rat retinas to result from stimulation of guanylate cyclase activity.[75]

Two types of marked regional differences in rod photoreceptor degeneration are observed in the retinas from lead-exposed rats. First, the selective loss of rod nuclei and the decrease in ROS length is more extensive in the inferior retinal hemisphere than in the superior retinal hemisphere. Similar regional differences in the response of rod photoreceptors to iodoacetate poisoning,[56] vitamin A deficiency,[38] and inherited retinal dystrophy[35,76] have been observed. Second, the rod loss exhibits a central-peripheral gradient of degeneration as evidenced by the greater percentage loss of rod nuclei in the superior and inferior posterior (-22.3%: 140.2 in controls vs. mean value of 109.0 in lead-exposed rats) than in the superior and inferior peripheral (-17.3%: 81.0 in controls vs. mean value of 67.0 in lead-exposed rats) temporal retina. A similar central peripheral gradient of photoreceptor cell degeneration has been noted following chemical, physical, metabolic, or genetic insult.[38,51,56,77-79]

The cellular mechanisms of action responsible for these regional differences in hooded rats are presently unknown. Several studies suggest, however, that they are not due to environmental lighting conditions but rather are intrinsic to the eye. Intrinsic morphological and biochemical regional differences in the retina of hooded rats have been reported. For example, mean ROS length is longer and rhodopsin content per eye is lower in the superior than in ther inferior retinal hemisphere.[31,26,80] On the other hand, more melanosomes are present in the inferior than superior retinal hemisphere.[79,81]

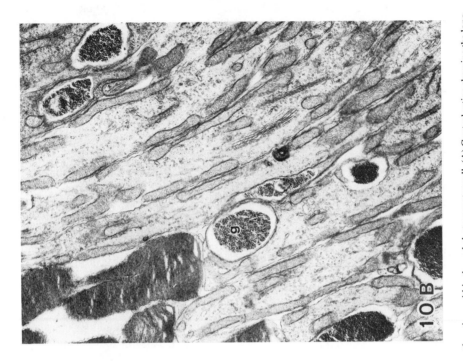

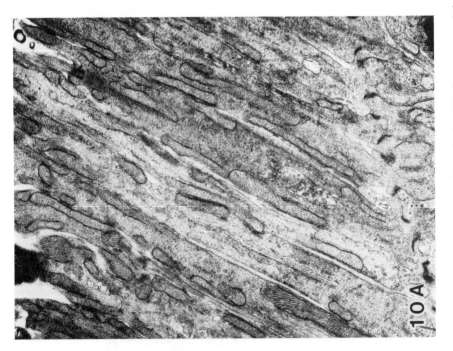

FIGURE 10.   Electron micrographs of intramitochondrial glycogen accumulation in different locations within the rod photoreceptor cell. (A) Control retina showing the long, slender rod inner segment mitochondria. × 17,000. (B) Lead-treated retina illustrating several rod inner segment mitochondria filled with beta-glycogen particles. × 17,000. (C) and (D), Various degrees of glycogen accumulation in rod synaptic terminal mitochondria. × 45,000. g, beta-glycogen particles. (Fox, D. A. and Chu, L. W.-F., *Exp. Eye Res.*, 46, 613, 1988. With permission.)

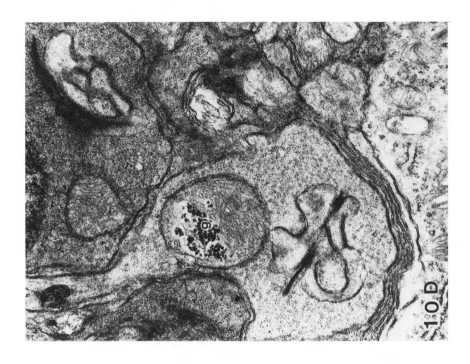

FIGURE 10D.

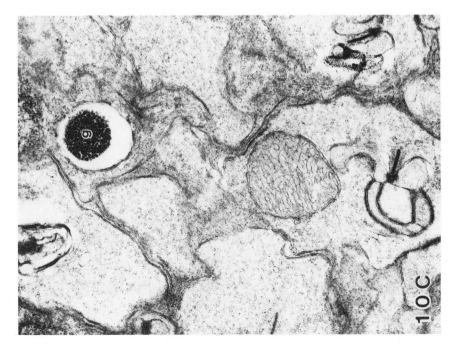

FIGURE 10C.

The exact mechanisms accounting for the loss of rod visual pigment, decreases in ROS length and the consequent decreases in retinal sensitivity are presently unknown. Our data suggests that these effects are the direct result of lead exposure since most of the changes occur by 30 d of age.[26] Possible cellular mechanisms of action responsible for these immediate and long-term lead-induced selective rod alterations include an inhibition of retinal protein synthesis, an alteration in phagocytosis, an alteration in vitamin A or zinc metabolism, an inhibition of retinal intermediary metabolism, and/or abnormal rhodopsin packing in the ROS. We believe, however, that no one mechanism will account for the diverse effects of lead intoxication. Most likely it will eventually be found that lead exerts its toxic effects by acting at several cellular and subcellular regulatory sites of action.

For example, the elevation of retinal cGMP by selective cGMP-PDE inhibitors has been found to selectively inhibit protein synthesis in rod photoreceptors.[73] This latter mechanism may be directly responsible for the decreased ROS length as well as the selective rod degeneration observed in our developmentally lead-exposed rats. Alternatively, the decrease in ROS length, the loss of rhodopsin and the decrease in retinal sensitivity could be due to an increased phagocytosis by the retinal pigment epithelium (RPE), as observed in patients with retinitis pigmentosa.[32] However, no morphological changes are observed in the RPE and no differences in the number of large phagosomes per number of rod cells are observed in similarly lead-exposed rats.[82] Decreased *de novo* synthesis of rhodopsin or incomplete regeneration of rhodopsin following bleaching in our lead-exposed rats, due to abnormal vitamin A and/or zinc metabolism, could result in a decrease in ROS length, a loss of rhodopsin and/or decreased retinal sensitivity.[27,33,83] The findings that a dietary zinc deficiency in developing rats causes a decrease in the activity of the metalloenzyme alcohol dehydrogenase which results in a decrease in the conversion of retinol to retinal,[84] coupled with our analytical data revealing a zinc deficiency in the brains of similarly lead-exposed rats,[85] suggests this as a possible mechanism accounting for our recent results. Finally, alterations in intermediary metabolism have been suggested by Noell[56] and Graymore and Tansley[86] to be the basis of selective rod degenerations. In support of this hypothesis, recently we have observed a lead-induced inhibition of energy metabolism in isolated whole rat retinas[87] and respiration in isolated rat retinal mitochondira.[88]

One notable structural alteration observed in the retinas from developmentally lead-exposed rats are the random patches of swollen and disrupted ROSs. We suggest these swellings, which primarily occur in the proximal one half to one third of the ROS, are treatment-related fixation artifacts and result from an altered rod intracellular ionic environment. Based on the factors controlling ROS permeability to sodium,[89] a lead-induced increase in retinal cGMP[20] or decrease in retinal calcium, as observed in other tissues,[64] may lead to a net influx of NaCl into the rods with subsequent swelling. An effect of this kind has been observed in X-ray diffraction measurements.[90]

Light microscopic, histochemical,[91] and quantitative electron microscopy[92] studies have revealed that under normal conditions the highly vascularized rat retina contains only a very small amount of glycogen beta-particles. The highest cencentration of glycogen is measured in the Muller cells and their processess, while a lower concentration is found in inner retinal neurons.[92] In marked contrast, large deposits of glycogen particles are observed in the distal retina (i.e., RIS, ONL, OPL) of lead-exposed rats. Interestingly, these accumulations appear to be exclusively localized to rod mitochondria: RIS, rod axon, and rod synaptic terminal mitochondria. This is probably followed by further accumulation in the cristae leading to an expansion of their internal space and eventually to the loss of all cristae.

Descriptions of intramitochondrial glycogen inclusions in the rat are relatively uncommon, except in aging control and diabetic animals.[92-94] The manner in which glycogen appears in the mitochondrion of the rat and the significance of this event are thus difficult to evaluate. The enzymes necessary for glyconeogenesis or glycogenolysis reside outside the mitochondria. However, it may be possible that glycogen is polymerized in the mitochondria from morpho-

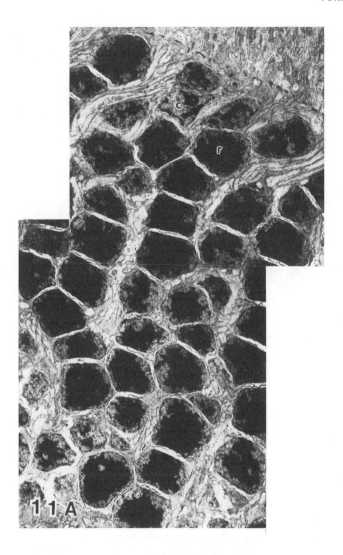

FIGURE 11.   Low power electron micrographs of the outer nuclear layer (ONL) from control (A) and lead-exposed (B) retinas. (A) Several cone photoreceptor nuclei are seen among the numerous rod photoreceptor nuclei. The distinctive chromatin pattern of the rod and cone cells seen by light microscopy are evident in this electron micrographic montage. The control ONL is 10 to 12 nuclei thick. × 5300. c, cone nuclei; r, rod nuclei. (B) Lead-treated retina illustrating the thinned (5 to 6 cells thick) ONL. Most of the remaining rod nuclei are shruken and/or pyknotic (p). The inter-nuclei spacing is greatly increased and there is extensive gliosis in this layer. The arrow points to a rod axon mitochondria containing a large accumulation of beta-glycogen particles (g). This is better visualized in the inset. × 13,000. (Fox, D. A. and Chu, L. W.-F., *Exp. Eye Res.*, 46, 613, 1988. With permission.)

logically undetectable monomer or dimer units taken up via mitochondrial membranes. This idea has been postulated to account for the intramitochondrial glycogen particles observed in the photoreceptor inner segments of aged rats.[93] Alternatively, it has been suggested that accumulation of membrane-bound glycogen in the retina may be related to trauma or ischemia.[91] The ischemic stimulus for this glycogen accumulation may be a function of a toxicant-induced (i.e.,

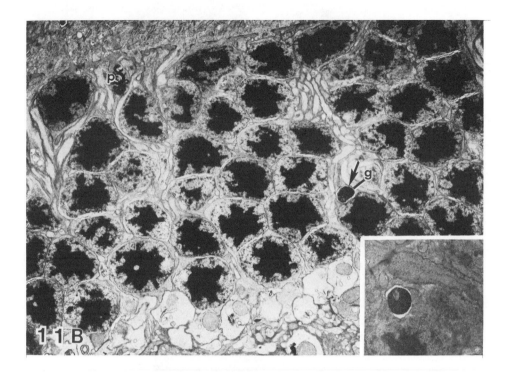

FIGURE 11B.

lead) or age-related change in the vascular supply. This raises the interesting and important possibility that low-level lead exposure during development, like other heavy metals,[95] accelerates the aging process.

In summary, recent studies in ERG, cyclic nucleotide metabolism, rhodopsin content, quantitative histology, and ultrastructure demonstrate that low-level lead exposure during early postnatal development produces long-term selective rod photoreceptor deficits and degeneration. In addition, general retinal damage as evidenced by mitochondrial glycogen accumulation is noted. The relevance and applicability of these data to subclinical pediatric lead poisoning has yet to be established. In addition, the data suggest that high levels of lead exposure in the mature organism may produce similar deficits. Furthermore, if rods and blue-sensitive cones in humans exhibit the same sensitivity to a lead-induced inhibition of cGMP-PDE as they do to the drug-induced inhibition,[60,96] color vision deficits, in addition to scotopic vision deficits, may result from lead exposure. The similarities between the effects produced by low-level lead exposure during development and those caused by iodoacetate,[86] vitamin A deficiency,[37,38] and isobutylmethylxanthine,[71,72] as compared to the inherited retinal dystrophies in man and animals,[24,35,78,97] establishes the need for further work in this environmentally relevant animal model of a selective rod photoreceptor deficit.

ACKNOWLEDGMENTS

I thank Drs. Debora B. Farber, Lena, W. -F. Chu, and Steve D. Rubinstein for their collaborative help and Dr. Andrew Alpar and Ms. Yvonne S. Blocker for their excellent technical assistance. Original work reported here was supported in part by an NIH Grant R01 ES 03183 from the National Institute of Environmental Health Sciences and a Sigma Xi Research Grant.

REFERENCES

1. **Grant, W. M.,** *Toxicology of the Eye,* 2nd ed., Charles C Thomas, Springfield, IL, 1974.
2. **Merigan, W. H. and Weiss, B.,** *Neurotoxicity of the Visual System,* Raven Press, New York, 1980.
3. **Williams, T. P. and Baker, B. N.,** *The Effects of Constant Light on Visual Processes,* Plenum Press, New York, 1980.
4. **Fraunfelder, F. T.,** *Drug-Induced Ocular Side Effects and Drug Interactions,* Lea & Febiger, Philadelphia, 1982.
5. **Dayhaw-Barker, P., Forbes, D., Fox, D. A., Lerman, S., McGinness, Waxler, M., and Felter, R.,** Drug photoxicity and visual health, in *Optical Radiation and Visual Health,* Waxler, M. and Hitchins, V. M., Eds., CRC Press, FL, 1986, 147.
6. **Massof, R. W., Sykes, S. M., Rapp, L. M., Robinson, W. G., Jr., Zwick, H., and Hochheimer, B.,** Optical radiation damage to the ocular photoreceptors, in *Optical Radiation and Visual Health,* Waxler, M. and Hitchins, V. M., Eds., CRC Press, FL, 1986, 69.
7. **Evans, H. L., Laties, V. G., and Weiss, B.,** Behavioral effects of mercury and methylmercury, *Fed. Proc.,* 34, 158, 1975.
8. **Plestina, R. and Piakovic-Plestina, M.,** Effect of anticholinesterase pesticide on the eye and vision, *CRC Crit. Rev. Toxicol.,* 6, 1, 1978.
9. **Bushnell, P. J., Bowman, R. E., Allen J. R., and Marlar, R. J.,** Scotopic vision deficits in young monkeys exposed to lead, *Science,* 196, 333, 1977.
10. **Cavalleri, A., Trimarchi, F. ,Gelmi, C., Baruffini, A., Minoia, C., Biscaldi, G., and Gallo, G.,** Effects of lead on the visual system of occupationally exposed subjects, *Scand. J. Work Environ. Hlth.,* 8 (Suppl. 1), 148, 1982.
11. **Bowman, R. E. and Bushnell, P. J.,** Scotopic visual deficits in young monkeys given chronic low levels of lead, in *Neurotoxicity of the Visual System,* Merigan, W. H. and Weiss, B., Eds., *Academic Press,* New York, 1980, 219.
12. **Fox, D. A., Wright, A. A., and Costa, L. G.,** Visual acuity deficits following neonatal lead exposure: Cholinergic interactions, *Neurobehav. Toxicol. Teratol.,* 4, 689, 1982.
13. **Fox, D. A.,** Psychophysically and electrophysiologically determined spatial vision deficits in developmentally lead-exposed rats have a cholinergic component, in *Cellular and Molecular Neurotoxicology,* Narahashi, T. Ed., Raven Press, New York, 1984, 123.
14. **Fox, D. A., Lewkowski, J. P., and Cooper, G. P.,** Persistent visual cortex excitability alterations produced by neonatal lead exposure, *Neurobehav. Toxicol.,* 1, 101, 1979.
15. **Costa, L. G. and Fox, D. A.,** A selective decrease of cholinergic muscarinic receptors in the visual cortex of adult rats following developmental lead exposure, *Brain Res.,* 276, 259, 1983.
16. **Guguchkova, P. T.,** Electroretinographic and electrooculographic examinations of persons occupationally exposed to lead, *Vestn. Oftalmol.,* 85, 60, 1972.
17. **Signorino, M., Scarpino, O., Provincialli, L., Marchesi, G. F., Valentino, M., and Governa, M.,** Modification of the electroretinogram and of different components of the visual evoked potentials in workers exposed to lead, *Ital. Electroenceph. J.,* 10, 51, 1983.
18. **Fox, D. A. and Sillman, A. J.,** Heavy metals affect rod, but not cone, photoreceptors, *Science,* 206, 78, 1979.
19. **Tessier-Lavigne, M., Mobbs, P., and Attwell, D.,** Lead and mercury toxicity and the rod light response, *Invest. Ophthalmol. Vis. Sci.,* 26, 1117, 1985.
20. **Fox, D. A. and Farber, D. B.,** Rods are selectively altered by lead. I. Electrophysiology and biochemistry, *Exp. Eye Res.,* 46, 597, 1988.
21. **Fox, D. A. and Chu, L. W. -F.,** Rods are selectively altered by lead. II. Ultrastructure and quantitative histology, *Exp. Eye Res.,* 46, 613, 1988.
22. **Santos-Anderson, R. M., Tso, M. O. M., Valdes, J. J., and Annau, Z.,** Chronic lead administration in neonatal rats: Electron microscopy of the retina, *J. Neuropathol. Exp. Neurol.,* 43, 175, 1984.
23. **Lolley, R. N., Farber, D. B., Rayborn, M. R., and Hollyfield, J. G.,** Cyclic-GMP accumulation causes degeneration of photoreceptor cells: simulation of an inherited disease, *Science,* 196, 664, 1974.
24. **Farber, D. B. and Shuster, T. A.,** Cyclic nucelotides in retinal function and degeneration, in *The Retina,* Part I, Adler, R. and Farber, D. B., Eds., Academic Press, New York, 1986, 239.
25. **Haynes, L. and Yau, K. -W.,** Cyclic GMP-sensitive conductance in outer segment membrane of catfish cones, *Nature,* 317, 61, 1985.
26. **Fox, D. A. and Rubinstein, S. D.,** Age-related changes in retinal sensitivity, rhodopsin content and rod outer-segment length in hooded rats following low-level lead exposure during development, *Exp. Eye Res.,* 48, 237, 1989.
27. **Dowling, J. E.,** Chemistry of visual adaptation in the rat, *Nature,* 188, 114, 1960.
28. **Rushton, W. A. H.,** Rhodopsin measurement and dark-adaptation in a subject deficient in cone vision, *J. Physiol.,* 156, 193, 1961.

29. **Fulton, A. B. and Baker, B. N.,** The relation of retinal sensitivity and rhodopsin in developing rat retina, *Invest. Ophthalmol., Vis. Sci.,* 25, 647, 1984.
30. **Bonting, S. L., Caravaggio, L. L., and Gouras, P.,** The rhodopsin cycle in the developing retina. I. Relation of rhodopsin content, electroretinogram and rod structure in the rat, *Exp. Eye Res.,* 1, 14, 1961.
31. **Battelle, B. A. and LaVail., M. M.,** Rhodopsin content and rod outer segment length in albino rat eyes: Modification by dark adaptation, *Exp. Eye Res.,* 26, 487, 1978.
32. **Ripps, H., Brin, K. P., and Weale, R. A.,** Rhodopsin and visual threshold in retinitis pigmentosa, *Invest. Ophthalmol. Vis. Sci.,* 17, 735, 1978.
33. **Dowling, J. E.,** Night blindness, dark adaptation, and the electroretinogram, *Am. J. Ophthamol.,* 50, 875, 1960.
34. **Dowling, J. E. and Sidman, R. L.,** Inherited retinal dystrophy in the rat, *J. Cell Biol.,* 14, 159, 1962.
35. **LaVail, M. M. and Battelle, B. A.,** Influence of eye pigmentation and light deprivation on inherited retinal dystrophy in the rat, *Exp. Eye Res.,* 21, 167, 1975.
36. **Rapp, L. M. and Williams, T. P.,** Rhodopsin content and electroretinographic sensitivity in light-adapted rat retina, *Nature,* 267, 835, 1977.
37. **Carter-Dawson, L., Kuwabara, T., O'Brien, P. J., and Bieri, J. G.,** Structural and biochemical changes in vitamin A-deficient rat retinas, *Invest. Ophthamol. Vis. Sci.,* 18, 437, 1979.
38. **Carter-Dawson, L., Kuwabara, R., and Bieri, J. C.,** Intrinsic, light-independent, regional differences in photoreceptor cell degeneration in vitamin A-deficient rat retinas, *Invest. Ophthamol. Vis. Sci.,* 22, 249, 1982.
39. **LaVail, M. M.,** Rod outer segment disc shedding in relation to cyclic lighting, *Exp. Eye Res.,* 23, 277, 1976.
40. **Bornschein, R. L., Fox, D. A., and Michaelson, I. A.,** Estimation of the daily exposure in neonatal rats receiving lead via dam's milk, *Toxicol. Appl. Pharmacol.,* 40, 577, 1977.
41. **Fox, D. A., Lewkowski, J. P., and Cooper, G. P.,** Acute and chronic effects of neonatal lead exposure on the development of the visual evoked response in rats, *Toxicol. Appl. Pharmacol.,* 40, 449, 1977.
42. **Dodt, E. and Echte, K.,** Dark and light adaptation in pigmented and white rat as measured by electroretinogram threshold, *J. Neurophysiol.,* 24, 427, 1961.
43. **Dowling, J.,** Visual adaptation: Its mechanism, *Science,* 157, 584, 1967.
44. **Green, D. G.,** Scotopic and photopic components of the rat electroretinogram, *J. Physiol.,* 228, 781, 1973.
45. **Farber, D. B. and Lolley, R. N.,** Measurement of cyclic nucleotides in retina, in *Methods of Enzymology,* Vol. 81, Packer, L., Ed., Academic Press, New York, 1982, 551.
46. **Lowry, O. H., Rosebrough, N. J., Farr, A. L., and Randall, R. J.,** Protein measurement with the Folin phenol reagent, *J. Biol. Chem.,* 193, 265, 1951.
47. **Farber, D. B. and Lolley, R. N.,** Enzymatic basis for cyclic GMP accumulation in degenerative photoreceptor cells of mice retina, *J. Cyclic Nucleot. Res.,* 2, 139, 1976.
48. **Lolley, R. N. and Farber, D. B.,** Cyclic nucleotide phosphodiesterases in dystrophic rat retinas: Guanosine 3′, 5′ cyclic monophosphate anomalies during photoreceptor cell degeneration, *Exp. Eye Res.,* 20, 585, 1975.
49. **Winkler, B. S.,** The electroretinogram of the isolated rat retina, *Vision Res.,* 12, 1183, 1972.
50. **Fulton, A. B., Manning, K. A., Baker, B. N., Schukar, S. E., and Bailey, C. J.,** Dark-adapted sensitivity, rhodopsin content, and background adaptation in pcd/pcd mice, *Invest. Ophthalmol. Vis.,* 22, 386, 1982.
51. **LaVail, M. M.,** Survival of some photoreceptor cells in albino rats following long-term exposure to continuous light, *Invest. Ophthalmol.,* 15, 64, 1976.
52. **Revel, J. P., Napolitano, L., and Fawcett, D. W.,** Identification of glycogen in electron micrographs of thin sections, *J. Biophys. Biochem. Cytol.,* 8, 575, 1960.
53. **Iwamasa, T., Fujisaki, A., and Takeuchi, T.,** Ultrastructure of glycogen particles, *J. Electron Microsc. (Tokyo),* 19, 371, 1970.
54. **Winer, B. J.,** *Statistical Principles in Experimental Design,* McGraw-Hill, New York, 1962.
55. **Kuwabara, T. and Gorn, R. A.,** Retinal damage by visible light, *Arch. Ophthalmol.,* 79, 69, 1968.
56. **Noell, W. K.,** The impairment of visual cell structure by iodoacetate, *J. Cell. Comp. Physiol.,* 40, 25, 1952.
57. **Miller, W. H.,** Physiological effects of cyclic GMP in the vertebrate retinal rod outer segment, in *Advances in Cyclic Nucleotide Research,* Vol. 15., Greengard, P. and Robison, G. A., Eds., Raven Press, New York, 1983, 495.
58. **Miller, W. H. and Nicol, G. D.,** Evidence that cyclic GMP regulates membrane potential in rod photoreceptors, *Nature,* 280, 64, 1979.
59. **Lipton, S. A., Rasmussen, H., and Dowling, J. E.,** Electrical and adaptive properties of rod photoreceptors in Bufo marinus. II. Effects of cyclic nucleotides and prostaglandins, *J. Gen. Physiol.,* 70, 771, 1977.
60. **Zrenner, E., Kramer, W., Bittner, Ch., Bopp, M., and Schlepper, M.,** Rapid effects on colour vision following intravenous injection of a new, non-glycoside positive inotropic substance (AR-L 115 BS), *Doc. Ophthalmol. Proc. Ser.,* 33, 493, 1982.
61. **Schneider, T. and Zrenner, E.,** The influence of phosphodiesterase inhibitors on ERG and optic nerve response of the cat, *Invest. Ophthalmol. Vis. Sci.,* 27, 1395, 1986.
62. **Yoshikami, S. and Hagins, W. A.,** Control of the dark current in vertebrate rods and cones, in *Biochemistry and Physiology of Visual Pigments,* Langer H., Ed., Springer-Verlag, New York, 1973, 245.

63. **Lipton, S. A., Ostroy, S. E., and Dowling, J. E.,** Electrical and adaptive properties of rod photoreceptors in Bufo marinus. I. Effects of altered extracellular Ca^{2+} levels, *J. Gen. Physiol.,* 709, 747, 1977.

64. **Pounds, J.,** Effect of lead intoxication on calcium homeostasis and calcium-mediated cell function: A review, *Neurotoxicology,* 5, 295, 1984.

65. **Audersirk, G.,** Effects of lead exposure on the physiology of neurons, *Prog. Neurobiol.,* 24, 199, 1985.

66. **Capovilla, M., Cervetto, L., and Torre, V.,** The effect of phosphodiesterase inhibitors on the electrical activity of toad rods, *J. Physiol.,* 343, 277, 1983.

67. **Bastian, B. L. and Fain, G. L.,** The effects of low calcium and background light on the sensitivity of toad rods, *J. Physiol.,* 330, 307, 1982.

68. **Goldberg, S. H., Frumkes, T. E., and Nygaard, R. W.,** Inhibitory influence of unstimulated rods in the human retina: Evidence provided by examining cone flicker, *Science,* 221, 180, 1983.

69. **Coletta, N. J. and Adams, A. J.,** Rod-cone interaction in flicker detection, *Vision Res.,* 24, 1333, 1984.

70. **Sillman, A. J., Bolnick, D. A., Bosetti, J. B., Haynes, L. W., and Walter, A. E.,** The effects of lead and of cadmium on the mass receptor potential: The dose-response relationship, *Neurotoxicology,* 3, 179, 1982.

71. **Lolley, R. N., Farber, D. B., Rayborn, M. E., and Hollyfield, J. G.,** Cyclic GMP accumulation causes degeneration of photoreceptor cells: simulation of an inherited disease, *Science,* 196, 664, 1977.

72. **Ulshafer, R. J., Garcia, C. A., ad Hollyfield, J. G.,** Sensitivity of photoreceptors to elevated levels of cGMP in the human retina, *Investig. Ophthalmol. Vis. Sci.,* 19, 1236, 1980.

73. **Ulshafer, R. J., Fliesler, S.J., and Hollyfield, J. G.,** Differential sensitivity of protein synthesis in human retina to a phosphodiesterase inhibitor and cyclic nucleotides, *Curr. Eye Res.,* 3, 383, 1984.

74. **Cohen, A. I., Hall, I. A., and Ferrendelli, J. A.,** Calcium and cyclic nucleotide regulation in inucbated mouse retinas, *J. Gen. Physiol.,* 71, 595, 1978.

75. **Lolley, R. N. and Racz, E.,** Calcium modulation of cyclic GMP synthesis in rat visual cells, *Vision Res.,* 22, 1481, 1982.

76. **Krill, A. E., Archer, D., and Martin, D.,** Sector retinitis pigmentosa, *Am. J. Ophthalmol.,* 69, 977, 1970.

77. **Bourne, M. C., Campbell, D. A., and Tansley, K.,** Hereditary degeneration of the rat retina, *Br. J. Ophthalmol.,* 22, 613, 1938.

78. **Carter-Dawson, L. D., LaVail, M. M., and Sidman, R. L.** Differential effect of the rd mutation on rods and cones in the mouse retina, *Investig. Ophthalmol. Vis. Sci.,* 17, 489, 1978.

79. **Howell, W. L., Rapp, L. M., and Williams, T. P.,** Distribution of melanosomes across the retinal pigment epithelium of a hooded rat: implications for light damages, *Invest. Ophthalmol. Vis Sci.,* 22, 139, 1982.

80. **Rapp, L. M., Naash, M. I., Wiegand, R. D. Joel, C. D., Nielsen, J. C., and Anderson, R. E.,** Morphological and biochemical comparisons between retinal regions having differing susceptibility to photoreceptor degeneration, in *Retinal Degeneration: Experimental and Clinical Studies,* LaVail, M. M., Hollyfield, J. G., and Anderson, R. E., Eds., Alan R. Liss, New York, 1985, 421.

81. **LaVail, M. M.,** Eye pigmentation and constant light damage in the rat retina, in *The Effects of Constant Light on Visual Processes,* Williams, T. P. and Baker, B. N., Eds., Plenum Press, New York, 1980, 357.

82. **Rubinstein, S. D., Fox, D. A., and Wilson, R. D.,** Rhodopsin content, rod retinal sensitivity and rod outer segment length decrease following developmental lead exposure, *Invest. Ophthalmol. Vis. Sci. Suppl.,* 27, 234, 1986.

83. **Kraft, S. P., Parker, J. A., Mutak, Y., and Rao, A. V.,** The rat electroretinogram in combined zinc and vitamin A deficiency, *Invest. Ophthalmol. Vis. Sci.,* 28, 975, 1987.

84. **Huber, A. M. and Gerschoff, S. N.,** Effects of zinc deficiency on the oxidation of retinol and ethanol in rats, *J. Nutr.,* 105, 1486, 1975.

85. **Fox, D. A. and Ku, R.,** Effects of inorganic lead on carbonic anhydrase activity and regional brain lead and zinc levels, *Toxicologist,* 2, 82, 1982.

86. **Graymore, C. and Tansley, K.,** Iodoacetate poisoning of the rat retina, *Br. J. Ophthalmol.,* 43, 486, 1959.

87. **Fox, D. A. and Rubinstein, S. D.,** Oxygen consumption in isolated rat retina: Relative inhibitory and excitatory effects of potassium and lead, *Invest. Ophthalmol. Vis. Sci Suppl.,* 28, 249, 1987.

88. **Fox, D. A., Medrano, C. J., and Rubinstein, S. D.,** Isolated rat retinal mitochondrial respiration: In vitro and in vivo lead studies, *Toxicologist,* 8, 44, 1988.

89. **Miller, W. H.,** *Current Topics in Membrane Transport,* Vol. 15., Academic Press, New York, 1981.

90. **Chabre, M. and Cavaggioni, A.,** Light induced changes of ionic flux in the retinal rod, *Nature New Biol.,* 224, 118, 1973.

91. **Kuwabara, T. and Cogan, D. G.,** Retinal glycogen, *Arch. Ophthalmol.,* 66, 96, 1961.

92. **Sosula, L., Beaumont, P., Hollows, F. C., Jonson, K. C., and Regtop, H. L.,** Glycogen accumulation in retinal neurons and glial cells of streptozotocin-diabetic rats, *Diabetes,* 23, 221, 1974.

93. **Ishikawa, T. and Pei, Y. F.,** Intramitochondrial glycogen particles in rat retinal receptor cells, *J. Cell. Biol.,* 25, 402, 1965.

94. **Moore, S. A., Peterson, R. G., Felten, D. L., and O'Connor, B. L.,** Glycogen accumulation in tibial nerves of experimentally diabetic and aging control rats, *J. Neurol. Sci.,* 52, 289, 1981.

95. **Spyker, J. M.,** Assessing the impact of low level chemicals on development: Behavioral and latent effects, *Fed. Proc.,* 34, 1835, 1975.
96. **Zrenner, E. and Gouras, P.,** Blue-sensitive cones of the cat produce a rodlike electroretinogram, *Investig. Ophthalmol. Vis. Sci.,* 18, 1076, 1979.
97. **Duke-Elder, S.,** Diseases of the retina, in *Systems of Ophthalmology,* Vol. 10., Duke-Elder, S., Ed., C. V. Mosby, St. Louis, 1967, 577.

Chapter 7

BEHAVORIAL EFFECTS OF HEAVY METAL EXPOSURE

Robert L. Bornschein and Shou-Ren Kuang

TABLE OF CONTENTS

I. INTRODUCTION

The topic of this chapter, like those that precede it, involves the toxic consequences of metal exposure. However the level of analysis and the methods of analysis are quite different. In this chapter we will discuss the effects of metal exposure on the behavioral development of children and the performance of adults. The methods used are those of psychologists and epidemiologists. Epidemiologists are concerned with problems that entail a level of organization more complex than that of the individual person. By contrast, clinicians are more likely to focus on the individual, while physiologists might study separate organ systems and biochemists study intra and extracellular events (see for example the reviews by Fox[1] and Minnema and Cooper[2] in this volume). The varying levels of complexity in these disciplines reflect the extent to which the system can be isolated and controlled for the purpose of analysis. However, just as the effects of a toxicant on an isolated organ system often are not predictive of what occurs in the intact animal, the results from clinical or physiological studies cannot be used to predict the outcome of epidemiological investigations. Epidemiologists study the causes of disease and focus on factors which modify the incidence and severity of disease in the population. The methods used are not strictly experimental. Experimental control is achieved through the manipulation of experimental design and the use of sophisticated statistical procedures intended to control for the influence of confounding factors. Causality is inferred from an examination of the association between the presumed cause and its effect.[3] True causal associations are most readily inferred when they are (1) strong, (2) statistically significant, (3) specific to a particular disease, and (4) consistent with current theory. Further support for causality can be marshalled if the presumed cause precedes the effect in time and there is a dose-response relationship between the cause and the effects. Ultimately the final confirmation of an epidemiological conclusion rests on the results of an experimental intervention.

In this review, we will examine epidemiologic investigations which have as their primary objective the neurobehavioral evaluation of exposed populations. We will summarize the current status of the field and highlight some of the problems involved with the conduct of neurobehavioral epidemiologic investigations, precautions to be taken in interpreting the results of such studies, and questions which remain to be answered.

Toxicity arising from occupational or environmental exposure to heavy metals frequently manifests itself through changes in neurological functioning and behavior. A summary of metals reported to produce such neurological and behavioral effects is provided in Table 1 and 2, respectively. During the last decade, we have witnessed a steady progression in both the quantity and quality of studies of the neurobehavioral effects of metal exposures in man. The majority of studies have dealt with occupational and pediatric lead exposure. Neurobehavioral consequences of organic and inorganic mercurty exposure have also been studied in recent years, although the number of such studies is relatively limited. There have been no comprehensive epidemiological neurobehavioral studies of other toxic metals such as arsenic, cadmium, manganese or organic tin.

II. OCCUPATIONAL LEAD EXPOSURE AND ITS EFFECTS ON PERFORMANCE

An estimated one million U.S. workers are exposed to lead every day. Several industrial processes including smelting, refining, and storage battery manufacturing can result in excessive exposure to lead. More moderate exposures occur in mining and solder production and its use. The production of alloys and paint pigments can also result in lead exposure. Highly variable and often excessive lead exposure can also occur in non-lead industries, e.g., during the cutting and welding of lead-painted structures or during lead paint removal via sandblasting of building

TABLE 1
Neurological Effects of Occupational Metal Exposures

Symptom or sign	Metal
Ataxia	Manganese, methyl mercury
Headache	Lead, nickel
Myoclonus	Mercury
Nystagmus	Mercury
Paraplegia	Organic tin
Parkinsonism	Manganese
Seizures	Lead, methyl mercury, organic tin
Visual impairments	Mercury
Tremor	Manganese, mercury

TABLE 2
Behavioral Effects of Occupational Metal Exposure

Symptom or sign	Metal
Emotional lability	Manganese
Psychomotor dysfunction	Lead, mercury
Memory Impairment	Arsenic, lead, manganese
Fatigue	Arsenic, lead, mercury, organic tin

exteriors and bridges or sanding of interior painted surfaces. The internal lead burden resulting from any of these exposures is highly dependent not only on the absolute amount of lead inhaled or ingested, but also on the particle size and chemical form of lead.[4-6] It should also be remembered that occupational exposures pose a threat not only to the individual worker, but also to other family members in the event that lead is carried home on the worker's clothes.[7]

In today's highly regulated workplace, high level occupational lead exposure usually occurs as a result of an acute accidental exposure of one or at most a few individuals. These situations, which can result in overt symptomatology leading to encephalopathy and even death, do not usually lend themselves to systematic investigation beyond the level of clinical case report. Nor is it necessarily the case that symptoms associated with acute high level exposures resemble those observed under chronic, low-to-moderate exposure conditions. It is the latter case which is more common, more systematically studied, and which forms the bases for our current perception of adult neurobehavioral lead toxicity.

Mood and personality changes are the most frequently reported symptoms among lead workers. Symptoms include fatigue, irritability, depression, and difficulty in concentrating. Unfortunately, these effects are not specific to lead or to metal exposures in general. They are usually reported to appear after several months of exposure, at blood leads of 40 to 60 μg/dl,[8] and usually resolve several months after cessation of exposure. An investigator's ability to quantify these mood and personality changes and to clearly relate them to lead exposure is highly dependent upon the use of a standardized questionnaire or mood scale, standardized interview format, appropriate control group, consideration of potential confounding factors (e.g., alcohol use), age and education, and a study sample of sufficient size to yield the statistical power

TABLE 3
Mood and Personality Changes Among Workers with Lead Exposure

Number of subjects		PbB (µg/dl)[a]			
Exposed	Control	Exposed	Control	Findings	Ref.
99	61	21—66	21—25	Significant dose-response association between exposure indices and mood scales	10
49	27	27—69	<21	Increased depression and irritability and decreased libido	67
13	19	81	16	Increased incidence of reported symptoms in association with elevated PbB and ALAU[b]	68
49	24	32	12	Increased neuroticism	11
49	36	49	15	Poorer scores on mood scales and somatic complaints	69
403	305	30—56	10—22	Increased depression and hostility	70
288	181	>35	<35	Increased conflict in interpersonal relationships	71
85	55	46	18	No significant effects on personality	13
70	35	61	22	Increased symptom reporting but not controlled for covariates	16

[a]   Mean blood lead or range of PbB's for exposed and control population.
[b]   Amino levulinic acid in urine

necessary to detect an effect against a background of larger inter and intra-individual variation. A summary of recent field studies of lead effects on mood is provided in Table 3.

Changes in cognitive function have also been reported in numerous cross-sectional studies. Impairments in verbal ability,[9,10] memory,[9-12] and visual-motor functioning[9-15] are most frequently found (see Table 4). Significant advances in our understanding of lead's effect on cognition are being made as a result of the use of longitudinal[8,10,16] or prospective[17] study designs and standardized assessment batteries.[18-20] For example, longitudinal follow-up studies[8,16,21] suggest that some lead effects are reversible if exposure is reduced. Baker et al.[18] in a 3 year study of foundry workers observed a 13 to 27% improvement in mood scores when workers with initial blood leads greater than 50 µg/dl experienced a 13 µg/dl per year decline in blood lead concentrations. Comparable improvements were not seen in those workers whose initial blood lead was less than 50 µg/dl. Tests of cognition and neurobehavioral functioning showed little change in association with declining levels of lead exposure. It may be that lead-related decrements in sensory/motor functioning and cognition are irreversible or require a longer time to recover. Baker et al.,[8] using multiple indices of exposure, also found that the average blood lead over the 2 year exposure interval correlated most highly with neurobehavioral outcomes and was thus a better predictor variable than maximum blood lead, blood lead at time of testing, average blood lead above a specified threshold, or zinc protoporphyrin concentrations. It is not clear yet whether this observation has some significance for understanding lead's mechanism of action or whether using the average value merely improves the precision of the exposure estimate over that obtained by any single measure.

There are many questions that need to be resolved in the field of occupational neurobehavioral lead toxicity. These questions are summarized in the following paragraphs.

What is the best index of chronic lead exposure to use in conjunction with each of the neurobehavioral outcomes? There are several options, including blood lead at the time of testing,

TABLE 4
Behavioral Changes (IQ and Sensory/Motor) In Lead Exposed Workers

Number of subjects		PbB (μg/dl)[a]		Findings on IQ and sensory motor test	Ref.
Exposed	Control	Exposed	Control		
106	65	8—80	3—36	Decreased verbal concept formation, memory and visual/motor performance improvement reduction in exposure	10
49	27	27—69	<21	Decreased memory and learning, increased visual reaction time	67
42	22	12—88	10—27	Decreased long term memory, verbal and visuospatial abstraction, and psychomotor speed	9
49	24	32	12	Significant relationship between impaired psychological performances and exposure level, visual-intelligence decreased (not significantly), visual- motor functions decreased (not significantly)	11
49	36	49	15	Significantly poorer performance on Digit Symbol Test, Bourdon-Wiersma, trail making test, Santa Anna test, flicker fusion and simple reaction time tests	69
403	305	30—56	10—22	Choice reaction time increased	70
24	33	15—30	10.5	Impaired performance on Block Design, Digit Span, and Santa Anna test of manual dexterity	17
190	100	61	25	Hand-eye coordination decreased, reaction time increased	72
288	181	>35	<35	No effects on IQ or sensory/motor tests	71
85	55	46	18	Auditory impairment and increased visual reaction time	13
70	35	61	22	Impaired oculomotor function	16
389	191	32—51	<25[b]	Impaired performance on Block Design, Digit Symbol, and Embedded Figures tests	15
90	127	81% > 40	98% < 40	Impaired performance on Block Design, Digit Symbol and Embedded Figure test	14
59	59	24—82	[c]	Impaired performance on critical flicker fusion, simple reaction time, hand steadiness, sensory store memory, Sternberg memory, and short term memory tasks	73

[a] Mean PbB or range of PbB.
[b] PbB assumed to be <25 μg/dl since zinc protoporphyrin levels were <50 μg/dl.
[c] Not reported.

average historical blood lead, maximum historical blood lead, months of exposure above a specified blood lead concentration, and zinc protoporphyrin, which provides a combined measure of recent exposure and individual sensitivity to lead. Few studies have been designed to permit a comparison of exposure indices. No neurobehavioral studies have used bone lead concentration as a cumulative historical exposure index, although advances in X-ray fluorescence technology have improved the accuracy, precision, and sensitivity of this measure.[74]

Is there a threshold for any or all of lead's neurobehavioral effects? Study designs that only include a comparison between an exposed group and a matched control group can contribute little to this issue. Furthermore, the absence of dose-response data weakens the case for assertions of causality between exposure and effect (cf. Mausner and Kramer[3]). Therefore efforts should be made to use study designs which permit multiple exposure group comparisons or treat lead exposure as a continuous variable and use regression analysis techniques to examine the relationship between dose and effect.

Are any or all of lead's neurobehavioral effects reversible? The recent use of follow-up and/ or prospective study designs is a valuable and necessary design feature to answer this question. Mood changes appear to be reversible responses to concurrent blood lead concentrations. Longer follow-up studies are necessary to assess the relative persistence of other lead effects such as impaired memory or visual motor functioning.

Why do individuals vary so markedly in their response to lead? To what extent are these differences attributable to toxicokinetic differences in how individuals handle lead or to the validity of whole blood lead as a marker for lead at the site of action, and to what extent are they attributable to modification of effect by other factors such as initial I.Q., education, socioeconomic status, and age? Studies of lead's effects on child development have found such factors to be important modifiers of the magnitude of the deficit and the rate of recovery.

What are the mechanisms responsible for lead-induced neurobehavioral impairments? Neurobehavioral test batteries have been designed to provide broadly based assessments of behavioral function. As such, they provided very little, if any, insight into mechanism of action. Successful test performance usually requires the integrated functioning of numerous sub-systems, such as vision, proprioception, perception, and memory. Poor test performance provides limited insight into which sub-systems might be impaired and no insight into biochemical mechanisms. It is therefore likely that most progress in this area will continue to come from experimental animal research and *in vitro* studies.

What is the optimal configuration of a test battery specifically designed to detect lead-induced neurobehavioral toxicity? Much of the inconsistency in the occupational neurobehavioral literature can be directly attributed to the lack of standardization in the tests used to assess behavioral functioning. Investigators have long known that performance is very task specific. Recently there has been a concerted effort to develop a core battery of tests which could be validated for use by many investigators to study the behavioral toxicity of a wide range of compounds. Hanninen and Lindstrom[22] in Finland have developed such a battery to study toxicant exposures in industrial populations. Valciukas and Lilis[15] have developed a limited battery which they have applied to lead workers. Recently Baker and co-workers[18,19] have developed a computer based neurobehavioral evaluation system that is particularly well suited for field use and is relatively comprehensive in that it assesses memory, visuomotor function, affect, and verbal concept formation. Letz[23] has continued to refine the battery through the development of other tasks that can be added to the core battery to suit specific toxicant-exposure situations. One of the objectives in developing such batteries is to use tests that have high test-retest reliability. The latter is essential in order to permit the study of the onset of toxicity and/ or recovery of function in longitudinal follow-up studies.

Perhaps the most ambitious effort to develop a test battery has been that undertaken by the World Health Organization (WHO) and the National Insitutute for Occupational Safety and Health (NIOSH). They are in the process of collecting international normative data on a core set of neurobehavioral tests. This battery is the only one developed by an international group of researchers with experience in epidemiologic investigations of neurotoxicity.[20] Further guidelines and rationale used in the development of such batteries are provided by Johnson and Anger.[24]

In summary, occupational lead exposure appears to produce reversible changes in mood at concurrent blood lead levels of 40 to 60 µg/dl. Deficits in visual-motor functioning, memory, and verbal ability are also reported to occur within this blood lead range. However, this "threshold" is likely to be revised downward as investigators use increasingly more sophisticated study designs and behavioral test batteries. The relative persistence of these effects is not well defined and awaits clarification in future longitudinal studies. The epidemiological studies reviewed here provide little insight into lead's mechanism of action.

III. CHILDHOOD LEAD EXPOSURE
AND ITS EFFECTS ON DEVELOPMENT

Children are exposed to lead from several general environmental sources such as air, water, and food, which are generally well regulated. They also incur accidental exposures through the use of lead-glazed pottery or exposure to lead used in hobbies such as painting or soldering stained glass. The major source of unregulated lead exposure today is household dust which is contaminated with lead. While deliberate ingestion of lead-contaminated soil is a frequently cited route of childhood lead exposure, current research indicates that the inadvertent ingestion of lead-contaminated house dust is a greater problem.[25,26] This ingestion occurs during the developmentally normal and frequent mouthing of hands, toys, and food which have become contaminated with lead-containing surface dust. Soil and surface dust contaminated with lead from exterior paints and past automobile and industrial emissions, are a major reservoir of environmental lead which continues to contaminate residential interiors. Current estimates are that 42 million homes contain lead-based paint.[27] Many children live in or near these contaminated homes and are thus at risk of excessive lead exposure. Childhood lead exposure is one of the few, if only, environmental illnesses whose incidence is expressed as a rate per 100 rather than as a rate per 1000 or 10,000. Estimates by the Agency for Toxic Substances and Disease Registry[27] are that 200,000 children less than 6 years of age have blood lead levels greater than 25 µg/dl, and 17% have blood lead levels greater than 15 µg/dl. Currently, in high risk urban neighborhoods 30% of the children may exhibit blood leads above 25 µg/dl. In addition to childhood exposure there is a renewed concern for fetal lead exposure. The ATSDR estimates that 4 million fetuses will be exposed to maternal blood leads greater than 10 µg/dl over the next 10 years.

There is mounting evidence from epidemiological investigations that low to moderate environmental lead exposure delays and/or impairs child development. During the 1960s and 1970s numerous cross-sectional and case-control studies provided suggestive but inconclusive evidence of lead's impact on neuropsychological development. These early studies have been the subject of intensive critical review in an effort to resolve the inconsistencies (cf.[28,29]). Reviewers pointed out the inadequacies inherent in using single blood lead measures as indicators of lifetime childhood lead exposure. It was also apparent that the developmental test batteries were often poorly constructed, incomplete, or relied on non-validated measures. Inconsistent results could be attributed to a failure to consider and/or adequately measure the broad range of confounding factors which also influence child development, e.g., nutrition, home environment and family structure, parental I.Q., and socioeconomic status. Statistical treatment of the data was often inadequate, failed to appropriately adjust for the influence of covariates and confounders, and lacked the statistical power to detect small effects. Finally, as discussed in detail by Pearson and Dietrich,[30] most investigators failed to explicitly enunciate the analytical model of behavioral development used to guide study design, selection of measurements and analytic strategy. The implicit models, either *main-effects* models or *interactive* models, were not adequate to capture the complex process of child development. Most developmental psychologists now subscribe to a *transactional* developmental model wherein the child plays an active role in his or her own growth and development. This type of model is capable of specifying how constitutional, biomedical, social, and environmental factors mutually influence each other over a period of time, thereby modifying exposure and subsequent morbidity. Despite the limitations of these early studies, there was sufficient evidence to suggest that low to moderate lead exposure was associated with small, subtle, but important changes in the neurobehavioral development of children.

Beginning in 1979, a series of general population studies were undertaken in several countries.[31-35] These cross-sectional studies incorporated significant improvements in study design, sample size, test battery construction and measurement of covariates and confounders.

Once again these studies were subjected to extensive critical examination.[36-40] Most studies found 3 to 5 point I.Q. differences between high and low exposure groups after appropriate statistical control for covariates and confounders. Again, as in earlier studies, there was some evidence that lead effects were larger in more socially disadvantaged children. However, not all studies attained conventional levels of statistical significance ($p < 0.05$). Smith[34] concluded that any causal effect of low level lead exposure on I.Q. learning or behavior was likely to be small and difficult to detect against the background of much larger social influences. Needleman and Bellinger[41] cautioned against making a Type II error, that is rejecting a valid association between lead and child development. They pointed out that while several studies failed to attain the conventional level of statistical significance, i.e., $p < 0.05$, the vast majority of studies found effects in the hypothesized direction and attained p-values <0.10. Unfortunately, because of the cross-sectional nature of these study designs, accurate characterization of earlier lead exposure and other critical developmental covariates was still lacking and controversy concerning the existence of a cause-effect relationship between lead and development remained.

The lack of antecedent data pertaining to lead exposure and developmental covariates in cross-sectional studies could only be overcome through the use of prospective longitudinal designs. To this end, a number of such studies were initiated in several locations around the world. Study design and early progress in these studies was the subject of the Second International Conference on Prospective Lead Studies held in Cincinnati in April 1984.[42] Further results from these studies were presented at a meeting in Edinburgh, 1986.[40,43] These longitudinal studies provide comprehensive longitudinal documentation of blood lead histories during pregnancy, infancy and childhood through about seven years of age. In addition, they provide extensive measurement of important covariates and confounders throughout pregnancy and early childhood. Key features of these studies have been summarized by Starr.[44]

Although these studies are still underway, several important findings have emerged. First, low level fetal lead exposure (maternal PbB range = 1 to 25 µg/dl) appears to have a negative impact on fetal growth and maturation.[45-48] Furthermore, fetal lead exposure appears to be associated with delays in early child development.[49-51] Bellinger and co-workers[51] in Boston reported that performance on the Bayley Mental Development Index (MDI) was inversely related to cord blood lead levels. Three groups of infants of approximately 70 subjects each and average cord blood lead levels of 1.8, 6.5, and 14.6 µg/dl were tested at 6, 12, 18, and 24 months of age. Bellinger found that the average difference in MDI scores between the low lead and high lead group was about 6 points. In their most recent analysis of this data set,[52] they report that the effect is greater in children of low socio-economic status. This latter finding confirms the earlier report of Dietrich et al.[49-50] In this study of a lower SES cohort of 305 infants born in a lead hazardous area of Cincinnati during 1980 to 1985, prenatal (maternal) PbB ($\bar{x} = 8.0$ µg/dl; range = 1 to 27) was found to be negatively associated with infant MDI scores at 3, 6, and 12 months. MDI scores declined about 4 points for every 10 µg/dl increase in maternal PbB or about 10 points across the observed range of maternal PbB concentrations (1 to 27 µg/dl). Furthermore the effects were greater in lower SES infants and in male infants.

Interestingly, neither the Cincinnati nor Boston studies found evidence of an association between postnatal lead exposure during the first 2 years of life and behavioral development during the first 2 years. This could mean that low level postnatal lead exposure has no effect on early developmental processes or more likely, that the developmental abilities which we are able to assess in young infants are not particularly reflective of postnatal lead toxicity. If the latter is the case, then we might expect to see an association between early postnatal lead exposure and later, more comprehensive, measures of development which include and require language abilities. Some evidence for this has recently been reported by another of the prospective lead studies currently underway. McMichael et al.[53] have been studying the developmental progress of a cohort of 537 children born during 1979 to 1982 to women living within a lead smelter community in South Australia. Maternal blood lead samples were obtained prenatally, cord blood at delivery, and infant blood was sampled at 6, 15, and 24 months postnatally and yearly thereafter. An analysis of the children's development at 4 years of age indicated that a child with an average postnatal blood lead concentration of 30 µg/dl had a general cognitive score 7.2 points (7%) lower than a child with a lifetime average blood lead of 10 µg/dl.[53] Similar deficits were obtained on perceptual-performance scores and memory scores within the exposure range studied, 5 to 58 µg/dl. There was no evidence for a threshold. Unlike studies of 1 and 2 year olds, that showed an association between prenatal blood lead and developmental status, this work with 4 year olds found no effect of prenatal exposure, only postnatal exposure. The authors attribute these effects in 4 year olds to a long term developmental impairment rather than a delay in maturation.

Based on these earlier reports from prospective studies, it appears that the fetal lead exposure, at levels below 25 µg/dl, the current CDC definition of excessive childhood lead exposure,[54] can result in measurable effects in well controlled studies of 1 and 2 year olds. The effects are greater in the most socially disadvantaged groups. Some of these fetal effects such as reduced early growth in stature and reduced MDI scores may be reversible if the infant is not subjected to additional adverse levels of lead exposure at an early age. Another interpretation is that these effects are not disorders or deficits from which the infant recovers, but rather developmental delays which are subject to eventual "catch-up". It remains to be seen whether or not fetal exposures will be related to later measures of CNS and behavioral functioning.

Although considerable progress has been made in our understanding of pediatric lead toxicity, several issues remain unresolved. These issues and their current status are discussed below.

What is the lowest level of lead exposure which might result in measurable harm? The results from the Boston and Cincinnati prospective studies[46,51] suggest that measurable developmental delays are observed at levels of fetal exposure considerably below the current pediatric guideline of 25 µg/dl. Cross-sectional and longitudinal studies of the developmental consequences of lead exposure suggest that postnatal lead effects are also demonstrable at levels below 25 µg/dl. Hawk et al.[35] found effects across a range from 6 to 47 µg/dl, with a mean of 27 µg/dl. Fulton et al.[55] in Scotland report a 6 point decline in IQ across a blood lead range of 3 to 34 µg/dl, with a mean of 11 µg/dl, while Hatzakis et al.[56] studying 533 children in Greece, found effects across a blood lead range of 7 to 64 µg/dl, with a mean of 24 µg/dl. McMichael et al.[53] in the first report of assessments of 4 year olds from the longitudinal studies finds substantial decrements in general cognition across a blood lead range of 5 to 58 µg/dl, mean lifetime average of 19 µg/dl. It is apparent that consideration must be given to a downward revision in the current Centers for Disease Control definition of lead toxicity as PbB greater than or equal to 25 µg/dl, in association with an erythrocyte protoporphyrin greater than or equal to 50 µg/dl.[54] However, a clear distinction should be made between guidelines for maternal (fetal) exposure and child exposure since sources, pathways, and biological sensitivity are likely to be quite different.

Is there a critical period of exposure during development which is responsible for these deficits? This will remain a very difficult issue to address. Blood lead concentration, the best available index of internal lead burden, is a function of current lead intake and existing stores of lead in other body compartments. In a rapidly developing child, the blood lead pool never attains equilibrium with either environmental intake or existing body stores. Thus, blood lead does not permit accurate estimations of when and how much lead a child might have acquired. Behavioral measures of effect are equally inaccurate in the ability to indicate when an insult occurred. Developmental measures are limited to tasks that the child is capable of performing. For example, lead exposure at 9 months of age might damage a region of the brain essential for complex cognitive functioning e.g., reasoning by analogy. However, since the infant at 9 months of age has yet to develop language skills or concept formation skills it will not be possible to detect the deficit or disorder until several years later.

Are there identifiable causes for individual differences in susceptibility to the neurobehavioral consequences of lead exposure? Individual differences in response to lead exposure are quite large. For example, McMichael et al.[53] report an average decline in IQ of 7 points. However, the 95% confidence limits are 0.3 to 13.2. This large variation might be reduced if we used other indicators of lead at the "target site" than whole blood lead, e.g., serum lead. Bioavailable lead might also be regulated by a genetically controlled lead-binding protein which in turn might modify individual lead toxicity. Non-biological factors also contribute to this large variance. There are undoubtedly individual differences in the compensatory strategies that individuals use to deal with toxic insults, as well as a large variation in the range of other stressors concurrently impacting the child. These factors can exacerbate or ameliorate the apparent magnitude of the insult, as well as modify the rate of any recovery process.

Can the underlying causes of lead-related reductions in intellectual performance be more precisely defined? Is there a single deficit, disorder, or delay which underlies the more general reduction in performance, e.g., an attention deficit or auditory processing disorder? Most of the recent cross-sectional studies and current prospective studies have not emphasized tests of specific functions but rather have focused on standardized tests of general intelligence. However, the Cincinnati study[57] has incorporated other developmental measures such as attention, language development, auditory processing, specific motor abilities, balance, electrophysiological indices of sensory functioning, and information processing. Most of these measurements are obtained on children between 48 and 78 months of age and no data have yet been reported, with the exception of a pilot study of balance.[58]

Do children recover from the effects of low level lead exposure? There are relatively few data yet available to address this question. Bornschein et al.[45] reported that prenatal lead exposure resulted in fetal growth retardation. Subsequently, Shukla et al.[59] found that the most adversely affected infants were able to "catch-up" during the first 15 months of life if they were not subjected to additional postnatal lead exposure. Preliminary analyses of mental development in the Cincinnati cohort suggest that a similar "catch-up" might occur during the first 2 years of life. However, Bellinger et al.[51] report no "catch-up" by 2 years of age. It is still to early to discuss the time course of neurobehavioral deficits following postnatal lead exposure. However, the longitudinal studies will be able to address the issue in future years.

What are the long-term consequences of childhood lead exposure at low-to-moderate levels? This question is closely linked to the previous issue of persistence/recovery. Development is characterized by its sequential dependency, as well as by its plasticity. It is possible that early developmental delays or deficits might disappear at later ages. However, this does not mean that full recovery has occurred. Deficits and delays in the sequential processes of development can have long term effects which alter a child's academic and social competency even though the initial deficit is no longer measurable. Thus the issues of persistence of effect, recovery, and consequences of early childhood insults are extremely complex and difficult to address even in longitudinal studies that afford the opportunity of repeated functional assessment.

IV. OCCUPATIONAL MERCURY EXPOSURE AND ITS EFFECTS

It is estimated that 10,000 ton of mercury are produced for industrial use world wide every year. Major consumers of mercury include the chloralkali industry, electrical equipment manufacturers and producers of paints and scientific instruments. About 150,000 workers are occupationally exposed to mercury. In addition to occupational exposures from these industrial users, mercury is released into the environment by other kinds of human activity such as the combustion of fossil fuel and waste disposal. About 20,000 ton of metalic mercury or its inorganic compounds are released into the environment by these routes.[60] The mercury concentration in the general atmosphere varies from a few ng/m^3 in remote uncontaminated areas to around 20 ng/m^3 in urbanized areas. This is in contrast to the much higher levels of lead contamination which range from 0.01 to 5.0 µg/m^3. Additional sources of inorganic mercury are organic mercury compounds which, when discharged into the environment, can be converted to inorganic forms. Conversely, in upper sedimentary layers of seabeds or lakebeds elemental mercury or mercuric compounds can be transformed into methyl mercury by bacteria. A large portion of mercury found in seafood is in the form of methyl mercury.

Most reports of hazardous exposure to inorganic Hg come from mining and chloralkali industries. In these industries, mercury vapor, inorganic mercury aerosols and dusts can pollute the worksite. Inhalation of mercury vapor and its compounds is the most important route of exposure, although some inorganic compounds can be absorbed via skin. Mercury can be measured in blood and is excreted primarily in urine. Urinary mercury levels correlate with exposure in recent months and are used in industry to monitor exposure. Unfortunately, as is the case with lead, there is considerable variation between concentration of mercury at the worksite and urinary mercury levels, as well as variation between urinary levels and health effects. At the present time there is no completely satisfactory index of body burden, although various parameters extracted from serial urine levels are most frequently used, e.g., a time weighted-average or peak level of excretion.

The effects of mercury exposure are dependent on the chemical form of mercury. Inorganic and metallic mercury affect both the nervous system and kidney, whereas alkyl forms of mercury are far more toxic and have their greatest effects on the central nervous system. Excessive exposure to inorganic mercury is associated with non-specific mood changes consisting of irritability, anxiety, timidity, and self consciousness followed by profound personality changes and memory loss. Organic mercurial compounds produce irritability, accompanied by depression and memory impairment. Impaired cognition, ranging from mild disturbances in analytical ability to severe mental retardation, has also been associated with organic mercury exposure.

Mercury, unlike lead, has not been the subject of extensive behavioral epidemiological investigation. The major effects of mercury exposure were established early in the history of industrialization. The most common indicator of overexposure is hand and arm tremor. It is not surprising then that the early studies focused on the quantification of neruomuscular function and tremor in the exposed worker.[61-63] Wood et al.[61] undertook a detailed longitudinal follow-up study of arm and finger tremor in two occupational exposed women. When first tested, about 1 month after termination of exposure, the workers exhibited highly variable amplitude and frequency of tremor and an inability to maintain a constant pressure on a finger trough equipped with a strain gauge. Their performance improved gradually and equaled that of control subjects after about 3 months. A power spectrum analysis showed the dominant frequency to range between 2 and 7 Hz as compared to a very narrow range of 2 to 3 Hz in controls. Their improvement in performance coincided with a decline in plasma mercury levels.

This work was confirmed in a group of 77 chloralkali workers[62,64] with elevated urinary mercury levels (mean of 374 µg/l) compared with a group of 65 control workers (mean urinary mercury levels of 63 µg/l). Langolf et al.[63] undertook a detailed dose repsonse study of the relationship between several indices of exposure and quantitative measures of tremor and psychomotor function. The exposure indices included: (1) average urine mercury concentrations for the past 3, 6, or 12 months, (2) the number of peak concentrations greater thean 150, 250, and 500 µg/l in the previous 3, 6, or 12 months, (3) the highest concentration ever recorded, and (4) the total duration of mercury exposure. Response measures included a power spectrum analysis of forearm tremor, choice reaction time, finger tapping, Michigan maze, and critical tracking tests. They found a slight decrease in maximum finger tapping rate and an increase in rate variability, an increase in hole-to-hole times in the Michigan maze, and reduced performance on a visual tracking task. The best single predictor of impaired performance was the number of peaks each subject experienced in excess of 500 µg/l in the previous year. However, as was the case in lead research, they found considerable individual variation in sensitivity to mercury as indexed by urine mercury levels. In a follow-up examination of 10 employees with high urine mercury histories, they found an improvement in test performance and tremor. At the time of initial testing average urinary mercury was 660 µg/l, whereas 6 to 10 months later the average level had dropped to 100 µg/l. Thus neuromuscular effects, reflected in behavioral performance tasks, appear to occur at levels greater than 500 µg/l and improvement occurs upon cessation of exposure.

Piikivi and co-workers[65] have studied a group of 36 Finnish men employed in a chlorine-alkali plant and 36 controls matched for age, education, and occupational skill levels. Mercury exposure had occurred for a minimum of 10 years and urinary mercury determinations had been obtained at quarterly intervals. The long term indicators of exposure were (1) the number of quarterly intervals with urinary mercury levels greater than 300 µg/l, and (2) the time weighted average. In addition, blood and urine mercury levels were measured at the time of testing. The behavioral tests were drawn from the Finnish Institute of Occupational Health Test Battery.[22] These tests of cognitive performance included one verbal and one visual test from the Wechsler Adult Intelligence Scale (Similarities and Picture Completion), three memory tests from the Wechsler Memory Scales (Digit Span, Logical Memory, and Visual Reproduction), and one test of eye-hand coordination (Santa Anna Dexterity Test).

The average urine mercury levels at the time of testing were 58 µg/l (range = 7 to 133 µg/l), while the average blood mercury levels were 20 µg/l (range = 5 to 68 µg/l). The time weighted average urine mercury level (minimum of 10 years) was 120 µg/l (range = 7 to 214 µg/l). The results were consistent with other studies reporting deficits in cognitive abilities and memory among workers exposed to mercury vapors. The verbal concept formation test (Simularities) proved to be the most sensitive test in the battery. Deficits were most frequent among workers with concurrent blood and urine mercury levels above 15 and 56 µg/l, respectively. Memory disturbances correlated better with concurrent blood mercury level than with urinary mercury levels.

The issue of mercury's effects on memory has been intensively investigated by Williamson and co-workers.[66] Instead of assessing general ability, they used a test battery specifically designed to provide a comprehensive assessment of specific aspects of information processing. They sought to determine if mercury exposure disrupted the initial attentional stages of processing, iconic (sensory) memory, short-term or long-term memory, either separately or in combination. The battery also permitted an examination of mercury's effects on long term concentration, fatigue, and speed and accuracy of motor responses. The specific tests included: (1) critical flicker fusion; (2) visual pursuit; (3) hand steadiness; (4) short-term memory (paired-associates); (5) reaction time; (6) Sternberg memory task; (7) vigilance; (8) iconic (sensory) memory; and (9) long-term memory.

The subjects were 12 workers exposed to inorganic mercury. They were matched with 12 control subjects according to sex, age, level of education, and ethnicity. Urine mercury levels were available for the preceeding 3 years. Levels at the time of testing ranged from less than 10 to 670 µg/l. Duration of exposure ranged from 1 to 8 h/d for 3 months to 8 years. The total number of hours spent directly working with mercury ranged from 180 to 2532 h.

Simple reaction time and critical flicker fusion were unrelated to any of the several indices of exposure. Individuals actively working with mercury at the time of testing showed impaired performance on several measures including the Sternberg memory test, visual pursuit, and short-term memory. Williamson et al.[66] also found fatigue to be an important factor in tasks requiring muscular coordination. The finding of the tremor and fatigue in conjunction with normal reaction times suggests that mercury's effect on motor performance tasks may well be at the level of the muscle or peripheral nervous system. They also concluded that short term memory effects were not due to changes in arousal or attention, since critical flicker fusion performance and vigilance performance were normal.

Efforts to relate various indices of exposure to effect also resulted in interesting findings. Urine mercury, the traditional measure of exposure, was related to only one measure, hand steadiness. Regardless of the length of their exposure, workers who were actively using mercury at the time of testing and whose urinary mercury levels were increasing, showed the greatest performance deficits. Thus it may be that change in mercury status, rather than absolute exposure level, is the key factor in nervous system toxicity. This work awaits confirmation by other studies with larger samples and perhaps a longitudinal follow-up design.

The studies reviewed above report increased muscular tremor and decreased finger tapping speed in workers with periodic urine mercury levels greater than 500 µg/l. Deficits in memory and other tests of cognitive function have recently been reported at levels at or below 100 µg/l. However, given the large inter-individual variation in response to mercury exposure, it will be necessary to replicate these studies, utilizing more sophisticated data analytic strategies to control for possible confounders.

V. FETAL AND PEDIATRIC METHYL MERCURY EXPOSURE AND ITS EFFECTS

Early reports of organic mercury poisoning date back to the mid 19th century. Several dozen case reports document the effect of *direct* exposure to organic mercury by chemical workers or others using mercury treated products, e.g., lumbermen and farmers. In 1956, we learned that the general population could be *indirectly* exposed to methyl mercury through the consumption of fish that had become contaminated by mercury discharged into their water by a nearby chemical plant.[74] This poisoning of the food chain resulted in a massive poisoning of the human population in the region. Approximately 120 people living around Minamata Bay, Japan were poisoned between 1953 and 1960. Forty-six people died and twenty-two infants were poisoned prenatally. This environmentally mediated form of methyl mercury poisoning is now referred to as "Minamata Disease". It is characterized by sensory impairment of the extremeties, impaired gait, slurred speech, restricted visual field, and loss of hearing (see Fox[1] for further discussion of metal effects on sensory systems). Prenatal exposure can result in mental retardation and cerebral palsy. Subsequent to the reports from Minamata Bay, there have been other reports of mercury poisoning in the general population. On several occasions during the 1960s and 1970s there have been reports of large numbers of people poisoned as a result of eating bread made from wheat which had been treated with organic mercury-containing fungicides.[75,76] Of greater concern to the general public are reports that industrial and agricultural use of mercury is resulting in increasingly higher amounts of methyl mercury being found in shellfish, fish, and birds.[77]

The individuals most susceptable to organic mercury poisoning appear to be fetuses and infants. They are exposed to mercury *in utero* through their mother's blood supply and postnatally through mother's mercury-contaminated milk. The toxicokinetics of maternal mercury burden and its transfer to the fetus and clearance via milk are discussed by Bakir et al.[75]

Despite the large number of individuals suffering from methyl mercury poisoning and the increasing mercury burden of the general population there have been no detailed epidemiologic studies of the neurobehavioral consequences in infants and young children. However, there has been one study of neurologic effects in a cohort of 234 Cree Indian children, aged 12 to 30 months, living in Northern Quebec.[78] Mothers of these children were exposed to mercury by consuming a diet of contaminated fish and fish-eating mammals.[77] Segments of maternal hair coinciding with the period of pregnancy were analyzed for mercury content and used as a marker for fetal mercury exposure. Exposure levels were considerably lower than those reported in Japan and Iraq. The prevalence of neurologic abnormalities in this cohort was not related to exposure indices, with the exception of abnormal tendon reflexes observed in 11% of the male infants. However, the generally negative findings of this study should be viewed conservatively for several reasons. As the investigators point out, they lacked good normative neurological and developmental data for this cohort. Furthermore at these lower exposure levels, neurological deficits might by minimal. A high incidence of alcohol and cigarette consumption and poor prenatal care may have masked subtle effects. Finally, sensitive tests of neurobehavioral development were not used. Given the obvious neurotoxicity of mercury at high doses and the apparent high susceptibility of fetuses and infants, more rigorous assessment of similar pediatric cohorts is certainly called for.

The experience in recent pediatric lead toxicity studies has shown the utility of prospective longitudinal designs, rigorous documentation of exposure, and comprehensive behavioral assessment batteries. Toward this end, Marsh and co-workers[79] have initiated a prospective study of infants exposed to mercury *in utero*. The study is focusing on a cohort of women in the Seychelle Islands exposed to mercury through their diet of fish. Concentrations of mercury incorporated into maternal hair during pregnancy will be used as an index of dose to the fetus. Outcome measures will include repeated measures of growth, neurologic and general physical examinations, and results of repeat assessments with the Denver Developmental Screening Test (DDST). At the present time 525 infants have been recruited. The maternal hair mercury concentrations during pregnancy ranged from 0 to 30 ppm. This concentration range can be contrasted with that of the Iraqi study[80] which found evidence of adverse fetal effects at a concentration of 20 ppm, and the recent New Zealand study[81] which found an increased prevalence of developmental delays, according to the DDST, at maternal maximum hair mercury concentrations during pregnancy of 9 ppm. The large Seychelles cohort, which the investigators intend to follow through 1990, should provide a very good initial characterization of effects and dose-response relationships.

VI. CONCLUSION

The general population studies discussed in preceding sections are shedding more light on the toxic consequences of heavy metal exposure and the sensitivity of the nervous system to such insults. They raise new questions about metal toxicokinetics in adults, pregnant women, fetuses, and young children. Improved assessment batteries are revealing increasingly more subtle, yet not inconsequential, effects of metals on behavioral development and adult performance. More sophisticated data analysis strategies are revealing how metals interact with other developmental factors or performance factors. These studies not only require a reconsideration of current health guidelines, but also set the stage for further experimental investigations of mechanisms of action of metals.

REFERENCES

1. **Fox, D.,** Sensory system neurotoxicology of heavy metals, in *Focus on Biological Effects of Heavy Metals,* Foulkes, E., Ed., CRC Press, Boca Raton, FL, chap. 6.
2. **Minnema, D. and Cooper, G.,** Effects of heavy metals on neurotransmitter release, in *Focus on Biological Effects of Heavy Metals,* Foulkes, E., Ed., CRC Press, Boca Raton, FL, chap. 2.
3. **Mausner, J. and Kramer, S.,** *Epidemiology: An Introductory Text,* W. B. Saunders, Philadelphia, 1985, 180.
4. **Barltrop, D. and Meek, M.,** Effect of particle size on lead absorption from the gut, *Arch. Environ. Health,* July/ Aug., 280, 1979.
5. **Barltrop, D. and Meek, M.,** Absorption of different lead compounds, *Postgrad. Med. J.,* 51, 805, 1975.

6. **Roy, B. R.,** Effects of particle sizes and solubilities of lead sulfide dust on mill workers, *Am. Indust. Hyg. Assoc. J.,* 38, 327, 1977.

7. **Rice, C., Fischbein, A., Lilis, R., Sarkozi, L., Kon, S., and Selikoff, I. J.,** Lead contamination in the homes of employees of secondary lead smelters, *Environ. Res.,* 15, 375, 1978.

8. **Baker, E., White, R F., Pothier, L. J., Berkey, C. S., Dinse, J. P., Travers, P. H., Harley, J. P., and Feldman, R. G.,** Occupational lead neurotoxicity: improvement in behavioural effects after reduction of exposure, *Br. J. Ind. Med.,* 42, 507, 1985.

9. **Grandjean, P., Arnvig, E., and Beckmann, J.,** Psychological dysfunctions in lead-exposed workers: relation to biological parameters of exposure, *Scand. J. Work. Environ. Health,* 4, 295, 1978.

10. **Baker, E. L., Feldman, R. G., White, R. A., Harley, J. P., Niles, C. A., Dinse, G. E., and Berkey, C. S.,** Occupational lead neurotoxicity: a behavioural and electrophysiological evaluation. I. Study design and year one results, *Br. J. Ind. Med.,* 41, 352, 1984.

11. **Hanninen, H., Hernberg, S., Mantere, P., Vesnato, R., and Jalkanen, M.,** Psychological performance of subjects with low exposure to lead, *J. Occup. Med.,* 20, 683, 1978.

12. **Hogstedt, C., Hane, M., Agrell, A., and Bodin, L.,** Neuropsychological test results and symptoms among workers with well-defined long-term exposure to lead, *Br. J. Ind. Med.,* 40, 99, 1983.

13. **Repko, J. D., Corum, C. R., Jones, P. D., and Garcia, L. S.,** The effects of inorganic lead on behavioral and neurologic function. U.S. Department of Health, Education and Welfare, National Institute of Occupational Safety and Health, Cincinnati, Publication No. 78-128, 1978.

14. **Valciukas, J. A., Lilis, R., Eisinger, J., Blumberg, W. E., Fischbein, A., and Selikoff, J. F.,** Behavioral indicators of lead neurotoxicity: results of a clinical field survey, *Int. Arch. Occup. Environ. Health,* 41, 217, 1978.

15. **Valciukas, J. A. and Little, R.,** A composite index of lead effects, *Int. Arch. Occup. Environ. Health,* 51, 1, 1982.

16. **Spivey, G. H., Brown, C. P., Baloh, R. W., Campion, D. S., Valentine, J. L., Massey, F. J., Jr., Browdy, B. L., and Culver, B. D.,** Subclinical effects of chronic increased lead absorption—a prospective study. I. Study design and analysis of symptoms, *J. Occup. Med.,* 21(6), 423, 1979.

17. **Mantere, P., Hanninen, H., and Hernberg, S.,** Subclinical neurotoxic lead effects: two-year follow-up studies with psychological test methods, *Neurobehav. Toxicol. Teratol.,* 4, 725, 1982.

18. **Baker, E. L., Jr., Feldman, R. G., White, R. F., Harley, J. P., Dinse, G. E., and Berkey, C. S.,** Monitoring neurotoxins in industry: development of a neurobehavioral test battery, *J. Occup. Med.,* 25(2), 125, 1983.

19. **Baker, E. L., Letz, R. E., Fidler, A. T., Shalat, S., Plantamura, D., and Lyndon, M.,** A computer-based neurobehavioral evaluation system for occupational and environmental epidemiology: methodology and validation studies, *Neurobehav. Toxicol. Teratol.,* 7(4), 369, 1985.

20. **Johnson, B. L., Baker, E. L., El Batawi, M., Gilioli, R., Hanninen, H., Seppalainen, A. M., and Xintaras, C.,** *Prevention of Neurotoxic Illness in Working Populations,* John Wiley & Sons, New York, 1987.

22. **Hanninen, H. and Lindstrom, K.,** Behavioral test battery for toxicopsychological studies, *Inst. Occup. Health, Helsinki,* 1979.

23. **Letz, R. and Baker, E.,** Computerized neurobehavioral testing in occupational health, *Sem. Occup. Med.,* 1, 165, 1986.

24. **Johnson, B. L. and Anger, W. K.,** Behavioral toxicology, in *I. Environmental and Occupational Medicine,* Rom, W., Ed., Little Brown, Boston, 1983, 329.

25. **Bornschein, R., Succop, P., Dietrich, K., Clark, S., Que Hee, S., and Hammond, P.,** The influence of social and environmental factors on dust lead, hand lead, blood lead levels in young children, *Environ. Res.,* 38, 108, 1985.

26. **Bornschein, R., Succop, P., Krafft, K., Clark, C., Peace, B., and Hammond, P.,** Exterior surface dust lead, interior house dust lead and childhood lead exposure in an urban environment, in *Trace Substances in Environmental Health,* Hemphill, D., Ed., University of Missouri, Columbia, 1986, 322.

27. **Anon.,** The Nature and Extent of Lead Poisoning in Children in the United States: A Report to Congress, Agency for Toxic Substances and Disease Registry Public Health Service, U.S. Department of Health and Human Services, July, 1988.

28. **Bornschein, R. L., Pearson, D., and Reiter, L.,** Behavioral effects of moderate lead exposure in children and animal models, I. Clinical studies, *CRC Rev. Toxicol.,* Nov., 43, 1980.

29. **Rutter, M.,** Raised lead levels and impaired cognitive/behavioral functioning: a review of the evidence, *Dev. Med. Child Neurol. Suppl.,* 22, 800, 1980.

30. **Pearson, D. T. and Dietrich, K. N.,** The behavioral toxicology and teratology of childhood: models, methods, and implications for intervention, *Neurotoxicology,* 6(3), 165, 1985.

31. **Needleman, H. L., Gunoe, C., Leviton, A., Reed, R., Peresie, H., Maher, C., and Barrett, P.,** Deficits in psychologic and classroom performance of children with elevated dentine lead levels, *N. Engl. J. Med.,* 300, 689, 1979.

32. **Lansdown, R. Yule, W., Urbanowicz, M. -A., and Hunter, J.,** The relationship between blood lead concentrations, intelligence, attainment and behaviour in a school population: the second London study, *Int. Arch. Occup. Environ. Health,* 57, 225, 1986.

33. **Winneke, G., Kraemer, U., Brockhaus, A., Ewers, U., Kujanek, G. Lechner, H., and Janke, W.,** Neuropsychological studies in children with elevated tooth lead concentrations. II. Extended study, *Int. Arch. Occup. Environ. Health,* 51, 231, 1983.

34. **Smith, M., Delves, T., Lansdown, R., Clayton, B., and Graham, P.,** The effects of lead exposure on urban children: the institute of child health/Southampton study, *Dev. Med. Child Neurol. Suppl.,* 47, 1983.

35. **Hawk, B. A., Schroeder, S. R., Robinson, G., Otto, D., Mushak, P., Kleinbaum, D., and Dawson, G.,** Relation to lead and social factors to I.Q. of low SES children: A partial replication, *Am. J. Mental Def.,* 91(2), 178, 1986.

36. **Pocock, S. J. and Ashby, D.,** Environmental lead and children's intelligence: a review of recent epidemiological studies, *Statistician,* 34, 31, 1985.

37. **Smith, M.,** Recent work on low level lead exposure and its impact on behavior, intelligence, and learning: a review, *J. Am. Acad. of Child Psych.,* 1, 24, 1985.

38. **Yule, W. and Rutter, M.,** Effects of lead on children's behaviour and cognitive performance: a critical review, in *Dietary and Environmental Lead: Human Health Effects,* Mahaffey, K. R., Ed., Elsevier, Amsterdam, 1985.

39. **Smith, M.,** The effects of low level lead exposure on children: the state of the art in 1986, in *Lead Exposure and Child Development: An International Assessment,* Smith, M. Grant, L. D., and Sors, A., Eds., MTP Press, Lancaster, U.K., in press.

40. **Smith, M., Grant, L. D., and Sors, A., Eds.,** *Lead Exposure and Child Development: An International Assessment,* MTP Press, Lancaster, U.K., 1989, 293-305.

41. **Needleman, H.. L. and Bellinger, D. C.,** Type II fallacies in the study of childhood exposure to lead at low dose: a critical and quantitative review, in *Lead Exposure and Child Development: An International Assessment,* Smith, M., Grant, L. D., and Sors, A., Eds., MTP Press, Lancaster, U.K., in press.

42. **Bornschein, R. and Rabinowitz, M.,** Proceedings of the Second International Lead Conference, *Environ. Res.,* 38, 1985.

43. **Davis, J. M. and Svendsgaard, D. J.,** Low-level lead exposure and child development, *Nature,* 329, 297, 1986.

44. **Starr, R. H.,** Current research on the developmental ecology of lead exposure during childhood, *Environ. Res.,* 38, 197, 1985.

45. **Bornschein, R. L., Grote, J., Mitchell, T., Succop, P. A., Dietrich, K. N., Krafft, K. M., and Hammond, P. B.,** Effects of prenatal lead exposure on infant size at birth, in *Lead Exposure and Child Development: An International Assessment,* Smith, M., Grant, L. D., and Sors., Eds., MTP Press, Lancaster, U.K., 307, 1989.

46. **Dietrich, K. N., Krafft, K. M., Bier, M., Succop, P. A., Berger, O., and Bornschein, R. L.,** Early effects of fetal lead exposure: neurobehavioral findings at 6 months, *Int. J. Biosocial Res.,* 8(2), 151, 1986.

47. **McMichael, A. J., Vimpani, G. V., Robertson, E. F., Baghurst, P. A., and Clark, P. D.,** The Port Pirie cohort study: maternal blood lead and pregnancy outcome, *J. Epidemiol., Comm. Health,* 40, 18, 1986.

48. **Ernhart, C. B., Wolf, A. W., Kennerd, M. J., Erhard, P., Fillipovich, H. F., and Sokol, R. J.,** Intrauterine exposure to low levels of lead: the status of the neonate, *Arch. Environ. Health,* 41, 287, 1989.

49. **Dietrich, K. N., Krafft, K. M., Bornschein, R. L., Hammond, P. B., Berger, O., Succop, P. A., and Bier, M.,** Low-level fetal lead exposure effect on neurobehavioral development in early infancy, *Pediatrics,* 80, 721, 1987.

50. **Dietrich, K., Krafft, K., Bier, M., Berger, O., Succop, P., and Bornschein, R.,** Neurobehavioral effects of fetal lead exposure: the first year of life, in *Lead Exposure and Child Development: An International Assessment,* Smith, M., Grant, L. D., and Sors, A., Eds., MTP Press, Lancaster, U.K., 320, 1989.

51. **Bellinger, D., Leviton, A., Waternaux, C., Needleman, H., and Rabinowitz, M.,** Longitudinal analyses of pre- and postnatal lead exposure and early cognitive development, *New Engl. J. Med.,* 316, 1037, 1987.

52. **Bellinger, D., Leviton, A., Waternaux, C., Needleman, H., and Rabinowitz, M.,** Low-level lead exposure, social class, and infant development, *Neurotoxicol. Teratol.,* in press.

53. **McMichael, A., Baghurst, P., Wigg, N., Vimpani, Robertson, E., and Roberts, R.,** Port Pirie cohort study: environmental exposure to lead and children's abilities at age four years, *N. Engl. J. Med.,* 319, 469, 1988.

54. **Anon.,** Preventing lead poisoning in young children: A statement by the Centers for Disease Control, U.S. Department of Health and Human Services, Atlanta, Jan. 1985.

55. **Fulton, M., Thomson, G., Hunter, R., Raab, G., Laxen, D., and Hepburn, W.,** Influence of blood lead on the ability and attainment of children in Edinburgh, *Lancet,* 1, 1221, 1987.

56. **Hatzakis, A., Salaminios, F., Konev, A., Katsouyanni, K., Maravelias, K., Kalandidi, A., Koustelinis, A., Stefanis, K., and Trichopoulos, D.,** Blood lead and classroom behavior of children in two communities with different degree of lead exposure: evidence of a dose-related effect?, in *Heavy Metals in the Evnironment,* Lekkas, R. D., Ed., CEP Consultant, Edinburgh, 1985.

57. **Bornschein, R., Hammond, P., Dietrich, K., Succop, P., Krafft, K., Clark, S., Berger, O., Pearson, D., and Que Hee, S.,** The Cincinnati prospective study of low-level lead exposure and its effects on child development: protocol and status report, *Environ. Res.,* 38, 4, 1985.

58. **Bhattacharya, A., Shukla, R., Bornschein, R., Dietrich, K., and Kopke, J. E.,** Postural disequilibrium quantification in children with chronic lead exposure: a pilot study, *Neurotoxicology,* 9(3), 1, 1988.

59. **Shukla, R., Bornschein, R., Dietrich, K., Buncher, R., Berger, O., Hammond, P., and Succop, P.,** Effects of fetal and infant lead exposure on growth in stature, *Pediatrics,* in press.

60. **Berlin, M.,** Mercury, in *Handbook on The Toxicology of Metals,* Vol. II, Friberg, L., Nordberg, G. F., and Vouk, V. B., Eds., Elsevier, New York, 1986, 387.

61. **Wood, R. W., Weiss, A. B., and Weiss, B.,** Hand tremor by industrial exposure to inorganic mercury, *Arch. Environ. Health,* 26, 249, 1973.

62. **Miller, J. M., Chaffin, D. B., and Smith, R. G.,** Subclinical psychomotor and neuromuscular changes in workers exposed to inorganic mercury, *Am. Ind. Hyg. Assoc. J.,* Oct., 725, 1975.

63. **Langolf, G. D., Chaffin, D. B., Henderson, R., and Whittle, H. P.,** Evaluation of workers exposed to elemental mercury using quantitative tests of tremor and neuromuscular functions, *Am. Ind. Hyg. Assoc. J.,* 39, 976, 1978.

64. **Chaffin, D. B. and Miller, J. M.,** Behavioral and neurological evaluation of workers exposed to inorganic mercury, in *Behavioral Toxicology,* Xintraras, C., Johnson, B. L., and De Groot, I., Eds., U.S. Department of Health, Education and Welfare, Washington, D.C., 1974, 214.

65. **Piikivi, L., Hanninen, H., Martelin, R., and Mantere, P.,** Psychological performance and long-term exposure to mercury vapors, *Scand. J. Work Environ. Health,* 10, 35, 1984.

66. **Williamson, A., Teo, R., and Sanderson, J.,** Occupational mercury exposure and its consequences for behavior, *Int. Arch. Occup. Environ. Health,* 50, 273, 1982.

67. **Chogstedt, C. H.,** Neuropsychological test results and symptoms among workers with well-defined long-term exposure to lead, *Br. J. Ind. Med.,* 44, 99, 1983.

68. **Hammond, P. B., Lerner, S. I., Gartside, P. S., Henenson, I. B., Roda, S. B., Foulkes, E. C., Johnson, D. R., and Pesce, A. J.,** The relationship of biological indices of lead exposure to the health status of workers in a secondary lead smelter, *J. Occup. Med.,* 22(7), 475, 1980.

69. **Jeyaratnam, J., Boey, K. W., Ong, C. N., Chia, C. B., and Phoon, W. O.,** Neuropsychological studies on lead workers in Singapore, *Br. J. Ind. Med.,* 43, 626, 1986.

70. **Johnson, B. L., Berg, J., Xintaras, C., and Handke, J.,** A neurobehavioral examination of workers from a primary nonferrous smelter, *Neurotoxicology,* 1, 561, 1980.

71. **Parkinson, D. K., Ryan, C., Bromet, E. J., and Connell, M. M.,** A psychiatric epidmiologic study of occupational lead exposure, *Am. J. Epidemiol.,* 123(2), 261, 1986.

72. **Morgan, B. B. and Repko, J. D.,** Evaluation of behavioral functions in workers exposed to lead, in *Behavioral Toxicology,* Xintaras, C., Johnson, B. L., and De Groot, I., Eds., U.S. Department of Health, Education and Welfare, Atlanta, 1974, 248.

73. **Williamson, A. M. and Teo, R. K. C.,** Neurobehavioral effects of occupational exposure to lead, *Br. J. Ind. Med.,* 43, 374, 1986.

74. **Harada, H.,** Methylmercury Poisoning Due to Environmental Contamination (Minimata Disease), in *Toxicology of Heavy Metals in the Environment.,* Part 1., Oehme, W. F., Ed., Marcel Dekker, New York, 1978, 261.

75. **Bakir, F., Damluji, S. F., Amin-Zaki, L., Murtadha, M., Khalidi, A., Al-Rawi, N. Y., Tikriti, S., Dhahir, H. I., Clarkson, T. W., Smith, J. C., and Doherty, R. A.,** Methylmercury poisoning in Iraq: an inter-university report, *Science,* 181, 230, 1973.

76. **Marsh, D. O., Myers, G. J., Clarkson, T. W., Amin-Zaki, L., Tikriti, S., and Majeed, M. A.,** Fetal methylmercury poisoning: clinical and toxicological data on 29 cases, *Ann. Neurol.,* 7, 348, 1980.

77. **Mckeown, G. E. and Ruedy, J.,** Methylmercury exposure in northern Quebec, 2, Neurologic findings in adults, *Am. J. Epidemiol.,* 118, 461, 1983.

78. **Mckeown, G. E., Ruedy, J., and Neims, A.,** Methylmercury exposure in northern Quebec. II. Neurologic findings in children, *Am. J. Epidemiol.,* 118, 470, 1983.

79. **Marsh, D. O., Berlin, M., Cox, C., Meyers, G., and Clarkson, T.,** Seychelles: prospective study of infant-mother pairs exposed to methyl mercury in fish during pregnancy, in Environmental Health Sciences Center: 13th Annual Report, University of Rochester, Rochester, 11, 1988.

80. **Clarkson, T. W., Cox, C., and Marsh, D. O.,** Dose-response relationships for adult and prenatal exposures to methyl mercury, in *Measurements of Risk,* Berg, G. G. and Maillie, H. D., Eds., Plenum Press, New York, 1981.

81. **Kjellstrom, T. and Kennedy, P.,** Physical and mental development of children with prenatal exposure to mercury from fish. National Swedish Environmental Protection Board, Report 3080, 1986.

INDEX